American Capitals

UNIVERSITY OF CHICAGO GEOGRAPHY RESEARCH PAPER

NUMBER 247

Titles published in the Geography Research Papers series prior to 1992 and still in print are now distributed by the University of Chicago Press. The list of available titles follows. The University of Chicago Press commenced publication of the Geography Research Papers series in 1992 with number 233.

TITLES IN PRINT (ADDITIONAL TITLES LISTED AFTER INDEX)

246. PHILIP W. PORTER, *Challenging Nature: Local Knowledge, Agroscience, and Food Security in Tanga Region, Tanzania, 2006*

245. CHRISTIAN A. KULL, *Isle of Fire: The Political Ecology of Landscape Burning in Madagascar, 2004*

244. CHARLES M. GOOD, *The Steamer Parish: The Rise and Fall of Missionary Medicine on an African Frontier, 2003*

243. JOHN A. AGNEW, *Place and Politics in Modern Italy, 2002*

242. KLAUS FRANTZ, *Indian Reservations in the United States: Territory, Sovereignty, and Socioeconomic Change, 1999*

241. HUGH PRINCE, *Wetlands of the American Midwest: A Historical Geography of Changing Attitudes, 1997*

240. ANNE KELLY KNOWLES, *Calvinists Incorporated: Welsh Immigrants on Ohio's Industrial Frontier, 1996*

239. ALEX G. PAPADOPOULOS, *Urban Regimes and Strategies: Building Europe's Central Executive District in Brussels, 1996*

238. EDWARD T. PRICE, *Dividing the Land: Early American Beginnings of Our Private Property Mosaic, 1995*

237. CHAD F. EMMETT, *Beyond the Basilica: Christians and Muslims in Nazareth, 1995*

236. SHAUL EPHRAIM COHEN, *The Politics of Planting: Israeli-Palestinian Competition for Control of Land in the Jerusalem Periphery, 1993*

235. MICHAEL P. CONZEN, *Thomas A. Rumney and Graeme Wynn: A Scholar's Guide to Geographical Writing on the American and Canadian Past, 1993*

234. DAVID M. KUMMER, *Deforestation in the Postwar Philippines, 1992*

233. RISA PALM AND MICHAEL E. HODGSON, *After a California Earthquake: Attitude and Behavior Change, 1992*

230. CHRISTOPHER MUELLER-WILLE, *Natural Landscape Amenities and Suburban Growth: Metropolitan Chicago, 1970–1980, 1990*

228–29. BARRY C. BISHOP, *Kamali under Stress: Livelihood Strategies and Seasonal Rhythms in a Changing Nepal Himalaya, 1990*

226. JEFFREY A. GRITZNER, *The West African Sahel: Human Agency and Environmental Change, 1988*

225. GIL LATZ, *Agricultural Development in Japan: The Land Improvement District in Concept and Practice, 1989*

American Capitals

A Historical Geography

CHRISTIAN MONTÈS

THE UNIVERSITY OF CHICAGO PRESS • *Chicago and London*

CHRISTIAN MONTÈS is professor of geography at the Université Lumière Lyon 2.

The University of Chicago Press, Chicago 60637
The University of Chicago Press, Ltd., London

Printed in the United States of America

23 22 21 20 19 18 17 16 15 14 1 2 3 4 5

ISBN-13: 978-0-226-08048-2 (cloth)
ISBN-13: 978-0-226-08051-2 (e-book)
DOI: 10.7208/chicago/9780226080512.001.0001

Library of Congress Cataloging-in-Publication Data

Montès, Christian, author.
American capitals : a historical geography / Christian Montès.
pages ; cm
Includes bibliographical references and index.
ISBN 978-0-226-08048-2 (cloth : alkaline paper) — ISBN 978-0-226-08051-2 (e-book)
1. Capitals (Cities)—United States—History. I. Title.
E180.M66 2014
307.760973—dc23 2013031757

♾ This paper meets the requirements of ANSI/NISO Z39.48-1992 (Permanence of Paper).

FOR MY PARENTS

Contents

1

Capitals

A New Light on American Cities and Territorial Processes

Like most boys living in the west-end of Chillicothe [during the 1850s], my entry into school life was through the Western Building, and, naturally, by way of good old Miss Pierson's room. I have yet a lively recollection of her early discovery of my predilection for geography, and of her choosing me, when visitors were present, to name and point out the State-Capitals; this pleasure was always tempered by the fear that New Hampshire's capital would be called for, and pride had a fall when such was the case, for the nearest my infantile lips could pronounce it was "corn-cob," and I dreaded their laugh raised at my expense.

L. W. RENICK, *CHE-LE-CO-THE, GLIMPSES OF YESTERDAY* 1896

According to Michel de Certeau (1988), historical investigation has only one way to resolve the dilemma between the social status of contemporary historians—the fact that the past is a construction of the present—and the necessary "virginity" of researchers: surprise—surprise with regard to a text, a silence, or an absence of archives. This study originated in a personal surprise at the small size of most American state capitals. The second surprise was that nobody has written a global history of these cities, the only exceptions being a few popular or juvenile works.[1] It is perhaps because memorizing state capitals at school has left scholars—all of them former pupils like Mr. Renick, cited above—with the idea that these cities are to be taken for granted and not seen as possible subjects of intellectual investigation (figure 1.1 nevertheless reminds readers of their names and locations). More than forty years ago, an American geographer asked the same question I am asking today (Browning 1970). He did so in a journal article intended for schoolteachers rather than for the academic world, and since then nobody has taken his first approach to the question further than his perceptive but general remarks.

FIGURE 1.1 State capitals of the United States of America. Credit: Christian Montès and M. L. Trémélo.

A Very Diverse Corpus

Although capitals appear alike on national maps (a star most of the time), they form a highly heterogeneous corpus, reflecting the absence of any uniform "national mind" in the United States of America. According to the 2010 census, our state capitals ranged in population from 7,855 (Montpelier, VT) to 7,559,060 (Boston combined statistical area [CSA]), and their share of the state's population ranged from 0.66% (Annapolis) to 152.1% for Providence, Rhode Island (exceeds 100% because the metropolitan area of Providence extends beyond Rhode Island's boundaries). Nor does any regional convergence emerge. For instance, the once clear difference between New England, where towns were founded at first, and southern tidewater plantations, which were almost without towns, is no longer perceptible in the current size of state capitals. The reverse might even be true. In New England, which includes the six states of Maine, Vermont, New Hampshire, Massachusetts, Connecticut, and Rhode Island, three capitals are small cities, and three are large (Boston, Hartford, and Providence). In the Southeast, Atlanta, Nashville, Raleigh, and Richmond are conurbations of a million or more people; Montgomery, Jackson, Columbia, and Harrisburg also have metropolitan statistical areas (MSAs) between 0.3 and 1 million people.[2] Of the capitals in that region, only Frankfort, Annapolis, and Dover are small cities. The same opposition is present in the percentage of a state's population living in its capital: 1.26%, 9.20%, 11.12%, 33.92%, 115.4%, and 152.1% for New England, versus 0.66%, 1.63%, 1.95%, 7.84%, 11.86%, 15.73%, 17.24%, 18.08%, 25.05%, and 54.39% for the Southeast.

Table 1.1 underlines the fact that the study of capitals cannot be reduced to small-town analysis, a field that has undergone recent renewal, mostly from a cultural point of view.[3] Nor does it belong to medium-sized city analysis (Hollingsworth and Hollingsworth 1979). The situation has changed since the 1950s because many capitals have become metropolises, from Atlanta to Salt Lake City and Austin. Indeed, the 2010 census revealed that 24% of state capitals had fewer than fifty thousand inhabitants in their municipalities, 28% had more than a quarter of a million inhabitants, and 34% had a million or more in their CSAs or MSAs. The difficulty in classifying state capitals was well expressed in the appraisal of Columbia, South Carolina's seat of government: "It's a wonderful city. . . . It's still a big small town—it's small enough that you are always running into someone you know, but not so small that you know everyone."[4] Such diversity shows that the history of American state capitals is complex and needs some explanation.

Fundamental Questions

> By a Capital I mean a city which is not only the seat of political government, but is also by the size, wealth, and character of its population the head and centre [*sic*] of the country, a leading seat of commerce and industry, a reservoir of financial resources, the favored residence of the great and powerful, the spot in which the chiefs of the learned professions are to be found, where the most potent and widely read journals are published, whither men of literary and scientific capacity are drawn.
>
> JAMES BRYCE, *THE AMERICAN COMMONWEALTH* (1891)[5]

First we require a definition of the word *capital*. Following Bryce's definition would lead to the conclusion that there is no "real" capital in the country, echoing Alexis de Tocqueville, who thought that the decentralization of power brought by the absence of a real capital was the cause of the survival of democratic institutions in America. But to work on an a priori definition is not the best way to analyze the question. First, Bryce's definition refers to European centralized states rather than to the American federal system, in which urban primacy and politics follow a different logic (Glaeser 1999).[6] Second, his definition is based on the broadest interpretation of the word *capital*; the Latin word *caput* means "head," and he stated that a capital

Table 1.1 State capitals: Basic facts

State	Capital	Year founded	Year chosen	State pop., 2010	μSA/MSA, 2010*	Percentage of state pop.	Rank when chosen	Rank in 2010
Alabama	Montgomery	1819	1847	4,779,736	374,536	7.84	2	3
Alaska	Juneau	1880	1900	710,231	*31,275*	4.40	2 or 3	4
Arizona	Phoenix	1874	1889	6,392,017	4,192,887	65.60	2	1
Arkansas	Little Rock	1821	1820	2,915,978	699,757	24.00	1	1
California	Sacramento	1848	1854	37,253,956	2,149,127	5.77	2	4
Colorado	Denver	1858	1868	5,029,196	2,543,492	50.57	1 or 2	1
Connecticut	Hartford	1636	1873	3,574,097	1,212,381	33.92	1	1
Delaware	Dover	1683	1781	897,934	162,310	18.08	1 or 2	2
Florida	Tallahassee	1824	1823	18,801,310	367,413	1.95	3	12
Georgia	Atlanta	1843	1868	9,687,653	5,268,860	54.39	2	1
Hawaii	Honolulu	1100	1900	1,360,301	953,207	70.07	1	1
Idaho	Boise City	1863	1864	1,567,582	616,561	39.33	New town	1
Illinois	Springfield	1821	1837	12,830,632	210,170	1.64	2	4
Indiana	Indianapolis	1821	1825	6,483,802	1,756,241	27.09	2	1
Iowa	Des Moines	1843	1857	3,046,355	569,633	18.70	7	1
Kansas	Topeka	1854	1861	2,853,118	233,870	8.20	4	3
Kentucky	Frankfort	1786	1793	4,339,367	*70,706*	1.63	2	7
Louisiana	Baton Rouge	1719 fort	1882	4,533,372	802,484	17.70	2?	2
Maine	Augusta	1628	1832	1,328,361	*122,151*	9.20	2	4
Maryland	Annapolis	1648	1694	5,773,552	(38,394)†	0.66	?	4
Massachusetts	Boston	1630	1692	6,547,629	4,552,402	69.53	1	1
Michigan	Lansing	1847	1847	9,883,640	464,036	4.69	New town	3
Minnesota	St. Paul	1840s	1849	5,303,925	3,279,833	61.84	1	2
Mississippi	Jackson	1821	1826	2,967,297	539,057	18.17	3	1
Missouri	Jefferson City	1822	1822	5,988,927	149,807	2.50	New town	6
Montana	Helena	1864	1874	989,415	*74,801*	7.56	1 or 2	4
Nebraska	Lincoln	1867	1867	1,826,341	302,157	16.54	New town	2

Nevada	Carson City	1858	1861	2,700,551	55,274	2.05	2 or 3	3
New Hampshire	Concord	1726	1808	1,316,470	*146,445*	11.12	2	3
New Jersey	Trenton	1680	1790	8,791,894	366,513	4.17	1 ea.	4
New Mexico	Santa Fe	1610	1610	2,059,179	144,170	7.00	1	2
New York	Albany	1624	1797	19,378,102	870,716	4.49	2	4
North Carolina	Raleigh	1792	1792	9,535,483	1,130,490	11.86	New town	2
North Dakota	Bismarck	1873	1889	672,591	108,779	16.17	4	3
Ohio	Columbus	1812	1812	11,536,504	1,836,536	15.92	New town	3
Oklahoma	Oklahoma City	1889	1910	3,751,351	1,252,987	33.40	1	1
Oregon	Salem	1840	1860	3,831,074	390,738	10.20	3?	3
Pennsylvania	Harrisburg	1785	1812	12,702,379	549,475	4.33	5	3
Rhode Island	Providence	1636	1900	1,052,567	1,600,852	152.09	1	1
South Carolina	Columbia	1790	1790	4,625,364	797,598	17.24	New town	2
South Dakota	Pierre	1880	1889	814,180	*19,988*	2.45	c. 5	7
Tennessee	Nashville	1779	1843	6,346,105	1,589,934	25.05	1	1
Texas	Austin	1839	1839	25,145,561	1,716,289	6.83	New town	4
Utah	Salt Lake City	1848	1856	2,763,885	1,124,197	40.67	1	1
Vermont	Montpelier	1780	1808	625,741	(7,855)†	1.26	6 at most	12
Virginia	Richmond	1737	1779	8,001,024	1,258,251	15.73	Village	2
Washington	Olympia	1850	1853	6,724,540	252,264	3.75	Village	16
West Virginia	Charleston	1788	1885	1,852,994	304,282	16.42	5	1
Wisconsin	Madison	1836	1836	5,686,986	568,593	10.00	New town	2
Wyoming	Cheyenne	1867	1869	563,626	91,738	16.28	1	1

Sources: US Census Bureau, 2010; local histories.

Note: "New town" means the town was intended to become the capital when founded; "village" means the town was very small when designated as the capital.

* μSA: micropolitan statistical area; MSA: metropolitan statistical area.

† For Annapolis and Montpelier, the population is the municipal population.

must be at the head of the political, economic, social, and cultural spheres. This study takes the opposite course, studying the actual status of political capitals in the organization of the United States instead of the position they should or might have. I also analyze how American leaders or citizens looked at their capitals and how they envisioned them. The executive secretary of the American Civic Association asked some interesting questions about the essence of state capitals.

> Where should the State buildings be located? How many of them should be centralized in the State capitals? What provision should be made for future expansion? What is the desirable location of the business district of the capital city in relation to State buildings? . . . What do the State officials owe the capital city? How can the city fathers serve the State? . . . Have the citizens of the State a feeling of ownership and pride in the State capital? Should the State capital be located in a commercial or industrial city? Should it be near the center of the State? Should the State universities and other State institutions be located in the State capital? Should all of the executive departments be centralized in the capital city? (James 1925–26, 387–88)[7]

Paul W. Pollock answered these questions, perhaps a little too emphatically, when he argued that "the state capitals still form the nucleus of our country's vital nervous system. They are the life-force of our economic and political society. As such, they together exert incalculable influence on all areas of American life." (Pollock 1960, foreword). Putting these questions and ideas in a broader perspective yields the thesis that underlies this book: studying state capitals is a useful way to look at some fundamental questions regarding the United States of America.

There are three fundamental concerns: the building of the national territorial structure; how the American democracy worked during that period; and the evolution of the American city and the inhabitants' relationship to it. To address these concerns, three geographical levels must be taken into account: a state capital is a capitol (a building; both words are pronounced identically), a municipality (a fact usually overlooked), and the state it symbolizes. All levels have precise boundaries that are both exclusive and inclusive (a Texan is not an Oregonian, for instance).

The first issue arising from that framework is the status of state capitals in the national democratic balance. The fact that American democracy is based on the idea that power must be as decentralized as possible—the definition of *possible* varying according to the ebbs and flows of federalist ideas—has been thoroughly studied.[8] But the operation of this idea at the local level has been given far less attention, and such studies have mostly tended to scrutinize the municipal level. Capitals have only been treated from the political point of view, as places where decisions are taken at the state level, and almost never from the spatial point of view. Nor have they been looked at through the lens of the changing relationships between the private sector and the public sector, especially during current deregulation and privatization processes, although capitals epitomize such relationships.

The second issue concerns the status of capitals in the American urban system: Do they form a parallel system, disconnected from the classic system, largely based on economic criteria? Capitals were indeed part of the construction of the states and the urban system, but in a spatial framework that is quite varied. Does the capital of tiny Rhode Island have a status similar to the capital of Alaska? This leads us to inquire into the role of state boundaries in the building and workings of the United States, and still further into the relationships between the political and economic realms. State capitals, with some exceptions (such as Boston, Atlanta, and Indianapolis) developed much more slowly than most other American cities. The reason was not the will to be secluded from economics. On the contrary, capitals also dreamed of becoming economic metropolises. This is clearly put by a member of Michigan's House of Representatives in his comment on the victory of Lansing at the end of the 1847 capital contest: "To me and my constituents in Clinton County, it was the opening and building of roads from Pontiac and Ann Arbor and for seventy miles into the wilderness where we lived. It was opening for us a way to markets and bringing us again into connection with the civilization from which we had unwisely but voluntarily exiled ourselves" (Upton 1990, 402).

The capitals' developmental delay is often charged to a corrupt government on one hand, and to a relatively unenlightened citizenry on the other. Only in the most recent generation or two have these factors changed for the better. A more generous interpretation of state capitals reads them as the currently fading expression of the Jeffersonian ideal—a democracy based upon small but educated farmers. All but eleven of the state capitals were selected during the nineteenth century, thirty-five of them before 1861, an era of pioneer and idealized territorial vision. Washington, DC, the national capital city, could therefore have served as the model for state capitals, just as the national capitol building came to be virtually the standard design for state capitols. The Washington model removed the political capital from economic centers. The choice of Washington grew from a compromise between North and South. By contrast, state capitals grew each in their own way, out of the flux and hurly-burly that was American development. Local elites tried to win state capital status not only for economic advantage but also for the political stability that such a status might provide. Most of the time, however, stability proved elusive, and state capitals migrated—often westward, like the pioneers. Capitals shifted as political factions grabbed control and railway companies fought over new territories. The United States is "littered" with towns that once dreamed of the capital's crown—like Prince Charmings—but were either unable to catch it or unable to retain it.

Do state capitals consequently express an imbalance between form and function? A historian of Frankfort, Kentucky's capital, wrote that "as early as 1977, some of the city's business and political leaders were voicing concern over Frankfort's image as a 'company town' whose economy was excessively dominated by state government" (Kramer 1986, 386). Are state capitals akin to company towns, dominated by one function? Are they in essence public company towns? The difference is that in this case, the one function—government—never ceases to reinforce itself. Unlike business, government almost never goes bust. Are state capitals more than simply symbolic towns? Most Americans believed—and still do—that political power is the key to the solution of most problems and to the advancement of society. Capital cities were and are the embodiment of such power, which brings us back to the first issue, American democracy.

A Study in Complexity

What would be the best scientific approach to understanding the choice and subsequent evolution of colonial, territorial, and state capitals? The major question is how to encompass four centuries of history for about 180 cities throughout the United States (all former and current capitals). We must also take into account that this would be the first book on the subject. A multifocused approach seems the best way to convey the complexity of the processes and to take into account the variety of sources available. My intention is to lay a sound foundation for an intellectual debate on American state capitals, as Donald Meinig did (in a far broader way) for the historical geography of the United States. To achieve such an aim, following a single theoretical path would be too narrow an approach. This does not mean that there is no major point of entry into the subject. This study is based on a transversal approach close to Immanuel Wallerstein's notion of space-time (2004). The analysis is cultural (and not culturalist), because culture has to deal with the political and economic as well as the spatial processes at stake. This means that I try to go beyond linearity through the building of a model that reveals a certain permanence as well as rendering (sources allowing) multiple temporalities. I also try to go beyond general processes to study groups, individuals, and representations (such as the small town). My approach is fundamentally incremental and tries to build an explanatory model progressively, without imposing it a priori and without predetermining its components. It is based on possible hypotheses—coming from various disciplines—among which I try to determine the most probable ones. As always in social sciences, this model is seen only as a means, not an end.

The use of narration goes with that cultural approach: it provides chronological frameworks, suspense around moments of crisis rendered through "stories" that offer some part of individuality, and models that take into account explanations based on the long term. I refuse dogmatism and wish to shed various kinds of light on the richness of my subject. If we must find an academic slot for this study, historical geography would be the best one, since it has long been characterized by "liberal eclecticism" (Holdsworth 2002).

I fully agree with David Hamer, who wrote, "Why should historians feel inhibited about doing research just because they are unable to find some convenient category in which to slot their work?" (Hamer 1990, 3). Ted Margadant, in his study of urban rivalries in Revolutionary France, expressed how difficult it is for such an analysis to follow the well-trodden paths of historical research and how necessary it is to broaden the analytical scope. According to him—and I heartily concur—that difficulty is the very interest of the study:

> The social historians have assumed that conflicts within towns were more important than conflicts between towns; most institutional historians have overlooked the extraordinary efforts that townspeople made to gain the new directories and lawcourts; and cultural historians have ignored the fundamental beliefs that these townspeople shared about the economic interests at stake in the reorganization of the kingdom. The subject of ur-

> ban rivalries over the institutions of the state does not fit easily into any of these interpretative frameworks. Its anomalous position has the advantage of bringing a different angle of vision to bear on contentious issues within the field of revolutionary historiography. (Margadant 1992, 443–44)[9]

To bring to light the processes of the selection and evolution of capitals, I have therefore tried in this study to follow what Anne Kelly Knowles described as the best method for historical geographers to follow: "empirical research that digs for answers to historical questions, critical examination of landscape, and a desire to understand how the material world has been shaped by human action and ideas" (Knowles 2001, 469). To fulfill such aims, I have intertwined two methodological paths. This study is located within "modern academic and civic concerns" but without eliminating "Meinigian narratives" (Holdsworth 2002, 675). The first approach searches for models that throw light on the processes at work and for theories that explain the formation, evolution, and image of urban America, which help in understanding capitals. The second methodology owes much to the work of Tuan, Giddens, and others on the role of perceptions and emotions in the understanding of cities and urbanity. Capitals are an interesting field to study because their heavy symbolism enhances the affects and sentiments that people manifest toward them (Widdowfield 2000). I have therefore tried to revive the main actors in what is now studied as a process or an overall pattern and also to revive the places—to describe their landscapes and their amenities, as well as their shortcomings, all of which played a part in the final choice of a new capital. The examples given aim not only to entertain readers but to prove that history and geography have to be true to their categorization as "social" or, better, "human" sciences. Three themes have therefore been privileged.

1. Capitals as what French historian Pierre Nora called the places of memory (*lieux de mémoire*), "that condense the long-lasting time in the instantaneous one" (Debarbieux 1995, 105).[10] Capitals having evolved throughout several centuries—for the oldest ones—Fernand Braudel's concept of long-lasting time (*temps long*) is very useful.
2. The process of capital choice as a revealing moment of crisis. A crisis belongs to short-term processes, but it should not be opposed to long-term ones, both being the two faces of the same coin, that is, urban rhythms. Braudel's approach was mainly based on long-term processes (since he had been influenced by Paul Vidal de la Blache and structuralists, for whom permanence and durability were foremost). I prefer Marcel Roncayolo's

approach. Inverting Braudel's method, he asks how short-term phenomena create long-term structures (Roncayolo 2002). The processes of capital choice are revealing because their effects are often still at work today. Selecting capitals induced the incremental processes of economic evolution and the construction of an image. This refers to the question of heritage and identity, issues that have received recent academic scrutiny. Identity is built by three major elements: the human subject, society, and geographical space. Identity is crystallized by elements that, by becoming part of a common heritage, create a collective identity and a powerful social mobilization (Di Meo 2002). Capitals and capitols are very strong unifying forces in a nation that is more and more diverse, be it ethnically or socially. Contrary to the cultural relativism of the postmodernist approach, which artificially fabricates cultural objects, capitals and capitols are clear bases of the American identity. They also were and remain major public spaces in a country where the forces of privatization are very strong.

3. Capitals as symbolic places that participate in territorial construction and in interactive adjustments to other ways of territorial structuring.

From a Plurality of Sources to Multiple Points of View

I did the research for this study primarily in libraries, searching for local histories of capitals and states, as well as for general studies in American politics, urbanism, and so forth.[11] Numerous local historical journals are published in the United States; almost every state has produced one, and some have two journals that proved very useful. Second, research was also conducted in state archives. This often proved disappointing, as in the Alabama State Library and Archives, where no minutes of the legislative debates are preserved. The only state newspaper published in 1846 preserved there was Tuscaloosa's *Independent Monitor*, but there was only one issue, for April 8, 1846, which held nothing of interest for this study. In fact, Montgomery's first capitol burned in 1849, destroying the state library with its numerous documents about the early history of the state, although the public records were fortunately saved. Likewise, research in the state archives, the state library, and the municipal library of New Jersey did not yield many results. Trenton was chosen in 1790 as the permanent capital, but the legislative minutes only began in 1800. Before that date, only "dry" records of the decisions taken (with no record of the debates preceding the votes) are available. The state's newspapers (e.g., the *State Gazette*) between 1786 and 1792 are not preserved. The primary sources seem therefore hard to find. The only possible ones are Governor Paterson's papers, which are to be found at Paterson University, New Jersey. In 1990, the Trenton Historical Society held "a bicentennial program honoring Trenton's selection as the state capital," called "A Capital Place: New Jersey's Quest for a Permanent Seat of Government," but the records of the events—if they exist—were nowhere to be found. Fortunately, secondary sources exist (see Walker 1929; WWP New Jersey 1939; McCormick 1981).

Field trips, the foundation of geographical research, provided the third way to look at state capitals—a more impressionistic but very useful means of knowledge. I visited thirty-one state capitals, from Augusta to Sacramento and from Albany to Baton Rouge. To inhale the atmosphere of a city, to walk in and around its monuments and streets, and to converse with its inhabitants gives more "reality" to the sometimes cold analyses that one can find on each capital.

As a result, this study does not pretend to provide the definitive and complete answer to the "capital question," but it offers a first approach to the complex processes leading to the selection of capitals as well as a first assessment of their subsequent evolution and of their integration into the broader processes of territorial and urban development. Bear in mind that this work was written in 2012, and that it is impossible to bring forward any "true" revival of the past. This book is meant as a basis for future work that will correct the factual errors that are certainly present and modify or challenge its interpretations. Since social sciences are more social than scientific, scholars know that theses are only put forward to be corrected or discarded by further research and new theoretical frameworks. This book presents my truth, and my hope is that readers will find that it is not too far off the(ir) mark.

The ten chapters follow a long-term analysis through two lenses—one long-term, one short-term—from the choice of the first colonial capitals to the fate of the fifty present capitals. Chapter 2 emphasizes that capitals are singular in the American urban fabric, from physical and social perspectives, because of the monumentality and the public spaces they offer. This sheds light on their enduring symbolic power as places of memory. The next three chapters consider the processes of capital choice as moments of crisis, and they progressively build an explanatory model. First, chapter 3 looks at the spatial patterns of the migration of capitals that almost all states have experienced, taking a diachronic approach to the framework of American territorial construction. This study progressively uncovers the processes at work, as well as their interrelations. The search for criteria for the model begins with chapter 4: Puritanism, small town ideals, the booster model, and the gateway model are successively examined. Chapter 5 takes a closer look at the major factor of explanation: politics. It examines the balance of powers that resulted from the process of choice and assesses the degree of democracy present at the time by looking at the power of king, Congress, territories, legislatures, and the influence of the people through referenda. It then proposes a global model.

The next two chapters change the perspective. Having scrutinized the processes of selection, we examine the evolution of present capitals in order to explain their place in American territorial construction and their position in the American urban system. The developmental delay of many of them is also addressed. Chapter 6 deals with what could be called the "sleeping" or "purgatory years" of state capitals, ending in the 1950s, when a majority experienced a slow demographic and economic growth in a rapidly growing nation and suffered under a "Babylonian" image. Chapter 7 looks at the current revival—or late awakening—of most state capitals, which became open to more political purity and economic success in the so-called postindustrial age. However, the national urban hierarchy has not been fundamentally altered by their growth. Chapter 8 aims to validate all the previous models, while rendering them more concrete. It presents a deeper study of the evolution of three capitals, which serve as examples of three types of transformation—from the brightest (Columbus, Ohio) to the medium-sized but locally grand (Des Moines, Iowa) and through modest but real evolution (Frankfort, Kentucky). Chapter 9 sheds light on the fate of former capitals, showing that although hosting the capitol did not often induce significant growth, being the seat of government allowed capitals a better fate than befell cities that lost such a coveted status, at least until the current politics of heritage revival. Chapter 10 briefly concludes the study by considering the modification of the image of capitals. The book ends with an appendix that provides demographic and historical tables in order to sum up the evolution of capitals.

2

Capitals as Places of Memory

Capital cities differ from other cities from two points of view. First, they often depart from the supposed uniformity of American cities from coast to coast. Second, one could argue that capitals are the "most American" of the nation's cities—not because they have more characteristic features than other cities (in terms of location, frontier history, etc.), as such thinking suffers from a necessarily subjective and incomplete choice of attributes,[1] but because they are symbols of the United States of America. Capitals are places of memory and are studied here through the three rhetorical figures put forward by Bernard Debarbieux (1995):

1. The place as an attribute, a stereotype that has a constant meaning, as does, for instance, the Empire State Building for New York. For state capitals, capitol buildings play that role.
2. The generic place that materializes an element of the historical core of the nation and is an allegory to the social group that constitutes that very place. The national capitol, for instance, materializes American democracy as the state capitols and grounds materialize the entire state.
3. The place of condensation, as an image and an environment. Such places are "built and identified by a society that gives itself to be seen through them, [they] . . . narrate its history and anchor its values" (Debarbieux 1995, 100). In these places, individual and collective experiences take place that revive their reference to the social group and to its territory. Borrowing a phrase from linguistics, they are spatial and social synecdoches. Urbanism and monuments are conceived so as to render a visit to the capital edifying. They concentrate emblems and dignitaries; they are the strong points of the territory and welcome the significant events of the society (e.g., laying the capitol's cornerstone or the inauguration of a new capitol).

Capitals belong to the three figures, the third one being the most important. These figures function as a signifying continuum, translating three relationships to time: past events, perceived as the coming of a new order; perpetuation;[2] and instantaneity of experience. In other words, they take part in both the long-term and immediate processes that are the basis of this study's approach. From the geographical point of view, they offer a unique pattern, because no other country demonstrates a similar condensation from the national scale (Washington, DC, with the national capitol) to the regional scale (fifty state capitals, with capitols that often mimic the national one). Scale and social construction are strongly linked: capitals express a symbolic language paraphrasing the social space and paraphrasing Washington, DC.[3] These rhetorical figures are seen through four successive lenses (which are not always simultaneously present in all state capitals):

1. Capitals as generic places, through the naming of capitals, which enhances their strong symbolic meaning, distinguishing them from other cities in the state.
2. Condensation processes in capitals, through their platting, which stage and translate the polity in the physical pattern of capital cities.
3. Debarbieux's three figures are seen through the lens of the erection of monuments, the most impressive of which are the capitols. Once symbols of the reality of statehood, they have become symbols of permanence, linking the states with their history.
4. Finally, we emphasize that "condensation" does not always encompass the whole place and the entire population of state capitals. This point is studied through the unique relationship between the public and the private spheres. Capitals certainly host and stage the most important public spaces and places in their states, but they do not always create harmony and equality.

Naming or Renaming Capitals—a Political Action

A capital is first a name on a map—a name that owes nothing to chance but often a lot to politics. The names of American capitals reflect the leading group(s) that founded or ruled the state. They also reflect the history of the building of the United States of America; in this sense, they belong to Debarbieux's generic places. Tables 2.1 and 2.2 show two overlapping categories of capital names. The first one is linguistic (40% of the names are not English); the second one is thematic.

Table 2.1 Naming the capital: Language (English excluded)

Language	Names
Native American (6)	Cheyenne, Honolulu, Indianapolis, Oklahoma City, Tallahassee, Topeka
Spanish (2)	Sacramento, Santa Fe
French (6)	Baton Rouge, Boise City, Des Moines, Montpelier, Pierre, Juneau
German (2)	Bismarck, Frankfort
Greek (4)	Indianapolis, Annapolis, Atlanta, Olympia

Table 2.2 Naming the capital: Main themes

Theme	Names
Grandness (7)	Atlanta, Augusta, Columbia, Columbus, Concord, Olympia, Phoenix
Honorary: US presidents (4)	Jackson, Jefferson City, Lincoln, Madison
Honorary: lesser known (9)	Albany, Annapolis, Bismarck, Carson City, Denver, Lansing, Montgomery, Nashville, Raleigh, Trenton
Religion (4)	Providence, Salem, St. Paul, Santa Fe
Pioneers and founders (6)	Austin, Charleston, Harrisburg, Pierre, Juneau, Frankfort
Existing towns (7)	Boston, Dover, Richmond, Hartford (England), Montpelier (France), Helena, Lansing (U.S.)
Natural features (6)	Boise City, Des Moines, Little Rock, Sacramento, Springfield (rivers), Salt Lake City
Feminine forms (5)	Atlanta, Augusta, Columbia, Helena, Olympia

These names express the diverse origins of the national population. The four main groups behind the United States' colonization are represented in table 2.1: besides the ultimate "winners," the British and the first immigrants—Native Americans—are represented, along with the French and the Spanish. The presence of German names is a testimony to the largest single group of immigrants that came to the United States. But there is a telling exception: the African Americans were left aside when choosing capitals' names, as they had been left aside in early American history and were denied their original names.

French names come mostly from local landscapes. Louisiana's future capital was christened Baton Rouge. This name, given to the fort built in 1721, comes from the red pole with the heads of fish and bear stuck upon it that the French explorer Iberville had seen on the same spot in 1699 (Davis 1959, 30).[4] Idaho's capital is called Boise City, an inconspicuous, bucolic name in the midst of a mining boom. The name derives from the Boise River (Wooded River), named by French Canadians in the early 1810s (Arrington 1994, 108). The city lives up to its name; it is now nicknamed the "City of Trees." Des Moines also comes from the French name of the Des Moines River (Trappist Monks had lived there). The War Department preferred it to Fort Raccoon (the name of the town's other river), which was suggested by Captain Allen, the first head of the garrison. But some think that Des Moines could be derived from the Indian word *moingona*, meaning river of the mounds—burial mounds were located near the river (see the city's website). Nature (or lack of imagination) also inspired the founders of Springfield, on the tributary waters of Spring Creek (Krohe 1976, 4); Sacramento (the river); Little Rock, and Salt Lake City.[5]

Another discernible trend in the choice of a name was the attempt to bring success to the newborn settlement by placing it under the aegis of God or natives. The oldest American state capital (1609) was given the name of Santa Fe (Holy Faith).[6] The explorers of that quite unknown part of the Spanish Empire must have had great faith for them to imagine that their tiny settlement would outlive them or even the coming winter. Salem's name comes from the Hebrew word *shalom* (peace) and refers directly to Jerusalem. In a Christian vein, when Roger Williams fled Boston's overly strict Puritans in June, 1636, and founded a new town, he named it Providence. In a famous speech made for the occasion, he said, "I, having made covenantes of peaceable neighborhood with all the sachems and natives round about us, and having in a sense of God's merciful providence unto me in my distresse, called the place Providence. I desired it might be for a shelter for persons distressed of conscience." (quoted in McLoughlin 1978, 9). More than two centuries later, a similar quest for peace with the local Indians influenced the naming of Cheyenne. The Union Pacific Railroad Company decided to call the town "by its present cognomen in hopes of conciliating the interesting Savages" (Stelter 1967, 7). Vain hopes. To pay lip service, Iowa City and Indianapolis stressed the "native" names of those states. Topeka, which means "wild potato" in Shawnee, was so named because this "vegetable grew plentifully upon the rich bottom land along the river"[7] (Cutler 1883). *Tallahassee* is the Apalachee Indian word for "old town" or "abandoned fields," which was perhaps not a very propitious choice. Michigan had also an eponymous new capital for a few months in 1847. The legislators had at first approved the name Aloda, but changed it to Michigan, before again modifying it to Lansing in 1848 (Dunbar and May 1995, 238). In fact, the only capital bearing an autochthonous name is Honolulu ("protected bay"), for the Hawaiians had already named it before the archipelago was colonized by the United States.

The presence of Greek names is interesting. It was the cause of some merriment among the legislators of Indiana. The process involved a prolonged discussion by the House, in Committee of the Whole. Judge Jeremiah Sullivan of the Supreme Court, who was a member of the legislature at the time, related the circumstances of the naming as follows. After an entire day had been lost, and many names had been rejected (e.g., Tecumseh or Suwarrow), Sullivan proposed Indianapolis the next day. After a first laugh, Sullivan reckoned it was adopted because "the Greek termination would indicate to all the world the locality of the town" (Dunn 1910, 26–27). But not everyone was pleased with this manifestation of onomastic Greek revival. Vincennes's newspaper, the *Indiana Centinel*, favoring Tecumseh, published this incensed account in its January 15, 1821, edition: "Such a name, kind readers, you would never find by searching from Dan to Beersheba; nor in all the libraries, museums, and patent offices in the world. . . . For this title your future capital will be greatly indebted, either to some learned Hebraist, some venerable Grecian, some sage and sentimental Brahmin, or some profound and academic Pauttowattomie" (Dunn 1910, 27).

In fact, despite some incensed prose, Americans had already had recourse to "venerable Grecian" when naming Annapolis, Gallipolis, or Philadelphia. Missouri's capital, City of Jefferson (commonly called Jefferson City), narrowly escaped being named Missouriopolis. Minneapolis was first named Teutopolis, owing to its German population. The *Indiana Centinel* would have been horrified at the name of a new town founded in December 1788 by a New Jersey speculator and his Kentucky associates. It was called Losantiville, meaning "the city across the mouth of the Licking River." *L* stood for "Licking"; *os* was Latin for "mouth"; *anti* was "opposite or across from"; *ville* was for "city." English, Latin, Greek, and French had been used in a single word! But when on January 2, 1790, Governor Arthur St. Clair arrived to make the town the new capital of the Northwest Territory, he rechristened it Cincinnati, to honor the Society of the Cincinnati, formed by Revolutionary War officers.[8] Because of (or despite) its new name, Cincinnati immediately thrived (Knepper 1989, 66–67). Indeed, before the Greek revival of the 1820s, the classical education given to the youngsters of the American elite had accustomed them to Latin and Greek.[9] But Latin was not abandoned: Corvallis, briefly the capital of Oregon in 1855, took its name from two Latin words that mean "the heart of the valley."

Besides classics, the usual inspiration behind the names of capitals leaned toward grandness, before as well as after independence. Albany owes its name to the second title (Scottish) of the Duke of York that was given to the Dutch Fort Orange (founded 1624) in 1664, when it became English.[10] The town did not retain its Mohican name of Pempotowwuthut-Muhhcanneuw, meaning "the fireplace of the Mahikan nation" (Kennedy 1983, 22). Lansing's name follows the wishes of the first settlers, who came from Lansing, New York, a town that had been christened in honor of John Lansing Jr. (1754–1829), one of the nation's founding fathers and a New York delegate to the Constitutional Convention.

Grander—and more secure—were the names of great presidents, used for Jefferson City, Madison, Jackson,[11] and Lincoln.[12] The latter's case is more complex. Nebraska's capital, Lincoln, is indeed a tribute to the great president, but also the result of a political miscalculation. North and South Platte factions embodied different political conceptions, for and against slavery. When a bill was drawn in 1867 that empowered the governor, the secretary of state, and the state auditor to select a site within southern counties to be named Capital City, the North Platters thought they could still retain the capital.[13] Senator Patrick proposed to name the future capital Lincoln, certain that his South Platte opponent, Senator Reeves of Nebraska City, a strong Confederate, would refuse it, thus crushing southern hopes. But Senator Reeves agreed, and Lincoln became the capital (McKee and Duerschner 1976, 1). In the case of North Dakota, the afterthoughts with regard to grandness were economic. The capital was named Bismarck to thank and attract German investors in a newly built transcontinental railroad, of which the capital was the temporary terminus in 1873.[14]

In a simpler grand vein is Olympia, referring to the abode of the Greek gods (and the nearby Olympic Mountains), as well as Columbia and Columbus, singing the praise of the official discoverer of America and hinting clearly at the federal capital, Washington, District of Columbia.[15] Ohioans received the name of their new capital with great pleasure and even pride. The name was suggested by Joseph Foos, a representative from Franklin County, and won immediate approval: "Columbia was the gem of the ocean; Columbus was an emerald that glittered in the heartland setting of Ohio. The name was a natural" (Condon 1977, 10). But, as William Ball remarked, the WASP elite had forgotten one thing about Christopher Columbus when naming the new capital in 1786—Columbus was a Roman Catholic (Ball 1932, 49). That such names often bear feminine forms is absolutely not homage to women, but the result of the usual feminization of classical references, owing to the fact that a city is supposed to be feminine. The Greek word *polis* as well as its Latin equivalent, *urbs,* were feminine, and most people use *she* when speaking of a city. Atlanta is the feminine form of the Atlantic Ocean; Augusta, of Imperator Caesar et Augustus; Columbia, of Christopher Columbus; Olympia, of Mount Olympus. All grandees who gave their names to capitals were naturally men. Exceptions are very rare, and somehow unfortunate, as in the case of the initial name of Montana's second capital (1865–75). It was first called Varina, the name of the wife of Jefferson Davis, the Confederate president. It was soon changed to Virginia City, this time honoring either Elizabeth the First, the Virgin Queen or, more probably, the state that bears her name.

But all capitals do not have grand names. Trenton was settled in 1680 and named in 1720 in honor of William Trent, a Philadelphia merchant who became Chief Justice of New Jersey. Carson City was named to honor the explorer and scout Kit Carson; Pierre, to honor Pierre Dorion, one the first white men to live in Dakota.[16] On December 3, 1819, the first legislature of the state of Alabama incorporated Montgomery, the present capital of Alabama, and named it in honor of General Richard Montgomery of revolutionary fame. The county already bore the same name, but it had been created by the Mississippi Territory, December 6, 1816, and named for Major Lemuel Montgomery of Tennessee, who was killed in the Battle of Horseshoe Bend, Tallapoosa River, in the Creek Indian War, 1813–1814 (Owen 1949, 204). Revolution was certainly better than one more Indian war!

Some local leaders or plain settlers, certainly thinking that God and grandees did not need more publicity, refused to relinquish their rights to fame. Georgia had first followed the classical denominational pattern in 1796 by naming its new capital Louisville in honor of France's king and in gratitude for French aid during the Revolution. But Louisville became too upcountry when a new land distribution system was set up. This led to the choice of a new capital in 1803. Its founder, governor John Milledge, modestly gave the new town his name.[17] Milledgeville remained the capital from 1807 to 1868, and soon became an important commercial center. Another governor who gave his name to a capital was James W. Denver, governor of Kansas, of which Colorado was a county at the time. But he was not behind the baptism. The "father" of the new settlement, William Larimer, thought he would please the governor by renaming it Denver—the former name was St. Charles Town—and in this way gain the seat of the newly created Arapahoe County, Kansas, and then the capital of the future territory. It proved useless, the governor having resigned some weeks earlier (Leonard and Noel 1990, 8).

The probable origin of the name of Frankfort, Kentucky (which was still in Virginia when it was founded in 1786), is Stephen Franck, a settler killed on a spot that was known afterward as Frank's Ford. Later German immigrants replaced that name with the name of the German town of imperial fame, Frankfort, which sounded similar. Harrisburg, Pennsylvania was named in 1785 after its founder, John Harris Jr., son of the first settler and head of the new town company, against the wishes of state officials who wanted to name it Louisville, to honor the French. Likewise, Charleston, West Virginia (Charles Town from 1788 to 1818), bears the name of the founder's father, Charles Clendenin—not that of Britain's King Charles, for whom the other Charleston was named (Goodall 1968a, 102).

Such a propensity among local leaders for naming new towns after themselves or after their former homes was widespread but neither universal nor everlasting. First, it was rejected by at least one town founder. When Colonel Jacob Davis was granted two towns in Vermont in 1781, he decided to give them French names, Calais and Montpelier. The decision was certainly influenced by gratitude toward the French but also by the fact that the *mont* in Montpelier echoed Vermont's name (Morissey 1981). Second, as the light of lesser stars does not shine very brightly, local names could soon fade. Alaska's capital was named for one of the town's first settlers, Joseph Juneau, only after internal division had created some controversy during the town's first year of existence, 1880–81 (DeArmond 1967). The new settlement was first called Harrisburg(h), from the name of Richard T. Harris, who was the first—with Joseph Juneau—to enter a town claim on October 18, 1880. But on February 10, 1881, the name was changed to Rockwell. The probable cause (it has not been recorded) was dissension among the early miners. Rockwell got eighteen votes, Juneau fifteen, and Harrisburg only one. Rockwell's town plat was filed on March 26, 1881. But, while the army, which had jurisdiction over Alaska, immediately adopted the new name of Rockwell, the new post office, created in April 1881, called the town Harrisburgh. That name was therefore used until a new meeting, on December 12, 1881, changed it to Juneau City (by forty-seven votes against twenty-one for Harrisburgh and four for Rockwell). The Post Office Department dropped *City*, making the official name of the town Juneau. Such changes were quite common in early mining camps, stressing the changing tides of popularity among the town fathers, and the modesty of fame. The founder's name was sometimes even dropped, as in Montana. In 1864 a group known as the "Four Georgians" struck gold in what is now Helena's main street and called it Last Chance Gulch. A settlement grew, and was first given the name of Crabtown, after one of the Georgians, John Crab. But it did not please the other miners, who preferred the name of a Minnesota town, Saint Helena, "pronounced Saint Hel-E-na. To the miners HEL may have been spelled HELL, and from then on was called Helena. Saint was dropped from the name as it was deemed unnecessary" (Southwest Montana Gold West Country website: http://southwestmt.com), perhaps because it referred to a way of life far removed from the one they were leading.

When a town already existed before being selected as the capital, its name was often changed. Fame and the desire for higher status were the main reasons, especially when the town's name was plain or even unfortunate. Indeed, not all the future capitals had names as propitious as Colorado's second capital, Golden. Its origin is not what one might expect. It was named for the first pioneer who settled there in 1858, Tom Golden (Leonard and Noel, 1990, 298). When colonial Virginia's General Assembly decided in 1699 to move the capital, it also voted to alter its name. The name of Middle Plantation, which was perhaps a little too matter-of-fact, was changed to Williamsburg. The name honored King William and certainly, though indirectly, the man behind the founding of Middle Plantation's jewel, William and Mary College, Reverend James Blair (Dabney 1971, 75). Even less fortunate were the first names considered for Montana's capital, Squashtown and Pumpkinville, soon dropped in favor of Helena. But Pumpkinville, owing to local production, was the real first name of Phoenix, Arizona, whose new name was linked to the renaissance of an old Indian canal system. Terminus, owing to its railroad origins,[18] became Marthasville (named for Governor Milledge's daughter) and then Atlanta. Pig's Eye became St. Paul, thanks to the prescience of Father Lucien Galtier, who had been sent there by the Roman Catholic bishop of Dubuque to evangelize the newly settled land near Fort Snelling. Galtier founded the chapel of St. Paul in 1840 (because he came from St. Peter, nowadays Mendota). The spot thus became St. Paul's Landing and then St. Paul. As a later wit wrote,

Pig's Eye, converted thou shalt be, like SAUL.
Arise, and be, henceforth, SAINT PAUL.

(Lass 1998, 100)

Politics was a second cause of name modification, as in seventeenth-century Maryland (Brugger 1988, 41). After the change of the British royal family with the coming of William and Mary of Orange, a Protestant Association was created in Maryland that overturned the Catholic government. Maryland was created as a proprietary colony by a liberal Catholic of the Lords Baltimore in 1624. When it became a royal province in 1689, Maryland also became a Protestant province. The name of the new capital, initially Proctor's Landing (1694) and then Ann Arundel Town (the family name of Lord Baltimore's spouse, scion of another prominent Catholic family, the Earls of Arundel), could not be retained. It was renamed Annapolis in 1695, no more honoring Lady Baltimore, but Princess Anne, her name being "grecianized," who was soon to become queen (Reps 1969, 133). A similar transformation happened to Fort Nashborough, named after brigadier general Francis Nash of North Carolina, a revolutionary soldier who was wounded at the battle of Germantown, October 4, 1777, and died three days later. But, as Nashborough sounded too English, it was changed in 1784 to Nashville, "perhaps in deference to French influence and the first settler of Nashville, Timothy Demonbreun" (McRaven 1949, 8, 15).

All states did not display such revolutionary zeal. Virginia did not change the name of its capital, Richmond, although it had been named for the English town of Richmond-upon-Thames, home to the Kew Gardens and the former royal Kew and Richmond Palaces. The same was true for Dover, an English port opposite French Calais, and also for the most important of all state capitals, Boston, named in 1630 after Boston, Lincolnshire, the hometown of the Massachusetts Bay Colony's first governor, John Winthrop, and its first deputy-governor, Thomas Dudley. Likewise, Albany and Hartford kept their British names. People who in 1636 seceded from Massachusetts on religious grounds had founded Connecticut. They migrated from Newtowne, Watertown, and Dorchester, and gave their new settlements the same names. But the next year the court ordered them to change the names, and Newtowne became Hartford Towne (Osborn 1925, 103). It was named for Hertford, England, the birthplace of the Reverend Samuel Stone, an assistant to Reverend Hooker, who led the seceders. The best example of the denominational continuity between the old regime and the new one is provided by North Carolina, where, although the War of Independence had only recently ended, the new capital was christened Raleigh. "Cittie of Ralegh" was the name the English explorer sir Walter Raleigh had given in 1587 to the future capital of his (never built) colony (Powell 1989, 212). Except for this last example, the reason behind such continuity probably lies in the fact that the original reference was unknown to the new settlers and to most of the old ones. The people and towns that gave their names to the capitals had long been forgotten.

Other changes of name were more commonplace. Concord symbolizes the spirit of compromise, but not the one needed for its choice as capital. It was incorporated in 1733 by Massachusetts under the name of Rumford, to honor the Countess of Rumford, who lived there. Disputes between Massachusetts and New Hampshire about their boundaries resulted in overlapping land grants. Because the major part of the town of Rumford had been included in the new township of Bow, numerous conflicts arose between both towns that hindered the conduct of their affairs. Finally, on May 25, 1765, an act of incorporation was obtained. A new parish town was then created out of one part of the town of Bow and given the name of Concord. Tradition has it that the name expressed "the entire unanimity in purpose and action which had characterized the inhabitants of Rumford during the period of their controversy with the proprietors of Bow, and, indeed, from the first settlement of Penacook" (Bouton 1856, 242). In 1774, the inhabitants of Bow asked for land in Maine and created there the new town of Bow, thus definitively ending the controversy between Bow and Rumford.

A name, however, does not alone make a capital. First, because a beautiful or a grand name was not necessarily an asset: neither St. Peter in Minnesota nor Fayetteville in North Carolina ever succeeded in becoming the capital. Second, because the founders or leaders of state capitals were not alone in wanting grand names for their cities. Table 2.3 indicates all the places listed by the census that have the same names as current state capitals (sometimes with an adjunction as East, West, St., or Fort).[19]

Only five capitals have the privilege of unique names (a Native American one for four of them; a French one for Baton Rouge). Salt Lake City has three occurrences, including North and South Salt Lake City, but all are in Utah. Cheyenne, Olympia, Pierre, and Raleigh are the only places with those names that qualify as cities. Salem is the most common name, a powerful

Table 2.3 Number of places bearing capitals' names in the United States

Name	Number	Name	Number	Name	Number
Montgomery	17	Baton Rouge	1	Oklahoma City	1
Juneau	2	Augusta	12	Columbus	19
Phoenix	5	Annapolis	2	Salem	33
Little Rock	4	Boston	10	Harrisburg	9
Sacramento	4	Lansing	7	Providence	11
Denver	6	St. Paul	16	Columbia	20
Hartford	18	Jackson	21	Pierre	5
Dover	21	Jefferson City	28	Nashville	9
Tallahassee	1	Helena	17	Austin	6
Atlanta	10	Lincoln	24	Salt Lake City	3
Honolulu	1	Carson City	9	Montpelier	6
Boise City	2	Concord	14	Richmond	24
Springfield	23	Trenton	14	Olympia	3
Indianapolis	1	Santa Fe	5	Charleston	13
Des Moines	4	Albany	20	Madison	29
Topeka	3	Raleigh	4	Cheyenne	3
Frankfort	11	Bismarck	3		

Source: US Census Bureau, 2000.

reminder of the Puritan trend of the United States. As for presidents, Madison (29) surprisingly overtakes Jefferson (28), Lincoln (24), and Jackson (21). One reason may be that Jackson and Lincoln became presidents, and therefore famous, when the process of urbanization was already seriously under way and many places had already been named. While a name does not make a capital, it helps to emphasize its status, as does a specific plat.

Planning the Capital: Another Way to Enhance the Capital Status

The examples of Annapolis, Williamsburg, Savannah, Washington, and many of the 19th-century planned state capital cities remind us that public initiative and investment [in] the planning of cities once served to create an urban environment superior in quality to that of the present when measured against available financial and intellectual resources.
(REPS 1969, 429)

Studying the plan of a capital is more than just studying the physical fabric of the city. It is also analyzing a framework that more or less directly commands the distribution of built and undeveloped areas and, what is more important, the distribution of public and private areas (Roncayolo 1990, 94). Plans therefore also express power relations (Rotenberg and McDonogh 1993, 196–214). According to Debarbieux's definition of places of condensation, urbanism is conceived so as to render a visit to the capital edifying (Debarbieux 1995).

Foreigners perceive American cities mostly as dull grids, witnesses to a lack of imagination and the absence of real town planning. This view is misleading. Indeed, John Reps has argued that town planning techniques perfected throughout the centuries in Europe were seldom known in colonial America before Major L'Enfant's plan for Washington in 1791. "The strangeness of the environment, the slowness of communications, the absence of traditions, the lack of institutional patterns, and the necessity to create anew even the most elementary of urban services and facilities—all contributed to prevent the speedy and complete transfer to the New World of what had been learned about city planning in the old" (Reps 1969, 423–24). Reps contended that American cities were dull because town planning, which had been a community enterprise, had passed to individuals and corporations seeking private profit, and also because the grid was the "most economical to survey, quickest to build, and easiest to understand"

(Reps 1969, 426–27).[20] The grid might be dull, but it is a testimony to "the openness and democracy of the American city, its accessibility to all" (Elazar 1987, xii). Stanley Schultz went further when he stated that Americans, like Europeans, wanted to settle in a lively urban environment. That meant, for the town jobbers, to give "some evidence that the promised settlement had been established or at least carefully planned. . . . In most cases of community boosterism, therefore, some sort of city design preceded land sales and construction" (Schultz 1989, 12). Among the most unusual plats were those of Circleville, Ohio, in the 1830s, and Octagon City, Kansas, twenty years later. According to Schultz, America was seized during and after the 1830s by a "town-designing and city-making mania."

In such a context, capitals were often planned even more carefully. Ohio illustrates the importance of the plan in the choice of a capital. In 1810, the legislature designated a commission to select a new capital near the center of the state. The centrally located city of Franklinton seemed a natural choice: as the county seat, it had become an important trading place. But the five commissioners discarded it. Among the reasons were Franklinton's low situation on the Scioto River—prone to floods—and its plan, which was deemed "objectionable" (Studer 1873, 14). The site that was eventually chosen two years later was located nearly opposite Franklinton, but on the high bank. It was to be called Columbus.

One of the major reasons behind such care is that capitals have been planned under public initiative and investment. The state often acquired the entire site of the new capital, or already owned it; this was true for Columbus as well as for Columbia, South Carolina; Jefferson City, Missouri; Lincoln, Nebraska; Raleigh, North Carolina; Tallahassee, Florida; Jackson, Mississippi; Austin, Texas; and Indianapolis, Indiana. Such a process derived directly from the procedure used for the national capital, which itself followed a pattern already used in colonial Virginia and Maryland for capitals, county seats, or port towns, such as Williamsburg, Annapolis, and Norfolk (Reps 1969, 307–8). The money to build the city would come from the sale of the town's lots, a pattern largely used for the colonization of the American West, the only (and major) difference being that the proceeds were used to build public buildings and parks and not a speculator's fortune.

Capitals have thus set the pace for urban planning. Colonial capitals' plats already reflected an urban philosophy, stemming from a baroque influence that viewed the city as a seat of power and culture (Fries 1977, xvi–xvii).[21] In the late 1630s, New Haven's plat—nine squares of equal dimensions—strictly followed the Old Testament's urban descriptions (Fries 1977, 66–67), as Boston was to embody the Puritan ideal of "the citty upon a hill" described by its first governor, John Winthrop. A city was a religious utopia with a covenant, as was the case between mankind and God (Lingeman 1980, 26). Another early example of careful planning is Philadelphia, the plat of which was later copied by other new capitals like Lancaster and Tallahassee. The founder-proprietor of Pennsylvania, William Penn (1644–1718), laid out Philadelphia using the latest developments in town planning, perfecting it in the city's rectangular gridiron with four symmetrical parks (Cochran 1978, 6). The grid was witness to the equality between all the heads of families that formed the basis of Philadelphia's citizenry. The parks were public spaces. But, as was the case more than fifty years earlier in Boston, his dreams of a "greene countrie towne" did not materialize; they gave way to a compact city, even if the basic plan survives. Later, Governor Francis Nicholson built Annapolis according to Wren's London. Annapolis's plan introduced "a new concept of civic design in colonial America," with two circles, a large square, and radiating diagonal streets (Reps 1969, 134). Nicholson later designed Williamsburg in 1705, putting the same great "emphasis upon a celebrative civic aesthetic" (Fries 1977, 30). Here too, "civic" meant that the plan provided not only for private needs but also for public ones.

A new phase was reached after the Revolution. A third dimension was added to the planning process with the inclusion of public buildings. Virginia's 1779 act to move the capital to Richmond provided for the first time for the inclusion of the three branches of government into a plat, as well as for the planning of specific government buildings (Reps 1972, 270). There had been earlier attempts, as in 1758 for George City, which was to be the colonial capital of North Carolina, in which Union Square in the center was reserved for the statehouse. Although it was never built, William Christmas, the surveyor who laid out North Carolina's new capital, Raleigh in 1792, followed George City's plat precisely. Besides Union Square, "four main streets, each 99 feet wide, led off in the four directions. All other streets were 66 feet wide, and a 4-acre square was reserved in each of the four quarters of the new city—an excellent example of city planning, be it 1758 or 1792" (Powell 1989, 212). This plat was rather close to Philadelphia's.

This example underlines that in the case of new towns, even if the grid had been selected, it was often altered to enhance the capital's status.

Columbus, Ohio's 1812 layout was a grid, but with some wider streets: 100 feet for High Street, 120 feet for Broad Street, versus the 82.5 feet for the other major streets and 33 feet for alleys (Condon 1977, 12). In 1839, Austin, the new national capital of Texas, was also platted with one 120-foot-wide avenue from the Colorado River to the capitol and a perpendicular avenue from the academy to the university. Fifteen years later, in 1854, Omaha, Nebraska, was platted by a private company intending to make it Nebraska's territorial capital. The surveyor therefore laid out 320 blocks, with 100-foot-wide streets, except for Capitol Avenue, which was 120 feet wide, as in Columbus and Austin (Olson and Naugle 1997, 80). The same state's permanent capital, Lincoln, also provided for streets of extra width that terminated at the capitol, the university, and the park. But in the case of Columbia, South Carolina's new capital, created in 1796, the unusually wide streets (up to 150 feet wide) were platted not for greatness but in order to prevent the spread of epidemics in this fairly humid state. Despite this, Colonel Thomas Taylor, whose plantation had been selected for the site of the new capital, reputedly commented, "They spoiled a damned fine plantation to make a damned poor town" (WWP South Carolina 1941, 215).

In planning a new capital city, grandness went with the search for beauty. This is why platters of new capitals often departed from the plain grid. To attain beauty, there were existing examples, such as Washington, DC, the grand design of which inspired the designers of some new capital cities.[22] Milledgeville was specifically founded to serve as Georgia's capital in 1803 (until 1868). The plat drew its inspiration from Savannah and Washington and included four public squares with streets laid out in checkerboard fashion. Likewise, Judge Woodward, having visited Washington, wanted to rebuild Detroit in 1807 according to L'Enfant's plan, with avenues two hundred feet wide radiating from parks.[23] But local opposition in 1817 resulted in the undoing of that great plan by narrowing the streets to 66 feet and cutting off others. Today's Detroit presents the common grid, with few exceptions, as for Grand Circus Park and the streets radiating out from it (Dunbar and May 1995, 114). Such a plan was perhaps too grand for a town of nine hundred inhabitants. This was not the case in Indiana. In January 1821, Indiana's legislature, after ratifying the selection of the capital's site, elected three commissioners to lay out a town on the site and an agent for the sale of the lots. They were to proceed "on such plan as they may conceive will be advantageous to the state and to the prosperity of said town, having specially in view the health, utility and beauty of the place" (quoted in Dunn 1910, 26). Although utility here comes before beauty, the latter was by no means forgotten. Indianapolis was thus laid out according to Washington's plan. The fact that one of the two town platters, Alexander Ralston, had assisted in surveying Washington certainly helped. Describing that first plan in 1910, Jacob Dunn was proud of its French inspiration, emphasizing that Indianapolis was, like the federal capital, modeled on the French royal city of Versailles (Dunn 1910, 29).[24] But this was not entirely true, as Indianapolis's plat combined the square and spider-web ideas, covering one square mile divided into one hundred squares. In the center was placed a circle, surrounded by a street 80 feet in width, designed for the governor's residence—though it is now Monument Place.[25] From the circle, four diagonal streets (ninety feet wide except for Washington Street, which was 120 feet wide) ran to the four corners of the plat, each of which cut four of the primary squares into two triangles. Pogue's Run, the river that ran through the southwestern part of the city, caused the only departure from the regularity of the plan.

But Washington only influenced a handful of capitals, unlike its capitol building. Local conditions were also important. They were topographical for Madison. The unusual plan of Wisconsin's capital (1836) enhanced its location on an isthmus between two lakes, Mendota and Winona. The capitol is located on the highest ground; from the square before it, four streets radiate, with the university at the end of one of them. Science and politics were clearly linked from the beginning in the pattern of the city, following Jefferson's ideal of learned democracy. In Washington Territory local conditions stemmed from the history of the founders. In 1850, when one of Olympia's founders, Edmund Sylvester platted the town site he had claimed in 1846 with Levi Lathrop Smith, he remembered his youth in Maine. He platted a New England–style town with a town square, tree-lined streets, land for schools, a Masonic Hall, and capitol grounds. For Topeka, Kansas, local conditions were more political. Long before the official designation of the capital, a private body—the Topeka Association, founded in 1855—had thought of a "capital" plan, amid the slave-state/free-state controversy. It provided for broad streets (up to 130 feet wide), and for two eighty-acre parks in the heart of the future city, one for an educational institution and the other for the capitol that was presented to the state in 1862 by the association (Adams 1903–4, 348).[26] A clear intention was thus at work to reflect the status of a capital in the town platting. But such an intention was not universal, especially for towns that became capitals after their foundation.

The Persistence of Plain and Unusual Plats

When Vandalia was designated as the new capital of Illinois in 1819, Ferdinand Ernst, a wealthy German emigrant, gave a precise description showing that uniformity was the motto: "The plan of the town is a square subdivided into 64 squares, and the space of two of these squares in the middle is intended for public use. Every square, having eight building lots, contains 320 square rods; each building lot is 80 feet wide and 152 feet deep. Each square is cut from south to north by a 16-foot alley; and the large, regular and straight streets, 80 feet wide, intersect each other at right angles" (quoted in Burtschi 1954, 24).

The plain grid was later evident in railroad cities, such as Bismarck, where the Northern Pacific Railroad organized the grid on both sides of the railroad in 1873. A mushroom town like Oklahoma City followed the same pattern. These three cities—like many others, such as Phoenix—only became capitals later. This explains why their founders had not thought of any specific planning requirements outside the grid. In some cases there was not even a proper grid. In St. Paul, Minnesota, the street pattern beyond the business district looks like a "crazy quilt" because landowners had no legal framework to follow (Reps 1998, 72). Pierre, South Dakota, though it is one of the smallest capitals, is very spread out and lacks a clear plat. This is the result of rivalry between speculators: one group was located north of the railroad, on the highest point, and the other preferred the south, so that the town expanded between the two areas (WWP South Dakota 1952, 129). Likewise, the plat of Denver, Colorado, lacks a central focus, parks, and civic spaces, being the result of five initial competing townsite companies (Reps 1981, 61).

Oklahoma City is the best example of fancy platting. Two factions each platted their own town, with amusing results. The first faction, the Seminole Land and Improvement Company of Topeka, Kansas, were secretly at work on their preferred spot before the official opening of Oklahoma Territory to settlers, at noon on April 22, 1889, and before the arrival by train of the official settlers. The latter organized themselves in the Kickapoos party, dedicated to law and order, in opposition to the gamblers and speculators of the other party.[27] After a fistfight that failed to produce a solution, each group decided to plat its own town. "The Seminole laid out their town in relation to the railroad track, which pointed slightly to the northeast. The Kickapoos laid out their town in relation to due north. Consequently, the main streets that were supposed to connect the two groups failed to meet.

(Even today, Oklahoma City motorists have to contend with a jog in the streets which connect the north and south)" (Thompson 1986, 58–59).[28]

The Evolution of the Capitals' Plats

Once platted, a capital was far from complete. Subsequent growth and annexations often required cities to rethink the first plat. Planning was therefore still ongoing in most capitals during the twentieth century—not always for the better. Cheyenne, Wyoming's first plat was plain enough, following the Union Pacific Railroad tracks that made the streets diagonal to the main compass points. However, as the twentieth-century additions to the town were platted with regular compass directions, "where these additions join the original townsite, streets are a maze of pointed intersections, short courts, and blind avenues" (WWP Wyoming 1941, 184). The architectural history of each state capital often shows that there were several attempts to create a beautiful capital at the beginning of the twentieth century. Hartford was the first city to benefit from a permanent city planning commission, in March 1907, and city planning was based on the work of Nolen in Little Rock (Scott 1971, 78–80). Commentators insisted on the aesthetic side of the movement, but beauty also had an economic value in the minds of the "City Beautiful" proponents: "Beauty has always paid better than any other commodity," remarked Daniel Burnham (quoted in Hancock 1967, 293). Harrisburg is a perfect example of this trend. Among the reasons for beautifying the city was Philadelphia's attempt to recover the capitol. Proposing a handsome capital with all the modern amenities was a way to ward off Philadelphia's claim. Economics and aesthetics went hand in hand, as the movement was launched by the town's economic elite in 1901, who set up a private committee that welcomed city officials and experts. The population approved the process (Wilson 1980, 213). Indianapolis took also part in the City Beautiful movement (Bodenhamer and Barrows 1994, 433). In 1908, the Board of Parks hired George Kessler, a landscape architect, who designed a system of boulevards linking the city's main parks that was largely built during the following decade. In 1921, the Indianapolis City Planning Commission was created. In the same decade, the huge Indiana World War Memorial Plaza was built north of the central business district.[29]

"City Beautiful" ideas were prompted by the building of a new capitol in Providence, Rhode Island. In his January 1890 message, the governor asked for a new capitol, the old one being too small. A committee was then formed, including business leaders, to select "a prominent architect firm to build a high-quality, highly visible capitol" (Conley et al. 1988, 66).[30] The committee introduced city planning to Providence through the selection of the capitol site.[31] The color of the capitol, white, is a direct reflection of Chicago's White City,[32] with which the firm of McKim, Mead, and White had been deeply involved. At the same time, New Union Station was built. But the beautification spirit was soon lost. In fact, when a City Plan Commission for the City of Providence was created in 1913, it was expected to prepare "a comprehensive plan for the systematic and harmonious development of the city, based primarily on practical utility and the public convenience and health" (*Annual Reports* 1921, 3). Several highways and parks were planned, as well as zoning regulations, in 1918. The principle that form follows function is well illustrated in the following statement: "The use of public funds solely for beautification of the city is not sought by this commission" (*Annual Reports* 1932, 16). Indeed, Exchange Place Mall, called "the most notable City Plan achievement in Providence," was highly practical. Today called Kennedy Plaza, it is a vast, open-air bus station between the town hall and the federal court.

Natural disasters sometimes prompted action. In Columbus, Ohio, after the great flood of 1913, the levees and bridges that were built to prevent future disasters served as the starting point of the city's civic center (Bureau of Business Research 1966, 14). Other cities prepared City Beautiful plans that could not be translated into reality. This was the case in Lansing. The capital of Michigan had commissioned an urban engineer from St. Louis, Harland Bartholomew, to develop a "plan for the Lansing of the future." But the Depression, World War II, and poor public finances brought that plan to a halt.

A second period of urban planning began after World War II. Faced with urban blight, central cities began a renewal movement that never really stopped. It touched the largest capitals (Denver, Honolulu) as well as smaller ones (Helena), and altered the initial plans. In Indianapolis, "modernity" (or barbarism?) caused some changes in the city's grand design. The plat's diagonals, once active business streets that shortened the distance to the city center (Dunn 1910, 29), have partly disappeared. A current map of Indianapolis shows that Washington Street has become an intraurban motorway (Interstate 40) and that the river has been covered to accommodate railroad tracks. Only two diagonals remain intact (Indiana and Virginia Avenues, from northwest to southeast), whereas Massachusetts and Kentucky Avenues have been cut near the center in favor of a rigid grid. The state capitol has also been slightly moved (1878–88), and occupies one block toward the north. Likewise, Richmond, Virginia, has suffered aesthetically from "modernization" processes since 1865: "Thoughtless additions of buildings in and encroachments to Capitol Square have seriously detracted from its beauty. Equally disastrous, tall office and business buildings in the lower portions of the town have now almost completely obliterated distant views of the Capitol" (Reps 1972, 281). This does not mean that capitols have ceased being distinctive buildings. During the last century, despite very visible modernization, the symbolism of the capitol always proved strong enough to prevent its demolition and replacement by a new building. There have been new constructions, but the beloved old capitol has always been preserved.

The Capitol and Political Monumentality

> Citizens who are proud of their state's unique locality and heritage in a vast and varied country seek tangible ways to affirm and validate their pride. The consequence is often building policies whose underlying aim is to create the "best" statehouse in the nation—or, at a minimum, a capitol equal to those of rival states.
>
> (GOODSELL 2001, 4)

Statehouses—or capitols—embody the three meanings of Debarbieux's places of memory: they act as stereotypes, they materialize democracy, they "narrate its history and anchor its values" (Debarbieux 1995, 110). Indeed, a capital is first a capitol. The first official building bearing the name of *capitol* was constructed in Williamsburg, Virginia, in 1701–1705, authorized by Governor Nicholson.[33] Capitols are places of symbolic condensation, as they represent a certain conception of the nation—federalism in this case. They are now American symbols and are considered as America's contribution to the world's monumental architecture, along with skyscrapers (Hitchcock and Seale 1976). Some analysts even liken the capitol to a civic church or a temple of democracy. As was the case for medieval cathedrals, the construction of the capitol was often the most expensive public project of the state.[34] While the construction of cathedrals amounted to a religious ritual, the erection of capitols was "a founding ritual of republican governance" (Goodsell 2001, 84). It is therefore quite natural for citizens to display great pride in "their" capitol—paid for with their money. Jacob H. Studer wrote as follows about Columbus, Ohio, in 1873: "The object that first strikes the eye of the visitor on entering the heart of the city is the Capitol, or, in familiar language, the State House. No other building of the kin on the continent rivals it in size, except the National Capitol in Washington" (Studer 1873).

Capitols also symbolize the state. In 1819, New Hampshire citizens were greatly pleased with the completion of the first capitol the state had ever built, seeing that building as testimony to the highest political unity ever attained in that state (Turner 1983, 200–201). Another good example of the powerful symbolic status of a capitol is evident in the wide array of items that were deposited inside the cornerstone of the new capitol in Kansas. On October 17, 1866, an imposing procession marched to the capitol grounds and deposited, besides the Holy Bible, business cards, "postal stamps and revenue stamps of all descriptions, . . . [and] 'one-piece sheet music' contributed by the Topeka Brass Band."[35] Everything is present—at least everything official—that makes a state: faith, laws, civil servants, newsmen, fraternities (Masons and veterans), businessmen, and even culture and education. "Mechanics" and "citizens" were among the three thousand people in the procession (almost Topeka's entire population). Such pride can still be felt one century later. Robert Richmond remarked that the capitol dome of Kansas was higher than Washington's and concluded with the following proud assertion: "On December 28, 1900, the *Topeka Mail and Breeze* carried a brief letter from a Kansan visiting in the nation's capital, which said in part: 'but Kansas legislators can congratulate themselves on meeting in neater and more elegant quarters than congress.' The words of *the Mail and Breeze,* in comment on the letter, might well apply in the 1970s—'A Kansan Need Not Blush for His State House'" (Richmond 1972, 266–67).

The case of Kansas also shows that what the population deemed worthy of being symbolized did not always include the whole citizenry of the state. Although it was built just after the Civil War, discussion of the capitol does not acknowledge former slaves, who were probably among the "citizens" who attended the ceremony. The same was true for Rhode Island. At the opening of the 1901 legislative session, the governor's message expressed great pride in the new statehouse, but he spoke mostly from aesthetic and practical perspectives: "Meeting in our splendid State House for the first time, on the birthday of the century, our New Year greetings one to another seem more than commonly cheerful and happily inspired. The noble proportions and dignified lines of this monumental building express the sovereignty of the State, while its harmony, beauty, and significance of architecture comprehend also abundant accommodation and ample facilities for the exercise of the functions of government and the performance of the public business in its several departments" (State of Rhode Island 1901, 3).

In reality, Rhode Island's legislators in the 1890s wanted, by building a new statehouse, to maintain the social status quo in a period of major economic and social changes, during the biggest increase in population that Providence would ever know. The growth stemmed from industrialization, and thousands of Italian immigrants arrived who challenged the old Protestant stock:

> The State House project was one significantly involved with the imaging of state government as a newly centralized institution to serve as mediator and ultimate authority in all socio-cultural as well as legal matters. With this end implicitly in mind, the State House project was fashioned as a sort of "reclamation" project of civic space by a reconsolidation of older political forces (namely, State ones) to challenge a newer near-majority, the diverse ethnic and religious newcomers. The carefully resurrected historical persona of Roger Williams was made to figure prominently in the State's project, seeming to encourage both the building of the State House and an assertion through that building of a visibly and functionally revivified social status quo. (University of Rhode Island 1996, 9)

In fact, the old status quo—that is, the power of the Yankee Protestants—was maintained though electoral fraud by Nelson Wilmarth Aldrich, a Rhode Island Republican senator who controlled the state at the time (McLoughlin 1978, 148–49). Naturalized citizens had to wait until 1928 to get full political equality.

A Symbolic Architecture

The symbolism of capitols is apparent in their very architecture. At the beginning, it varied greatly: during colonial or territorial times, county seats, churches, and even log houses were used. When specific buildings began to be erected, their architecture exhibited a variety of influences. The dome of the statehouse of Annapolis, Maryland, constructed between 1769 and 1788, is supposed to have been inspired by the tower of Karl-Wilhelm's palace in Karlsruhe.[36] This German influence came from the area's numerous German immigrants, one of whom might have brought a drawing of the palace (Gordon and Gordon 1972, 294–97).

But since the end of the eighteenth century, a common pattern has emerged, based on classical majesty. Capitols were "consciously designed to proclaim national distinctiveness" (Meinig 1986, 437).[37] Until then, the official buildings had mostly been of Georgian inspiration—red bricks and a cupola. They now seem rather small, although they seemed quite large to the relatively small American population—less than four million people in 1790. Although each capitol is unique in some respects, almost all are grand, domed buildings with a classic temple front. The seminal capitol is Boston's New State House (designed by Charles Bullfinch, 1795–98), and for newer postwar buildings, the standard is its Washingtonian avatar, along with City Hall in New York City. Wilbur Zelinski described the nation's capitol in 1988 as "the symbolic anchor of the American Union" and the "transcendent national icon" (quoted in Robertson 1996, 21). Reconstructed between 1814 and 1829 by Latrobe and Bullfinch (the dome was rebuilt in 1863 by Walter), it is symbolically the first one, although its architecture was borrowed from earlier buildings (fig. 2.1).

Frankfort, Kentucky; Des Moines, Iowa; and Concord, New Hampshire (for its 1864 enlargement) preferred perhaps the Hôtel des Invalides in Paris, but the very idea of a domed neoclassical building came directly from Washington. The other possible model, Jefferson's design for Richmond's capitol in Virginia, based on the Gallo-Roman Maison Carrée of Nîmes, France, remained unique. It nevertheless set the standard for buildings based on direct Greco-Roman influence. In fact, current capitols are

FIGURE 2.1 The national capitol, Washington, DC, the first model. Rebuilt according to classical tradition (1814–63), symbolizing Greek democracy and the American union, this impressive domed building has become the standard followed by most states when building their definitive statehouses after the Civil War. Photo by Christian Montès, April 2010.

not—with very few exceptions—the original ones, but their second, third or fourth incarnations. As they were mostly built after 1865 (most of them between 1880 and 1910), the influence of the Washington model was strong. The relative similarity of the capitols built during that period was enhanced by the links among their various architects, who had received similar training and sometimes worked together. The successful architectural firm of McKim, Mead, and White later adapted the model. They recreated the Washingtonian design for the capitol of Rhode Island (1904, fig. 2.2). Considered a highly successful early expression of American classicism, it succeeded the national capitol as a model (University of Rhode Island 1996, 7). Among the numerous capitols based on its design are Minnesota's (the architect, Cass Gilbert, had been a draftsman for McKim, Mead, and White) and Mississippi's. McKim, Mead, and White also worked on enlarging the capitols of Virginia, Florida, and Alabama. Other capitols followed suit in Arkansas, Kentucky, Wisconsin, Washington, Montana, South Dakota,[38] Pennsylvania, Idaho, and Utah.

The choice of the classical design—instead of any vernacular new one—was influenced by romanticism and independence: "The recent surge of interest in things Roman or Greek, the Greek struggle for independence from Turkey, and the new[ly] acquired autonomic rule for our own country from Great Britain, a 54-year-old heritage in 1837, all combined to make the Greek Revival an appropriate form of architecture" (Anderson 1974, 7). Standing capitols were even "actualized" (i.e., transformed according to contemporary architectural fashion). In Raleigh, North Carolina, "branded ugly and 'misshapen,' the capitol was not an architecturally imposing 'edifice' by any standards and would have to be glamorized with a new dome, false porticos, and layer of stucco by 1822" (Waug 1967, 5). Greek architecture also symbolizes democracy, as ex-Governor Ladd, president of the Board of State House Commissioners, emphasized in his opening address at the laying of the cornerstone for Rhode Island's new statehouse: "In going back to the stately and monumental style of the Greek[s] for our design, we are but adopting the model that our fathers chose for their system of government" (University of Rhode Island 1996, 19). The idea of democracy was already influential when Jefferson designed Richmond's capitol. Jefferson thus influenced the political significance of capitols more than their architecture. "Marcus Whiffen has suggested that the very term "capitol" invoked the Temple of Jupiter Capitolinus and thus signified a link to the civic life of ancient Rome. Standing on the summit of Shockoe Hill, Jefferson's capitol bore a striking resemblance to the Temple of Jupiter, which

FIGURE 2.2 An American classicist capitol: Providence, Rhode Island. Built between 1895 and 1904 with white Georgia marble, this statehouse is modeled on the national one and conveys the democratic and industrial pride of the leading community (therefore excluding minorities), focusing on the figure of Roger Williams. Photo by Christian Montès, July 2010.

looked down on Rome from atop the Capitoline Hill and was itself visible from many places within the city. There can be little doubt that Jefferson appreciated this similarity and all it suggested about the relevance of ancient tradition to American life" (Wenger 1993, 90).[39]

Exceptions do exist to the classical style, as in Albany, New York, where a

Second Empire style was preferred (i.e., neo-Gothic-Italian-Renaissance!), or Hartford, Connecticut, where Victorian Gothic was chosen. The post–World War I period produced only a few capitols. Three of them followed for the first time an "American style": Baton Rouge, Louisiana (the first capitol, built in 1882, was a mock-Gothic castle); Lincoln, Nebraska, and Bismarck, North Dakota. They all exhibit a 1920s–1930s skyscraper style that unites both of Hitchcock and Seale's "American contributions to world architecture," skyscrapers and capitols, in another example of the dual scale (nation and state) of condensation present in state capitals. Although they do not have a dome, they continue the tradition of the capitol as the tallest building in town. At 450 feet, Baton Rouge's capitol is the tallest in the nation. Nebraska's new capitol, built from 1920 through 1935 for just less than $10 million, was aptly nicknamed the "Tower of the Plains." According to McKee and Duerschner, the building is "considered the fourth architectural wonder of the world" (1976, 78). Another critic described it more modestly as "the foremost example of "Modernism" in the United States" (McCready 1974, 325). Even foreign tourist guides are in awe; the *Guide Bleu* (the oldest collection of tourist guides in France) states that "the strength of its building places it at the same rank as the Parthenon and Chartres's cathedral" (Collective authors 1989, 439). Likewise, Bismarck's capitol, almost 242 feet tall, is nicknamed "the skyscraper on the Prairie." It was built at the low cost of just under $2 million after a fire destroyed the old capitol in 1930. Due to the economic crisis of the 1930s, all exterior embellishments had to be abandoned, linking its Art Deco roots with the International style. Alaska was even more modest in designing its first real capitol, completed in 1931. It was a rectangular, five-storied building of brick-faced reinforced concrete. The only departure from the appearance of any other office building was a marble portico with four columns.[40] But old traditions were deep-rooted. When Delaware replaced its old statehouse at the beginning of the 1930s, it chose to go back to the Georgian model. Likewise, when the capitol in Charleston, West Virginia, burned in 1921, the architect chosen to replace it was Cass Gilbert, who gave the new capitol a neoclassical design and built it for slightly less than $10 million. Its dome is 293 feet high, 5 feet higher than that of the national capitol, making it the largest dome in the country. Likewise, when Oregon's capitol, modeled on Washington's, burned in 1935, the one built to replace it, completed in 1938 for only $2.5 million, was a four-story "modern Greek" building (fig. 2.3) that included a rotunda.

These highly visible domes are meant to convey the greatness of the

FIGURE 2.3 Salem, Oregon's "modern Greek" capitol. The design of the new capitol shows the permanence of the Greek ideal, while taking into account Art Deco modernism. Photo by Christian Montès, August 2010.

state: they are sometimes gilded, as in Georgia,[41] Colorado, and New Jersey; coppered, as in Arizona[42] and Utah (both nicknamed the Copper State), and in Montana (also a big copper producer); or silvered, as in Nevada (the Silver State). Denver's capitol has a 24-carat gilded dome, but the capitol's real treasure lies in its inside wainscoting, which used the entire world's supply of Colorado onyx, quarried in the state near Beulah.

Height races were not unheard of: Providence changed the blueprints of its capitol in 1899, in order to surpass St. Paul's, which was then under construction. Olympia, Washington, boasts the fourth biggest dome in the world. The capitol of Columbus, Ohio, underlines the importance of a glorious topping. It was built between 1839 and 1861, and cost $1,350,000, a huge sum at the time. Although it is locally said to be superseded in the United States only by the national capitol (which is not true), a controversy arose about its truncated dome. Some deemed it classy, but others compared it unfavorably to the Monitor, the first ironclad navy vessel, and to a mammoth cheesebox, according to the *Ohio State Journal* in 1882. Since then, the state has consistently denied that it was an economy measure (Condon 1977, 21–22). The lack of a dome on Oklahoma City's capitol did indeed result from a lack of money. When completed in 1917, however, it was the largest reinforced concrete building in the world (Robertson 1996, 21), and a dome was finally added in 2002, according to the original design. Money was a problem, because state legislatures usually had to allocate successive additional funding for the construction of their grand capitols. When future president Theodore Roosevelt became governor of the state of New York, he decided to stop all work on the capitol (begun about twenty-five years earlier) because $25 million had already been spent instead of the scheduled $4 million in four years. Visitors to Albany's capitol are therefore surprised to see unfinished carvings amid this otherwise magnificent—and composite—building. New York's legislators should have had the same idea as the ones from Tennessee. In 1844 they passed "An Act to Appoint Commissioners to superintend the construction of the State House, to direct the labor of the Penitentiary to the erection of the same, and thereby save the people of the State from Taxation" (*Acts of Tennessee*, 1843–44, chapter 205, 235). But not every state capital was lucky enough to have been designated as the site of the penitentiary, as Nashville was. Columbus and Concord also used prison labor.

New Mexico presents an exception to the quest for visibility, perhaps because Santa Fe has been the capital since the seventeenth century. Here, symbols of faith—the cathedral and many other churches—long overshadowed the gubernatorial palace (then the capitol, a low but centrally located building), as they still do the current rotund and low capitol built in the 1960s, which represents a sun symbol from the Zia Pueblo.[43] New Jersey's capitol building offers another kind of exception.[44] Amid the quasi-unanimous and overdone praise, we find the following 1939 description:

> The Statehouse is an unsatisfactory composite of additions and alterations, occupying a landscaped plot between State Street and Delaware River. What remains of the original structure, built c. 1792, is now a part of the present building, although exactly what part is uncertain. Subsequent growth has been without regard to any foresighted plan. . . . The ill-lighted main entrance is hung with indistinguishable portraits of early Jersey statesmen and patriots. . . . The second and third floors are labyrinths of gloomy corridors and passageways in and among erratically placed offices. (WWP New Jersey 1939, 404–5)

To summarize the aims and styles found in US capitols, the one in Harrisburg, Pennsylvania, is a good choice. When the capitol had to be rebuilt after an 1897 fire destroyed the gracious, domed, brick colonial-style structure erected in 1822, the city leaders planned a new modern building to celebrate the state's economic prosperity. The Chicago architect, Henry Ives Cobb, chose in 1897 an early American Renaissance design, close to that of the US capitol. But the project was poorly funded (only half a million dollars), and it failed. The Pennsylvanian Joseph Huston (later imprisoned for corruption) completed the project, with a design almost identical to Cobb's. During the 1906 dedication ceremony, President Theodore Roosevelt was deeply impressed by its six hundred opulent rooms (McInerney 1994, 30). The numerous grand comparisons made by the writers of the Works Progress Administration in 1940 help to explain the president's use of superlatives:

> An E-shaped granite structure of Italian Renaissance design, it rests on a basement of squared, dressed stone. A green dome, modeled on that of St. Peter's in Rome, rises to a height of 272 feet and is visible from almost every point in the city. . . . The decorations of the bronze doors depicting historic events in the State's life, recall Ghiberti's doors in the Baptistery at Florence. . . . A marble stairway within the rotunda, designed on that of the Great Opera house in Paris . . . (WWP Pennsylvania 1940, 240)

The image and location of American capitols indeed follow a grand civic vision, even if the planning and construction of these architectural masterpieces sometimes involved corruption, crude lobbying, factionalism, and pork-barrel policies. The contemporary capitol, be it branded a marvel or a horror, symbolizes the capital and performs the "encapsulation of history" (Goodsell 2001, 14) and sometimes also of geography: in Honolulu, Hawaii, the open atrium of the 1969 modern, rectangular capitol was designed to represent the archipelago's volcanoes. These symbolic functions may explain why it seems so difficult to build new capitols, the states' citizens mostly preferring to expand the old ones. New buildings flank the capitols in Phoenix (1960s) and Tallahassee (1978), for state offices as well as for associations and lobbyists.

Nevada built a new capitol near the old one (in the 1970s, enlarged in the 1990s), the architecture of which supposedly retained the old characteristics. Although the simplicity of the old democratic functioning of the state tends to be reduced by those new capitol "complexes," the social and symbolic meanings of the capitol still stand out. However, citizens mostly prefer inconspicuous—even if quite costly—modernization and addition processes, as in the case of Austin or Lansing. They also want the capitol to have a grand location within the capital city. Among the six elements of the American statehouse's "ideal type" stressed by Goodsell, four are linked to architecture: cruciform massing, a central dome, a temple front, and a grand central space. The last two are linked to the local geography of the building: a prominent site and parklike grounds.[45]

Capitols: A Prominent Site

The symbolism and the monumentality of state capitols are associated with their location: they have to be visible:

> Sir, when this building was originally completed, it was doubtless the most prominent and attractive feature of Albany. Dwarfed by his surroundings and depreciated by contrast, it has long since become an offense to the eye and a reproach to the State. . . . The State of New York is rich, and can afford to have a decent Capitol. One that shall fairly meet her wants, and correspond with her rank and power. (quoted in Roseberry 1964, 15)[46]

They were built on hills in eighteen capital cities (e.g., Annapolis, Montgomery, Jackson, Augusta, Springfield (on a slope), Austin, Albany, Denver, Olympia, Nashville, Salt Lake City, Harrisburg, and Hartford; see fig. 2.4).[47] They are also often situated in the center of the city, as in Raleigh, Columbus, Carson City, Concord, Lansing, Pierre, and Madison, and more rarely outside the city (Bismarck).

In the days when rivers were the primary (and sometimes the only)

FIGURE 2.4 A capitol upon the Hill, Hartford, Connecticut. This eclectic building expresses the visible power of democracy in its high and gilded dome and its location atop a hill. Photo by Christian Montès, April 2008.

means of transportation, capitols were usually located not far from a riverfront. Two-thirds of all capitols are so situated. The Augusta, Maine, capitol commission of 1822 selected Weston's Hill. "It was then a conical elevation, higher than at present, east of the then traveled road to Hallowell, and separated from the village of Augusta by a deep ravine across which State street now passes. This was a beautiful spot, of commanding prospect, pleasing to the eye, and every way suitable for the erection of public buildings" (North 1981, 470). The capitol was thus close to the center of the city, but kept some distance, owing to the deep ravine. The hilly location was not always the best one from a purely urban point of view, as it could be windy or less central and therefore given to "lower" land uses. Providence citizens manifested some hindsight when selecting the site of the future capitol at the end of nineteenth century. In fact, since the middle of the century, Smith Hill had been "a strange mixture of a factory district, a prison site, and a shanty town (derisively nicknamed 'Snowtown'). The area was within easy walking distance of downtown Providence, although separated by both the railroad station and track area, and the cove, which was fed by the Woonasquatucket (and largely filled in by the end of the century)" (University of Rhode Island 1996, 24). When the city's Public Park Association published in 1892 a booklet lobbying persuasively and successfully for the Smith's Hill site, it did forecast its potential to host a "magnificent boulevard." As noted earlier, the location and construction of the capitol were often part of an urban beautification and planning process. Even in Salt Lake City, headquarters of the Mormon Church, the capitol long remained the most visible building. According to a 1940 description, "From most approaches the copper-domed State Capitol, rising from the elevated North Bench, is the most conspicuous building in Salt Lake City. South of it, and lower down, is the six-spired Temple of the Church of Jesus Christ of Latter-day Saints, with the seclusion of high-walled Temple Square" (WWP Utah 1941, 226).

Besides dominating the city topographically, the capitol could dominate all aspects of the city's life, even its vegetation, as does the green-domed statehouse of Columbia, South Carolina.

> As it dominates the skyline, so politics dominates Columbia. The legislature meets the second Tuesday in January, and local tradition runs that the Japanese magnolias on the capitol grounds bloom when the assembly convenes and the Magnolia grandiflora blossoms in the late spring before it adjourns. When the solons are in town, hotels are jammed, and the buzz

> of politics in lobbies and cafés as well as in the State House, ends only when the sword of State in the senate and the mace in the house are put away, the palmetto flag of the State is hauled down from the State House, and the lawmakers at last go home. (WWP South Carolina 1941, 213)

As cities grew, however, the early impressive location of the capitol sometimes lost its power. To compare two lithographs of Saint Paul is enlightening (Reps 1994, 299, 303). The first, made in 1856, depicts a small town on the banks of the river; its focus is on the capitol, which is situated on the slope of a hill at the edge of a wood. In the second one (1874), the lithographer had to add a flag on top of the dome to make the capitol stand out among the many church spires. The building is almost hidden behind big commercial houses. Likewise, a contemporary description of Columbia, South Carolina, states that a viewer's eye would be caught by the Palmetto Center, the AT&T tower, and other Main (Richardson) Street structures, "that all but blot out the capitol dome" (Moore 1993, 465). As skyscrapers rose, some capitals, such as Sacramento, tried to retain the majesty of their grand capitols. The California legislature established the Capitol Building and Planning Commission in 1959. It delivered a master plan that expanded the capitol grounds from 69 to 138 acres, created a plaza and malls, and sought "architectural harmony." All new buildings around the forty-acre Capitol Park are limited to seventy-five feet in height, and the ones farther away to sixteen or twenty-six stories (Hansen 1967, 253). Later, the city set up the Capital Area Development Authority, the aim of which is to develop and rehabilitate the downtown area. The city of Lansing, Michigan, set up a Capital City Revitalization Task Force and helped private developers to refurbish more than five hundred residences near the capitol through a HUD loan (Bromley 1990, 17). Providence, Rhode Island, planned to beautify the capitol, situated on a hill, by providing it with a grand approach from the city's center. But the process took time. As the City Plan Commission remarked in 1930, "The State Capitol approach has been awaiting development for over 30 years. In the late nineties the old Cove [basin] was filled in, and the reclaimed area was dedicated to the people for park purposes" (*Annual Reports* 1932, 7). The 1914 plan had not been implemented, but the commission knew exactly what it wanted.

> One fact must not be lost sight of, namely, that at some future time the Union Station may be moved farther to the west and the tracks carried in a tunnel under Capitol Hill. With the removal of the viaducts and the freight yards, there is an opportunity for a continuation of the present open areas as an approach to the State House. It requires little imagination to visual-

> ize the opportunity then offered, a wide esplanade, half a mile long, with the State Capitol at one end, a possible water gate at the other and the war memorial in the center. (*Annual Reports* 1932, 13)

It was only at the end of the 1980s that a new station was built, and only during the 1990s that the old canal and the surrounding area were reclaimed and beautified. Old Union Station now hosts a microbrewery and offices, and artistic walkways grace the canal, where water-fires greet the evening stroller from spring until fall. At the end of a vast open park, the white capitol stands atop the hill. It seems that here the City Beautiful movement was born again.

Capitols convey far more than aesthetics, because they encapsulate the workings of local democracy. In 1925, the executive secretary of the American Civic Association concluded a brief sketch of seven southern capitals with an assessment that still rings true:

> New relationships to the people are being established by State governments. Whether the capital city exists mainly because of its capitalship or whether it is also a commercial metropolis there is every reason for the State to cooperate closely with the city government to develop a capital worthy of the commonwealth. There are also many opportunities for helpful contacts with the people of the State which State officials have somewhat neglected. The people of the States need better organized representation in the interests of their corporate public good during legislative sessions. They need more frequent contacts with the executive officials on questions of public welfare. There is a great opportunity for State officials to develop in the people a justifiable pride in their State Capitol building and grounds and to make them more familiar with the working of their State government. (James 1925–26, 395)

For James, careful planning of the capitol grounds and capital city should clearly express the smooth functioning of the local democracy. But she also insists on the need for cooperation and shared knowledge between citizens of the state and capital and their state government, in order to foster the "public good" and "public welfare." This leads to the question of the tension between public and private spheres in capitals. Capitals are most likely to see such tension, because they enhance the relationships between both spheres, be they good or bad. Since they are the locus of the expression of local democracy, state capitals can be described as fundamentally public places. Our concern now is to see how they are evolving to cope with the powerful contemporary forces of deregulation and privatization,[48] as cash-strapped municipalities increasingly use the privatization of public spaces to try to recover from their current financial woes.[49]

Public and Private Life and Places in State Capitals

They're parasites, they're as much a burden and added expense as they are an attribute.
(LEMOV 1993, 46)

Anthropologists, (micro)sociologists, and historians have emphasized the importance of individuals, who have specific ways to use and think about cities.[50] It is impossible here to render the full richness of urban life in state capitals, because city life involves a huge memory of multiple practices that do not follow the official "rules" set by city halls and theoreticians (Certeau 1980). I discuss the question from three perspectives. The first is social, employing the concept of fragmentation. The second is the political relationship between the two worlds of the capitol and its legislators, and the municipality. The last perspective has been put forward by Michel de Certeau and others: the importance of heritage and spaces of heritage as "archetypal" public places. This approach allows us to see that although capitals are indeed places of condensation, such condensation does not always reach the entire population.

A capital city is both a municipality and the seat of state government. The two do not always go hand in hand: capitals host citizens living a "normal" life in a nation based on freedom and liberal capitalism, as well as the "small world" of politicians and lobbyists concentrated in a part of the city where they tend to stay together.[51] On the one hand, we have the sphere of private concerns and life; on the other, we have the public sphere and places. In a country where the private world is foremost in life and landscapes, capitals are exceptions. In most of the quotations this book offers, capital cities are referred to with the pronoun *she* rather than the more neutral *it* in translating the appropriation of their space by society. Such an appropriation goes through schools, in the study of the state's history and visits to the capitol,[52] as well as through news coverage. The capitol's dome is today used as a symbol of the legislature by television: when journalists release political news, they speak either in front of a photograph of the building or with the real dome in the background. Goodsell rightly stated that "because of this enormous mass exposure, the building's familiar façade is presented for cultural absorption to a degree infinitely greater than seeing it in person could ever generate" (Goodsell 2001, 174). However, the building and the dome are not precisely located in the minds of most viewers: they are pure symbols. This explains why, on a local scale, capital cities are not always spoken of respectfully, as in Lemov's quotation, which refers to state government and was uttered by the mayor of Harrisburg, Stephen Reed. This clearly emphasizes the exasperation that mayors—as well as citizens—can experience in a capital city. There are often two worlds living apart in the same municipal boundaries. Washington, DC, can once again be taken as a bad model. The national capital epitomizes the evils of urban America: high poverty rates as well as crumbling public utilities and services are to be found not far from neighborhoods that rank among the richest in the nation. Slums are not far from the White House. The mayor of Washington could thus share the grievances put forward by the mayors of many state capitals: "The design of state buildings creates barriers within the city; the dominance of state government as an industry fragments civic leadership" (Lemov 1993, 46).[53]

Social Fragmentation in Capitals

The same sectionalism that divides parts of a state is at work in the cities themselves, on similar geographic and social lines. Montpelier is the only case—to my knowledge—where fragmentation was so high that it led to the definitive division of the capital into two municipalities.[54] Paradoxically, this occurred in the nation's least populated capital. In 1848, East Montpelier seceded from Montpelier. Since the designation of Montpelier as Vermont's capital, in 1805, the town had divided itself in two distinct groups, the "governmental seaters" and the "rurals." The "seaters," seeing in Montpelier a prosperous, growing commercial and political center, suffered from their subjection to a rural town government. The "rurals" saw the capitol area as a separate place that disregarded them. "The inconvenience of having to travel up the hill to the Center for town meetings cited by division advocates was a matter of civic pride. Meetings had taken place on the hill since 1792, and the location symbolized the control over town affairs exercised by the rural district" (Hill and Blackwell 1983, 130–31). Legislators therefore quickly agreed to sit in a "city" severed from its rural surroundings. As a result, the population of Montpelier fell from 3,725 inhabitants in 1840 to 2,310 in 1850 in an area of 10.2 square miles.[55]

Fragmentation was sometimes at work from the very beginning, but in a less definitive fashion. The precise location of the capitol sometimes ignited local feuds between citizens of the chosen new capital. In Concord, New Hampshire, North End residents, of old stock, were opposed to those of new stock in the South End (Anderson 1981, 327). The latter eventually won, after long (1815–16) and fierce debates that were intensified and directed by the local press and businessmen.

> The arguments for and against the respective locations in dispute were: That Stickney's land [in the north] was dry, elevated and airy; a building erected thereon could be seen far and wide; it was near the Town House, where the Legislature had been accustomed to meet, and had been recommended by a committee of that body. The other location [south] was said to be more central, and less difficult of access: to which it was objected that it was low and wet, and contemptuously it was called a "frog-pond." . . . [Opponents argued] that the expense of laying a foundation in such a spot would swell the cost far beyond the estimate. (Bouton 1856, 363–65)

The South Enders had to contribute $4,000 for the $86,000 statehouse. They also had to pay for the splitting of granite from the Rattlesnake Hill quarry as well as for its transportation—first to the state prison, where inmates cut the blocks, and then to the building site. Although the capitol has been enlarged twice (1864 and 1909), it still stands on the initial site. Today, Concord boasts an elegant granite capitol in the center of the small city, surrounded by well-tended grounds where no frogs are to be found (except me on the day I visited!).

In Des Moines, the location of the capitol became the main theme in the governor's election campaign. The committee appointed in 1854 to choose the site of the capitol in the newly designated capital, Fort Des Moines, had to choose between the eastern and western sides of the river. The fort was located on the west side and competed commercially and culturally with a village built on the opposite bank.[56] To win the capitol, east-siders donated forty acres, gave parties for the committee members, and were helped by the new governor, who had purchased lots (and votes?) there. The capitol was eventually built on the east side, to the indignation of the other side, "but harmony was finally restored."[57] The city later grew on both sides of the river, and Drake University was located on its western side, close to the "literary fellows."

Harmony was once again in danger when capitol neighborhoods were included in urban renewal policies in the 1950s and after. Creating or enhancing the above-mentioned "barriers within the city," renewal often destroyed the old neighborhoods surrounding capitols and replaced them with freeways and huge governmental complexes, as in Lansing or Indianapolis. Built around the capitol, they almost annihilate it. In St. Paul, Minnesota, the state capitol area formed the core of the city's downtown. Using funds available under the Federal Urban Renewal Act of 1949,[58] the State Veteran Service Building Commission launched in 1962 the forty-three-acre Capital Center renewal project, which destroyed almost the entire area, which included mostly pre-1920 buildings (Abler, Adams, and Borchert 1976, 47). The common tendency of the richer people to move to the suburbs further distended the links between local populations and their capitols. Paradoxically, a majority of the citizens who remain in the inner cities are the poorer ones, mostly belonging to the very minorities that were excluded from the opening celebrations.

That was the case in Albany, which undertook what was undoubtedly the most celebrated and controversial capital-planning operation since 1945. In the 1960s, the city's permanent population was at its lowest point, and its downtown had been badly battered by the suburbanization process. Governor Rockefeller decided to reverse the trend after he felt shamed during a visit by Queen Juliana of the Netherlands. She had come for the celebration of New Jersey's tercentenary in 1964. Fort Orange—Albany's first name—had held its tercentenary in 1924 (City of Albany 1924). "The Queen was riding with Mayor Corning and myself. . . . I could see the way the city was running down and what this lady might think. Here was a great Dutch city built in the New World, and then she comes to look at it, never having seen it before. My God!" (quoted in Kennedy 1983, 305). To wash out that shame, Rockefeller wanted to build "the most spectacularly beautiful seat of government in the world." The principal architect, Wallace Harrison, had already worked for the family on the Rockefeller Center in New York. He drew his inspiration from the Dalai Lama's palace in Lhasa. "He remembered that at Lhasa there was an approach across a low plain to the hilltop palace and to the cliff supporting the palace, and that one could enter that complex at the base of the cliff through a portal within a high wall. This is precisely how the South Mall was conceived" (Kennedy 1983, 305). In March 1962, 98.5 acres were expropriated, 1,150 structures destroyed, and nine thousand people (17% black) moved. The original project should have lasted four years and cost $250 million, but the total cost was estimated to be $1.9 billion in 1982. The financing involved complex negotiations between the mayor and the governor. An "inner-center renaissance" was happening, owing to Governor Rockefeller's Urban Development Corporation. It involved the coming of the State University of New York to Albany, the completion of a residential-commercial-hotel (Hilton) project, and much rehabilitation, as for Union Station. The result is impressive: the forty-four-story building dwarfs the statehouse, but it is perhaps a little "cold." The complex is not often used as a public place: this should not be a surprise, since the model was the Potala at Lhasa, which owed its origins to a strong theocracy that was far removed from democracy.[59]

A less-known pharaonic project was proposed in Michigan at about the same time (Chartkoff 1990, 28–31). In 1963, George Romney, Michigan's new governor, wanted to modernize state government and build a new capitol, the 1879 one being inefficient. The state had already acquired land and cleared it in the 1940s and 1950s. The plan proposed in 1966 comprised a new capitol that would symbolize "Michigan's position in the Atomic Age." In 1969, the people of Michigan were asked to vote on a new capitol to replace the old one. Seventeen designs were presented for their consideration, but none of them created the same enthusiasm that had greeted the old capitol. Finally, a downturn in the national economy stopped everything. Instead of a brand new "atomic" capitol, citizens and legislators had to content themselves with new flooring in the old capitol, a solution that proved not entirely satisfying. The capitol is now "adjacent to a smaller platform complex derisively called Fort Romney after its gubernatorial progenitor" (Goodsell 2001, 20).

While such civic centers may have been divisive, they also illustrate a specific conception of public life. When Washington State legislators voted in 1910 for a whole civic center instead of a single building (Washington was the first state to do so), the decision stemmed not only from practical or grand aims, but also from their high conception of civic power (Johnson 1988, 13).[60] Civic centers aided in municipal reform by undermining social fragmentation: "Grouping public buildings was one way to outwit 'local rings of interests' who wanted to scatter public structures in order to divide the benefit of their presence among various sections of the city" (Scott 1971, 63).

Smaller capitals are less likely to experience distension processes, for they have neither real suburbs, nor white flight to speak of, nor huge civic center projects. In Montpelier, the least populated state capital, as legislators have neither offices nor secretaries, they are easy to meet in the corridors of the statehouse. "Citizens can probably know their state representative—or the representative knows where constituents work, or live, or who their children married" (Morrissey 1981, 43). Capitals, their citizens, and the state's population are also brought together through conventions. Many public and private organizations use capitals' luxurious facilities to lodge politicians and reserve their meeting rooms for electoral jamborees. Most public and private South Dakotan state associations, for instance, are headquartered in small Pierre, owing to the capital's central location. With only 13,646 inhabitants in 2010, the capital city has thirty meeting rooms in six facilities, a convention center of three thousand square meters, and fifteen motels with 974 rooms to serve the 124 clubs and organizations of the town, as well as the ones from out of town. To strengthen the links between town and legislature, Florida established "major social events to welcome the legislature to the capital region at the beginning of each legislature, and to bid them farewell at the end of each session" (Bromley 1990, 15).

This shows that all capitals are not like Jefferson City, where a true schizophrenia seems to be at work. The city clerk's web page posted the following in 2003: "Because we are located *in the same city as the Missouri state capital,* we receive many misdirected calls on a daily basis for Missouri state government [my emphasis]." It is as if the capital city were distinct from the "normal" city, when both are naturally the same, although their politicians most of the time differ.

Town and State Governments: Worlds Apart?

Unlike Washington, DC, state capitals are "normal" cities within counties. Some are city-county, such as Honolulu and Nashville,[61] or city-borough, as is the case for Juneau. But this does not always induce close relationships between mayors and governors.[62] Larry Sabato asserts that being a mayor is not helpful for becoming governor or congressman and vice versa.[63] The following table shows the lack of connection between being mayor and winning election to other political positions.[64]

Out of fifty mayors, only four (counting Karen Hasara, mayor of Springfield, 1995–2003) served in their state's house or senate, and none became governor or served in the US Congress. This does not mean that some capital mayors lacked higher ambitions. Typical cases are those of Franck F. Fasi and Richard Harmon Fulton. Franck Fasi was mayor of Honolulu from 1969 to 1981 and again from 1985 to 1994. But he was defeated when he ran for the Hawaii House in 1950, for the US Congress in 1962, and for governor in 1974. He was also defeated in the 1996 and 2000 mayoral elections

Table 2.4 The political experience of state capitals' mayors in 2003

Capital	Mayor in 2003	Other political positions
Montgomery, AL	Bobby Bright	—
Juneau, AK	Sally Smith	—
Phoenix, AZ	Skip Rimsza	—
Little Rock, AR	Jim Dailey	—
Sacramento, CA	Heather Fargo	—
Denver, CO	Wellington E. Webb (since 1991)	—
Hartford, CT	Eddie Perez	Defeated candidate for California's 34th District
Dover, DE	James L. Hutchinson	—
Tallahassee, FL	Scott Maddox	—
Atlanta, GA	Shirley Franklin	—
Honolulu, HI	Jeremy Harris	D
Boise City, ID	Carolyn Terteling-Payne (since 2003)	—
Springfield, IL	Karen Hasara (1995–2003) Timothy J. Davlin (April 16, 2003)	Illinois House (1986–93), Senate (1993–95) —
Indianapolis, IN	Bart Peterson	D
Des Moines, IA	Preston Daniels (since 1997)	—
Topeka, KS	Butch Felker	—
Frankfort, KY	William I. May Jr.	—
Baton Rouge, LA	Bobby Simpson	—
Augusta, ME	William E. Dowling	—
Annapolis, MD	Ellen O. Moyer	—
Boston, MA	Thomas M. Menino	D
Lansing, MI	David C. Hollister (since 1993)	Michigan House (1975–93)
St. Paul, MN	Randy Kelly	—
Jackson, MS	Harvey Johnson Jr.	—
Jefferson City, MO	Thomas P. Rackers	—
Helena, MT	James E. Smith	—
Lincoln, NE	Don Wesely	D
Carson City, NV	Ray Masayko	—
Concord, NH	Michael L. Donovan	—
Trenton, NJ	Douglas H. Palmer	D
Santa Fe, NM	Larry A. Delgado	—
Albany, NY	Gerald D. Jennings	D
Raleigh, NC	Charles Meeker	—
Bismarck, ND	John Warford	—
Columbus, OH	Michael B. Coleman	D
Oklahoma City, OK	Kirk Humphries (since 1998)	—
Salem, OR	Janet Taylor	—
Harrisburg, PA	Stephen K. Reed (since 1982)	See the following chapter 2 text.
Providence, RI	David N. Cicilline (since 2003)	Rhode Island House (1995—2003)
Pierre, SD	Dennis Eisnach	—
Columbia, SC	Bob Coble	D
Nashville, TN	Bill Purcell	D
Austin, TX	Gus Garcia	—
Salt Lake City, UT	Rocky Anderson	—
Montpelier, VT	Chuck Karparis	—
Richmond, VA	Rudolph C. Mc Collum	—
Olympia, WA	Stan Biles	—
Charleston, WV	Jay Goldman	—
Madison, WI	Susan J. M. Bauman	D
Cheyenne, WY	Jack R. Spiker	—

Source: Capitals' websites and http://politicalgraveyard.com; http: //bioguide.congress.gov.

Note: An em dash (—) means none recorded; *D* means the person was a delegate to the Democratic National Convention in 2000.

(Johnson 1991). Richard H. Fulton was first a member of the Tennessee Senate from 1958 to 1960, and went on to become a US representative for Tennessee from 1963 to 1975, when he became mayor of Nashville and held that office until 1987. During his time as mayor, he was twice defeated in the gubernatorial primaries, in 1978 and in 1986.[65]

There are naturally exceptions to that general trend.[66] James A. Rhodes, mayor of Columbus, Ohio, from 1944 to January 1953, became state auditor and later governor (Condon 1977, 119). Erastus Corning II (1909–1983), before he became Albany's "king" from 1942 until his death, was a member of the New York senate from 1937 to 1941.[67] Stephen R. Reed, before being elected as mayor of Harrisburg in 1982 (serving until 2010), served three terms as a member of Pennsylvania's House (from 1974, when he was only twenty-five years old, to 1980) and was then Dauphin County commissioner (1980–81). To become governor, it seems best to have served as mayor of a large city. On the other hand, between 1950 and 1980, only 2 out of 290 ex-governors became mayors. One of them, J. Bracken Lee, was elected mayor of Salt Lake City, the capital of Utah (Sabato 1983, 46, 49).

The relationships between mayor and governor depend on several factors, for instance, the party to which they belong, although intergovernmental relations tend to supersede political ones.[68] In answer to my e-mail inquiry (in March 2003, about the advantages and disadvantages of being a state capital), the Honorable Stan Biles, mayor of Olympia, commented, "Meetings with the governor or lieutenant governor and senators and representatives come easier." An Oklahoma City representative wrote that the main benefit of state capital status was that it brought "people from across the state here (state representatives and senators)," and that the spatial and cultural relationships between "legislative hill" and the rest of the capital city were "good and improving with increased awareness of the various venues available in the community." Topeka's mayor listed among the main advantages of being state capital, "the stature as seat of government." As for Madison, "being the state capital has instilled a greater attitude of stewardship among many Madisonians that's reflected in our respect for our natural resources and our commitment to make Madison the best place it can be to live, learn, work, and play." The case of Madison rests indeed on a progressive history. Governor Robert LaFollette's "Wisconsin idea"—that is, the social progress he wanted—was "symbolized by State Street: at one end of the street stood the State University, at the other end, the State Capitol. LaFollette and the Progressives regularly—almost ostentatiously—drew on the university community for assistance with public matters" (Morone

1990, 117). Springfield provides another example of close ties between town and state. In a 1970 study, Daniel J. Elazar argued that "a politically conscious society" was set up,

> in which every articulate element of the local community, from the downtown businessmen to the local pursuers of "culture," is visibly attuned to the political life, a situation unique among the cities of the prairie and apparently unique in American society generally. . . . The community leadership of Springfield—the archetypal political community of the group—reflects this dual commitment [nation/state loyalties] most clearly, accepting the national role of their community's development and their role as a national shrine (the home of Abraham Lincoln), while remaining deeply immersed in the interests and operations of the State of Illinois. (Elazar 1970, 85–86, 143)

To sum up, although home rule still remains incomplete, and local borrowings are limited by the states, the relationships between the states and their municipalities have greatly improved since the 1960s. State aid to local governments has improved, commissions on intergovernmental cooperation have been created, and the State and Local Coalition was created in 1972 by states and localities to lobby the national government (Sabato 1983, 177–79).

Capitol and Capitol Grounds as "Ideal" Public Places

Opposing views exist about public spaces. Historians like Jane Jacobs and Lewis Mumford have compared them to the sites of ballet and dramatic productions, focal points of urban life, where social interactions flourish. Radical researchers have since disagreed. They argue that public space is considered a threat, inducing what Mike Davis calls "defensible space," creating a clear demarcation between public and private space. Contemporary geographers bemoan the current trend of privatization of public spaces, because they consider them as elements of collective identity (although the effect of the increasing polarization of the society must be taken into account). Likewise, scholars of "supermodernity" study the rise of "non-places" in modern cities, meaning places that have no identity, no history, and that induce no relations, contrary to the very definition of the word *place* (Augé 1995).[69]

Bearing in mind the fact that public places are not as cohesive and unitary as they might appear and that they reflect the tensions of urban societies, they nevertheless are more unifying than divisive.[70] Public spaces are not the foremost features of American cities. They are not utterly absent: one can cite streets and sidewalks,[71] public squares, spaces for memory or spectacle, and public parks. All these places are supposed to be open to everyone, but they are seldom used for truly collective gatherings. This certainly stems from the roles of individuality (meaning privacy) and community (meaning a particular group, excluding others) in America. It is perhaps also why private places are used in a quasi-public way, such as churches (for community or fund-raising events), hotels (for marriages, political rallies, and conventions), or malls (a running association uses the Mall of America, south of Minneapolis).

Capitols stand as an exception. They are among the few American buildings that are out of the real estate market: they belong to everyone. Everyone can freely walk in and look at all the state memorabilia displayed in their corridors and their rotunda or attend a session in their public galleries.[72] As was the case for Michigan, citizens would rather modernize old capitols than build new ones. Justin Kestenbaum's conclusion, that the capitol "was and is–a direct link to Michigan's past and a potent symbol of its future" (Kestenbaum 1986, 51), testifies to the beginning of a new period in the history of capitols: after urban renewal came the period of heritage policies. Capitols have largely benefited from the rise of the preservation movement since the 1960s. Two-thirds of them have undergone massive restoration. If they are no longer used as capitols, they are not sold or torn down. They are transformed into museums, free of charge, thus allowing everyone to see and learn the history and geography of the state.

Besides the capitol building, capitals also provide citizens with a unique public space, the capitol grounds. The grounds are more or less impressive, but they are always present to emphasize the building. They are one of the main sources of local pride in state capitals. A common feature is the lack of any wall (there are sometimes low fences), to make clear that, although the capitol might seem impressive, it is open to all citizens as a temple dedicated to democracy.[73] Ceremonies are often held on the grounds, particularly in the park, on the steps, or in the vast rotundas. In April 2000, for instance, in and around Baton Rouge's old capitol (now a museum dedicated to Louisiana's history), a local school held its annual celebration. Children and parents barbecued on the grounds or rehearsed in the building, amid local and foreign visitors. This was a simple but efficient appropriation of a public space. In a similar manner, Madison's capitol grounds "are used as the venue for free community orchestra concerts each summer, and are used to host one of the best farmers markets in the country every Saturday through the growing season," according to Ryan Mulcahy, assistant to the mayor (personal communication, March 2, 2003).

Capitols were therefore from the beginning enshrined in well-tended parks. In Augusta, Maine, the 1827 legislature, during the process of selecting the permanent seat of government, decided "that five hundred dollars be placed at the disposal of the governor to enable him to cause to be fenced, improved and ornamented with trees, such lot as may be conveyed to the State" (North 1981, 476). In Concord, New Hampshire, the legislature seized the opportunity of the 1864 debates concerning the location of the new capitol to warn the "winner," Concord, that the city had to create a new street south of the capitol within six weeks; otherwise, the capitol would go to Manchester. This area included the legislature's outdoor toilets, as well as unsightly stables, sheds, and backyards (Anderson 1981, 336). Even with the Civil War shortage of manpower, Concord complied on time.

In addition to the capitols and their grounds, there are the governor's mansions, the state's miniature White Houses. Like the White House, they are often open to the public,[74] but unlike it, they do not stand out isolated in the city. In Carson City, Nevada, the governor's mansion is just one among other large houses in a quiet and shaded neighborhood. In Pierre, South Dakota, the (former) residence was built in 1937 by the Works Progress Administration workers, at no cost to the state, thus enhancing the bonds between local people (who found jobs), the state, and the federal government. In 2003, a humorous (or ridiculous?) situation occurred in Columbia, South Carolina. Faced with a budget shortfall of $340 million, the governor—already known as parsimonious—proposed to close down his mansion until July and live with his family in an upstairs flat. Fortunately for him (and interestingly for this study), local businessmen, "convinced that the mansion [was] a powerful attraction in luring investors to the state," offered $100,000 to keep the mansion open ("The Homeless Governor" 2003, 50).

As places of condensation, capitols are places where individual and collective experiences take place that revive their reference to the social group and to their territory. This process is best expressed when a state celebrates its own or its capitol's centennial. Besides renovating the building and grounds, states often sponsor traveling exhibitions. They draw upon extensive research not only on the history of the capitol and on the reputations earned by early legislators who served in the capitol, but also

FIGURE 2.5 Capitol Centennial, 2010, Pierre, South Dakota. Capitols are more than grand buildings. In ceremonies such as centennial celebrations, they embody the link between a state and its citizens. Photo by Christian Montès, July 2010.

on local pride.[75] Citizens of South Dakota consider Pierre's 1910 capitol (fig. 2.5) to be "one of the more successful buildings from the period known as the American Renaissance, and it is certainly the grandest expression of revivalist architecture,"[76] forgetting that their capitol only mimics Rhode Island's.

The same aim of socialization motivated the building of New Mexico's new capitol in the 1960s. The modern regionalists "sought to develop a regionalism that progressed beyond façade symbolism to the use of traditional building and planning forms to enhance contemporary life" (Wilson 1997, 290). However, visitors to Santa Fe cannot help feeling the separation between the new capitol and the rest of the city, for it is located on the other side of the river in a rather isolated area. Capitols certainly benefited aesthetically from an elevated or isolated location, but this was not always the case from the social point of view. Denver's Capitol Hill was the best address in town from the 1890s to the 1930s before it suffered a sharp decline: "What once evoked visions of splendor and wealth from everyone now conjures up images of seediness and danger in the minds of some" (Shay 1983, 25). The ascendancy of posh suburbs and the advent of the Depression led to the conversion of Capitol Hill's mansions into bleak boardinghouses. Although postwar renewal policies replaced them with hotels, apartments, and offices, the decline continued. Two different trends have emerged since the 1960s: gentrification on the one hand (six historic districts have been created) and counterculture on the other (with the arrival of hippies and later of gays). These changes have induced a renewal of the neighborhood and a rise in its population (24,500 in 2002). Once the site of "millionaire row," Capitol Hill offered "the city's widest range in housing types and price ranges" according to Downtown Denver's website in 2003. Neighborhood organizations have been created, proud of the broad basis of their community.[77] Capitol Hill was described as the "most diverse, eclectic, and 'alive' neighborhood" in Denver.

Some Necessary Caveats

Caveats are first necessary because in some states, parts of the population still resent the capital city. When Harlean James traveled through southern capitals in 1925, she noticed how proud Richmonders were of their capitol, with its famed statue of Washington, modeled from life by Houdon, as well as the "affection for Richmond" among all Virginians, but she also remarked that there was an exception, the panhandle on the Kentucky-Tennessee border (James 1925–26, 389). Even today, unanimity does not exist in Richmond. There was a fuss about a new statue of President Lincoln that was to be unveiled on April 5, 2003, in a city where Monument Avenue already displays statues of Robert E. Lee and Thomas J. "Stonewall" Jackson. The commander of the Virginia division of the Sons of Confederate Veterans said it was "a not-so-subtle reminder of who won the war and who will dictate our monuments, history, heroes, education and culture" (quoted in "The South. Be Just Ain't Right" 2003, 48). Richmond is the capital of Virginia, but part of the population sees it primarily as the former capital of the Confederate South. Likewise, any American of African origin must see with very mixed feelings the Confederate flags and statues that are still evident in the other southern state capitals. Montgomery, Alabama, presents different aspects to visitors. On one hand, they can look for the traces left by the first Confederate capital and visit General Lee's house. On the other hand, they can search for the traces of the well-known 1955–56 bus boycott (begun on December 1, 1955, by Rosa Parks), visit the civil rights museum and research center, and meditate before the fountain where the names of all the African American civil rights advocates who died for their ideas are carved in black stone. Further west, Native Americans visiting Santa Fe are certainly happy to see that the capitol's architecture represents the Zia Pueblo symbol of the sun, but they also remember that the land on which it was built was stolen by white colonizers. Similar perceptions certainly prompted the study group led by Ray Bromley to urge capitals' leaders to foster a "capital region mentality" by bringing together all key groups in capital MSAs, emphasizing the fact that social unity has not yet been attained (Bromley 1990, 20).

It is also necessary to bear in mind that capitals were intended to become economic metropolises. Plats were therefore sometimes changed for economic reasons. Public good and public spaces had to yield to the power of big business, that is, private interests. Such an abdication could have positive results. John L. Hancock argued that the "desiccation" of Philadelphia's plat enabled the city to experience growth, as opposed to Savannah, where the rigidly maintained plat (residents were forbidden to sell or to profit from the land) inhibited the economy.[78] In 1900, all fifty MSAs, that is, all the major American cities, used plain grids (Hancock 1967).

Nowhere is the role of private interests more flagrant than in Oklahoma City, where Capitol No. 1 oil well was drilled in 1942 in a flower bed less than one hundred meters from the capitol's south entrance—hence its nickname, "Petunia Number 1" (Robertson 1996). The oil development after 1928 did greatly profit the city council. The gubernatorial creation in 1931 of a one-mile oil-free zone around the capitol ended in 1935 when the governor resigned, leaving the field open for his successor, Ernest W. Marland, an oilman. In fact, the petroleum industry contributed one-half of total state revenue, and in successive city referenda, the majority supported the pro-development party. Fifteen oil wells were drilled on the capitol grounds between 1936 and 1942 (out of a total of 4,850 oil and gas wells for the city). Topping out at 38 meters (123 feet), the derricks overshadowed the capitol building, and oil sometimes sprayed on its walls. Oil was produced until the mid-1980s and generated $8 million in revenue for the state, which was used to build government offices on the site. The derricks remain, at the demand of the petroleum companies, for which they are good publicity. Although Robertson contended that this situation resulted mainly from the decision of a single man, Governor Marland, it nevertheless symbolizes the sometimes very close relation of lobbies and politics.

In Phoenix the state has taken great care to "renew" the old capitol. Although two buildings were added in the 1960s, and the capitol grounds are perfectly tended, the capitol complex still seems lost amidst the new skyscrapers built in the capital to celebrate its economic success. Capitol buildings remain grand only in the smaller capitals, even though several larger capitals have set limits to the height of buildings surrounding the capitol.

This initial consideration of American state capitals has raised two paradoxes. First, while state capitals are undoubtedly places of memory in today's America, how can we explain why they have been the objects of conflicts, hostility, and seemingly endless debates? Second, while capitals are the most monumental and historical cities in the nation, which signifies permanence, why did they experience instability and mobility? This chapter has provided some preliminary answers, showing that state capitals are torn between two polarities: civic fragmentation on the one hand and the staging of democracy on the other. The next chapter intends to shed some light on the geographical pattern of the instability they experienced.

3

Geographical Patterns in the Migration of Capitals

Given the tension between models and individual histories, it seems useful to present both sides, beginning with the one that appears less nomothetic but has the advantage of informing readers about the complexity of the quest for causality and providing an image—far from perfect and distorted by the varying quality of the sources used—of life in the states considered. Moreover, this approach enables us to consider the question in the context of a chronology that models do not always follow (this indeed often constitutes their very strength). Chronology is not to be taken as an explanatory element: it does not provide any linear causal chain. However, to unfold a certain number of events according to a temporal scale has the merit of providing a known framework for the explanatory models that will be eventually proposed. The relation to time differs according to periods, groups, and individuals, but chronology has a heuristic value, particularly when the object of the research is the chronological sequence of capitals. Therefore, this chapter aims to provide an initial, broad view of the numerous processes at work along geographical patterns. The two quotations that follow summarize quite well the length of the process, which was similar before and after industrialization, urbanization, and the development of modern politics.

> The framers of the Michigan Constitution of 1835, unable to agree on where to place the state capital, let it remain temporarily in Detroit, but decreed that a permanent site be chosen in 1847. In the interim, communities all over the lower peninsula—seeking the enormous economic growth certain for the capital site and its hinterland—flooded the legislature with petitions and memorials offering free land and other inducements should the state locate its capital in their midst. (Kestenbaum 1990, 390)

> In fact, contention over the location of the seat of territorial and state government did not cease until the completion of the capitol building in 1927. During the interval of seventy-three years, many efforts were made to relocate the capital, and at some time in this period nearly every important city within the present boundaries of the state made plans or entertained hopes to become the capital. . . . The records show how closely several cities came to winning it, and how one city had the prize within its grasp, only to lose it through a legal technicality. (Beardsley 1941, 239)

A common factor in the history of American capitals was their instability. Only eight states never changed their capital. In fact, even after the capital was chosen, the story was often not over. Competitors soon arose to challenge the temporary capital with arguments and often cash. American states have had an average of 3.84 successive capitals, the median being 3.

The path to becoming a state capital was a hard one, especially in the early period of colonial and territorial America. Figures 3.1 and 3.2 give the five main spatial patterns of change that have been found (states may belong to two categories). The figures show that instability is present in all the spatial patterns of capital movements: even the first pattern—stability—often covered rough fights that led nowhere, but consumed time and money at the expense of more useful legislative work.

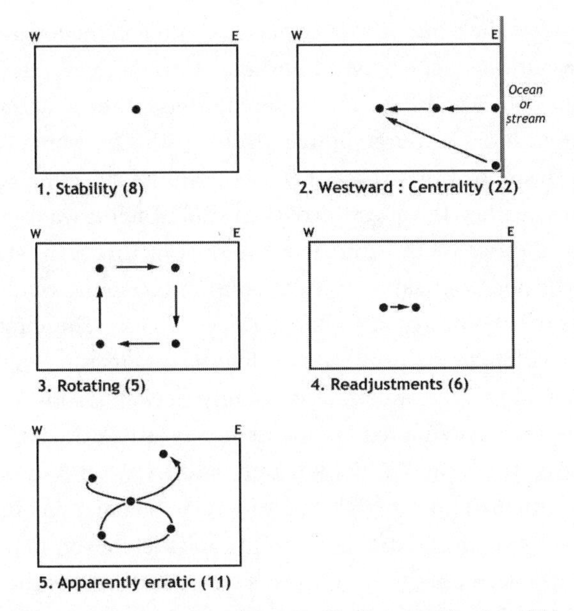

FIGURE 3.1 The five main types of instability. Credit: Christian Montès and M. L. Trémélo.

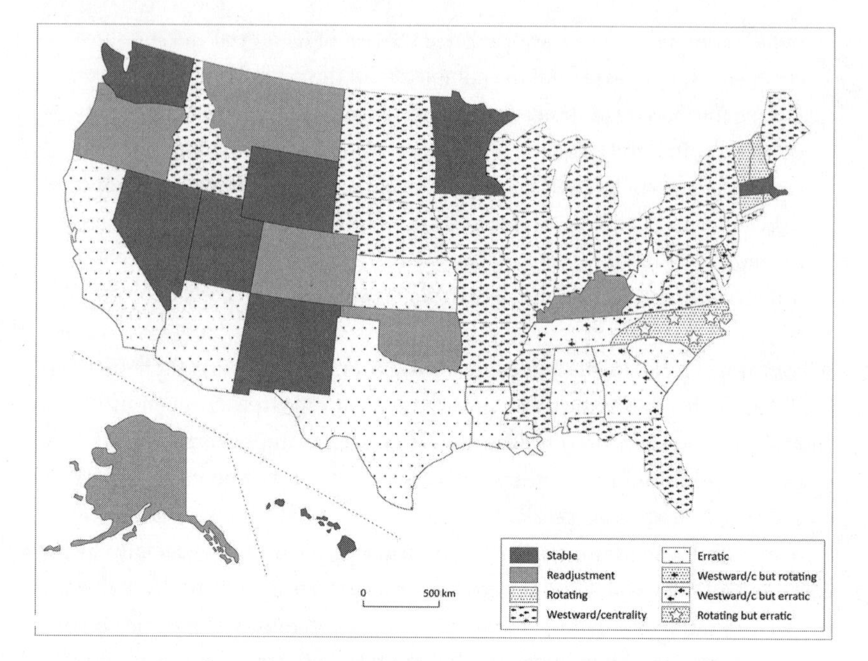

FIGURE 3.2 Spatial patterns of change in capitals' locations. Credit: Christian Montès.

Stability

Eight states (16%) have never experienced a change in the location of their capital: Hawaii, Massachusetts, Minnesota, Nevada, New Mexico, Utah, Washington, and Wyoming. The apparent stability of a unique capital, however, often hides a more complex history. The following cases show that, although some of these cities had been capitals since the seventeenth or eighteenth century, the question of their capital status was not definitely settled until the twentieth century. Numerous factors were at work: accessibility, shifting political majorities, shifting economic centrality, even monumentality (the power of building a new capitol). These factors were linked to the colonization processes during the nineteenth century, which long maintained an atmosphere of instability. Hawaii stands as an exception. The Royal Court adopted Honolulu in 1821 as its principal residence. It was declared the capital of the kingdom and a city on August 29, 1850. The capital remained unchanged after American colonization in 1898, and most Hawaiians are proud of their ancient capital (Johnson 1991).

Santa Fe has been the capital of today's New Mexico since the beginning of the seventeenth century. Faced with economic decline and the parallel growth of competing cities, its status was challenged during the whole nineteenth century by Albuquerque. Congress designated Santa Fe as the permanent capital only in 1897 (Wilson 1997, 79). Minnesota provides a fine example of apparent stability. The location of the capital remained on the legislature's agenda for almost half a century. Indeed, it took St. Paul from 1849 to 1893 to become secure as the state capital, insuring it by constructing Minnesota's third capitol. William Dean aptly called the legislative scheming upon the subject "a legislative foot-ball" (Dean 1908, 33). Since the convening of the first legislature in September 1849, the location of the temporary seat of the state government and then the threat of removing the capitol from St. Paul were used as bargaining chips to win St. Paul's (and Ramsey county's) votes. The following chronology tells the broad lines of the story.

ST. PAUL'S WAY OF THE CROSS

1849: The governor chose St. Paul, but legislators could not decide upon the matter.

1851: The second legislature decided at last to build a capitol in St. Paul.

1857: Driven by a speculative move, legislators relocated the capital to the new town of St. Peter. After a legal battle, the decision was overturned.

1858: An attempt to remove the capital to Nicollet Island was aborted.

1861: An attempt to locate the capitol on the Kandiyohi lands was aborted. The House agreed, but the proposal was defeated in the Senate.

1869: New attempt on the same location, sustained by rural members, combined with those from Winona, Stillwater, Minneapolis, and St. Anthony. The House and Senate concurred, but governor William R. Marshall vetoed the bill, "giving as his reasons for so doing, that there were no public sentiment in favor of the removal; that the question was not before the people at the last election; that the location was not central, and the time not opportune for the State to go into an expenditure of a million dollars or more" (Dean 1908, 17).

1872: An attempt to move the capitol to Stanton in Kandiyohi County was aborted.

1881: The first capitol, twice enlarged, burned. "A most vigorous move was set on foot, the morning after the destruction of the old building, to remove the capital from St. Paul, and great inducements were said to have been offered to members of the legislature to consider the proposition. . . . Governor Pillsbury, however, was inflexible in his refusal to entertain any consideration of the question" (Dean 1908, 22), and the bill was withdrawn.

1891: Minneapolis and Mennetaga (Kandiyohi County) asked for the capitol, because the capitol at St. Paul was too small and unhealthy. "It was the first time within the history of Minnesota legislature that the St. Paul delegation assumed an aggressive attitude on the capitol question. But the prize was worth the fight; for, if successful, it would forever settle the location of the seat of government" (Dean 1908, 32).

1893: On April 7, the assembly and the governor approved the building of a new capitol in St. Paul.

1905: Inauguration of the capitol, built by Cass Gilbert of St. Paul for $4.5 million.
The people never voted on the question, although the constitution required them to do so.

Our second example is the state of Washington, where the capital question has been a political issue for a century (Beardsley's introductory quotation was too optimistic, since the issue was not settled in 1941). When creating the Washington Territory, Congress provided for the capital choice in section 13 of the Act of March 2, 1853:

> That the legislative assembly of the Territory of Washington shall hold its first session at such time and place in said Territory as the Governor thereof shall appoint and direct; and at said first session, or as soon thereafter, as they shall deem expedient, the legislative assembly shall proceed to locate and establish the seat of government for said Territory, at such place as they may deem eligible; which place, however, shall thereafter be subject to be changed by said legislative assembly. And the sum of five thousand dollars, out of any money in the Treasury not otherwise appropriated, is hereby appropriated and granted to said Territory of Washington, to be there applied by the Governor to the erection of suitable buildings at the seat of government. (quoted in Beardsley 1941, 241)

Washington's first governor, Isaac Ingalls Stevens, selected Olympia as a capital because Puget Sound communities had been at the heart of the creation of the new territory; because Olympia had been the port of entry for the collection district of Puget Sound since 1851 and was the home of the first territorial newspaper, the *Columbian*; and because its county, Thurston, was the most populous on Puget Sound.[1] But Vancouver, head of the most populated county (Clark), immediately arose as a strong contender.[2] On January 10, 1855, after fierce debates (mostly between the "Olympia clique" and the "Vancouver clique"), Olympia became the permanent capital, owing to its accessibility and "grand scenery." Olympia's competitors, after losing the capital, united to prevent the building of the state capitol, thinking that this would allow them to bid later for it. Congress appropriated $30,000 in 1857 for a permanent capitol, but the money was never spent, and the territorial capitol, intended to be temporary, remained in service until 1901. During the 1859–60 session, the "Olympia clique" narrowly defeated a pro-Vancouver bill. But during next session, in December 1860, an act relocating the seat of government in Vancouver was passed, owing to the scheming of representatives from the Columbia River and northern Puget Sound, who divided between them the spoils of territorial government. Seattle would have received the university, Port Townsend the penitentiary, and Vancouver the capital, without debate (Nicandri and Valley 1980, 3). It seems that the people of Portland really directed the campaigns for removal for business reasons. The issue was nevertheless far from settled. Confusion reigned, because the date of passage and the enacting clause had been omitted, and because Olympia won at the referendum on the capital question held on July 8, 1861. The Supreme Court therefore repealed the relocation act in December 1861, and Vancouver was presented with the penitentiary.

When the territory became a state in 1889, the capital question was once again on the legislature's agenda during the constitutional convention, and conditions had changed. Olympia was no longer one of the territory's major towns (it had only 4,698 inhabitants in 1890), whereas Tacoma (terminus of the transcontinental railroad since 1872) and Seattle (a major port, linked to the coal mines) had become prime commercial centers. For Olympia, keeping the capital thus had a greater importance than during the 1850s, for its survival was at stake. The main question was whether the capital should be located near the state's demographic center (i.e., west of the mountains) or near its geographical center. The arguments against Olympia stressed the cost for legislators to travel there and its lack of good connections and good infrastructures. The arguments in favor of Olympia were that it was a moral city where the climate was mild and the cost of living was low. The two main contenders were North Yakima[3] (1,535 inhabitants in 1890) and Ellensburg (2,768), along with Pasco (a paper town), Centralia (2,026 inhabitants), Waterville, and Waitsburg. Seattle (42,837 inhabitants) and Tacoma (36,006 inhabitants) could not win the capitol, because their size would deter the legislators. Their businessmen nevertheless participated in the debates by supporting North Yakima and Ellensburg, where they had large real estate investments and commercial links (Dodds 1986, 174).

To avoid the fate of the 1860 decision, the legislators decided to tie the referendum on the capital to the vote on the constitution. It was held on October first, 1889; Olympia got 25,490 votes (48.06%), North Yakima got 14,711 (27.74%), and Ellensburg got 12,833 (24.2%). As the required majority was not attained, another referendum was held in 1890. This time, only Olympia campaigned, and it received 37,413 votes (72.77%); Ellensburg got 7,722, and North Yakima got 6,276.

The matter nevertheless remained unsettled, as the legislature kept waiting to build the permanent capitol (Johnson 1988, 19). In 1893, a State Capitol Commission was created, which was to fund construction of the capitol with funds from the sale of timber on the 132,000 acres given by the federal government for that purpose. But potential buyers considered the timberlands either useless or inaccessible, and no bids were made. As a result, opposition to Olympia continued, based on the argument that the city was inaccessible. Governor, John J. Rodgers, a Democrat, favored Tacoma as the capital and had the state purchase in 1901 the big Thurston County Courthouse (1892), located in downtown Olympia, instead of building a permanent capitol, which cast doubts on Olympia's permanence as capital. The legislature occupied the building from 1905 to 1927. In 1905, the House asked for another referendum on the capital question, but Mead, the new governor, vetoed it, seriously undermining the anti-Olympians. At last, in 1927, the new capitol was built, and Olympia felt secure. But as the state bureaucracy grew, many agencies located their main offices in the state's metropolis, Seattle, and the question of removal arose again. In August 1953, Olympia citizens and the Casco Company filed a suit in Thurston County Superior Court against Governor Arthur Langlie. The judge ordered the state to locate all agencies in Olympia. On appeal, the Washington Supreme Court noted in August 1954 that Seattle was unconstitutionally becoming the second seat of government, and ordered thirteen agencies to relocate in Olympia. They complied immediately, except for the Game Department, which complied in 1958, closing at last (?) the capital location question (Nicandri and Valley 1980, 33–34).

Westward/Centrality

When capitals moved, they often followed the general trends of the settlement of the United States. It is therefore not surprising to see the importance for the location of capitals of the "western pull factor" that initiated a movement inland from the coast and to the center of a territory from its border. Examples are numerous, such as Iowa City, Indianapolis, and Raleigh. But that movement was far from universal, affecting only twenty-one of the fifty capitals (42%). This is true for the East as well as for the West. In the East, Boston proved able to resist the westward movement, certainly because early religious partitions of the colony had left Massachusetts relatively small, unlike Philadelphia (where yellow fever played a great role) or New York. In the West too, St. Paul succeeded in remaining a capital, despite an off-center geographical—but not economic—location, when others had quickly to abdicate (St. Stephens, Alabama, after only two years). The process first occurred in the East, where coastal capitals had to yield to more centrally located cities, due to the westward expansion inside the state.

The movement was sometimes hastened by war. This was the case for New York, which lost its capital status because of the British occupation during the War of Independence. The British likewise controlled Savannah, Georgia, from 1778 to 1782, which led to the rise of the upcountry and its political importance. Apportionment favored the upcountry, enabling it to seize control of the state. Although the seaport of Savannah had been confirmed as the state capital in 1782, the assembly alternated its sessions between Savannah and Augusta, the upcountry's largest city, and the governor and the council resided part time in both cities. In 1786, Augusta became the sole capital and remained so until a new inland capital, situated on the Ogeechee River and called Louisville, was completed in 1795 (Coleman 1991, 91).

Population shifts can also partly explain Tennessee's wandering capitals. Called Southwest Territory upon its creation in 1790, the first seat of Tennessee's government was chosen by Governor Blount. He selected a mansion, Rocky Mount, at the juncture of the Holston and Watauga Rivers. Two years later, he founded a new capital, Knoxville (named for Blount's boss, Henry Knox), farther south but still in the northeastern part of the state, "so as to be more conveniently situated to deal with the Indians and to communicate with the Mero District" (Bergeron, Ash and Keith, 1999, 58). In 1812, Nashville, in middle Tennessee, became the capital, but it was moved back to Knoxville in 1817. In 1819 it was moved from east Tennessee to Murfreesboro, thirty-five miles southeast of Nashville. In 1826 at last, the government definitively settled in Nashville. Between 1810 and 1820, when east Tennessee had gained 34,000 inhabitants, middle Tennessee had grown by 125,000 and had also gained the political power (Bergeron, Ash, and Keith, 1999, 73).[4] This surge in its population was a consequence of the Treaty of San Lorenzo (1795), which ended both Spanish claims and the Indian threat, and induced an important westward migration (Billington 1974, 242).[5]

Tennessee illustrates the pattern of capital movement in the trans-Appalachian West outlined by Donald Meinig, who described a two- or three-stage process (Meinig 1993, 441). First, a capital was located near the border of a newly designated territory. Second, it was soon moved closer to the vanguard of settlement. Third, it was shifted to the approximate geographical center of the state. However, Meinig left aside another factor that was linked to the incremental way in which states were created. Contemporary states were often carved from vaster territories that sometimes bore the same name. Geographical centrality was therefore not stable. The case of Indiana is typical. When Indiana Territory was created in 1800, it was part of the Northwest Territory and included present-day Indiana, Wisconsin, the major part of Michigan, and a portion of Minnesota. The seat of government was Vincennes, located on the lower Wabash, which is now on the southeastern fringe of Indiana.[6] When Indiana Territory reached, more or less, its present boundaries after two more divisions, Vincennes seemed too far from the center. In 1812, the capital went to Corydon, near Kentucky's Louisville, in the southern part of the state, which was still the most populated part of Indiana because the north was not yet organized. The 1816 constitution gave the capital some respite, mandating that "Corydon, in Harrison County, shall be the seat of government for the State of Indiana, until the year 1825, and until removed by law." Corydon's citizens apparently encouraged the legislators to write that mandate with a bond for $1,000 (Carmony 1998, 107; see also Langdon 1916). But Corydon soon proved inaccessible to settlers who were moving into the northern and central areas of the state that had been opened to colonization. Legislators therefore passed the Act of January 6, 1821, "in order to locate the seat of government at the exact center of Indiana and, more important, to end the constant rivalry for the right to welcome the General Assembly every year" (Cayton 1996, 273). The governor also wanted to link the creation of the transportation system and the location of the capital. Geography, politics, and economy were therefore behind the quest for centrality.

Rotating Capitals

Previous to 1808, there had been 46 sessions of the General Assembly in 14 different towns; 23 sessions in the eastern side of the State, in or near the valley of the Connecticut river; 22 on the western side, 11 of which were in Bennington County, and 11 in or near the valley of Lake Champlain, and one session in the north-eastern part. These locations at extreme points entailed hardships of access, alternately on the one side of the Green Mountains and the other.

(HILL AND BLACKWELL 1983, 283)

Faced with a seemingly never-ending process of selection, some states yielded to a complex system of wandering capitals. Only five states—mostly small ones—experienced this: Delaware, New Hampshire, North Carolina, Rhode Island, and Vermont. Several more states thought of rotating capitals, but never implemented the idea, such as Tennessee in 1843 (Mahoney 1945, 101). According to Rosemarie Zagarri (1988), there were two main reasons for rotation: to avoid British troops during the War of Independence and to bring the capital to the people in smaller states. In fact, other forces were at work, sectionalism and factionalism being foremost. For North Carolina, the force that delayed the choice of a permanent capital for fifty years was clearly factionalism. First, because its agrarian economy was based on rural estates instead of cities, the Carolina colony did not have a town until 1706, with the founding of Bath. Even so, throughout the eighteenth century, candidates for the governmental seat were so numerous—Bath, Edenton, Halifax, Hillsborough, New Bern, and Wilmington—that the General Assembly was unable to reach a majority on the subject, wandering from one town to another with the official records. This system proved highly inefficient because of the frequent lack of a quorum and the loss of records, and a compromise was finally reached at the 1788 Convention, held in Hillsborough. Under this agreement, a permanent seat of government—a new town to be called Raleigh—was chosen three years later, on December 5, 1791, in New Bern (Waug 1967, 3).

In Delaware, although rotation occurred during the War of Independence, the British had little impact on the decision. As in North Carolina, rotation was rooted in political rivalries. In 1777, the fall session of the Delaware assembly was moved from New Castle to Dover. The move expressed the resentment of downstate counties—Kent and Sussex—for leading New Castle County. As was often the case, the downstate coalition did not last long, and in October 1779, a new temporary coalition between

New Castle and Sussex delegates resulted in an act that allowed the assembly to meet "anywhere in the state." Until October 1781, when Dover was made the permanent capital, the legislature wandered throughout the state's three counties, from Wilmington to Lewes, then to Dover, then New Castle, and to Lewes again. Dover's centrality made it an acceptable compromise (Munroe 1993, 75).[7]

Vermont experienced rotating capitals for seventeen years. The cause was this time geo-economic: "No place near the center is sufficiently settled to accommodate the [capital]," said the preamble of the legislature's Act of November 1, 1791. Despite the "great inconvenience and expence [*sic*]" of having no fixed place to hold its sessions, the legislature designated Rutland for the session of 1792; after that, sessions were held in Windsor and Rutland alternately for eight years. But on November 8, 1796, the House repealed the act of 1791, because it was "found to be inconvenient and expensive." Rutland, Windsor, Vergennes, Middlebury, Newbury, Burlington, Westminster, Danville, and Woodstock thus alternated until 1807. This solution was as inconvenient as the former: state records were lost, and access was difficult. After two years of debates (1803–1805), Montpelier was finally selected, on the grounds that it was "the most convenient place for the accommodation of the state at large" (*Records* 1877, 423–26).

The best example of rotating capitals is found in the smallest state, Rhode Island. It is also where rotation lasted the longest, because it is the only case where rotation was not the result of legislative deadlock. Rotation partly came from the way Rhode Island was created, as a religious offshoot of the Massachusetts Bay Colony. Its leader, Roger Williams, was granted by England a patent on March 14, 1644, based on the four existing towns (Providence, Portsmouth, Warwick, and Newport). The 1647 Legal Code—unchanged until 1843—created a "federal Commonwealth," with a rotating legislature (Conley, Jones, and Woodward 1988, 9–11). The 1663 Royal Charter changed the situation for a short period of time, designating Newport as the sole capital until 1681. Rotation then resumed, although it was irregular, between Newport, the main assembly site, Providence, Warwick (until 1741), Portsmouth (until 1739), and Kingstown (since 1698). After its division into North and South Kingstown in 1723, South Kingstown became, in 1733, a regular meeting place (every other October). Rotation was enhanced by the creation of five counties in Rhode Island between 1703 and 1750; the legislature rotated between all five county seats, where the county house would accommodate the General Assembly. An 1840 act of the legislature left four rotating capitals: Bristol (Bristol County), East

Greenwich (Kent County), Newport (Newport County) and Providence (Providence County), although South Kingstown (Washington County) was added again in 1843. Rotation was soon simplified, as some "capitals" remained small towns and were close to Providence, which was on its way to becoming an industrial stronghold. In 1850 Providence had 41,512 inhabitants, and Newport had 9,563, while East Greenwich had only 800 inhabitants, and South Kingstown had 50 (Bristol had 3,669 in 1865).

In November 1854, an article of the Third Amendment to the state constitution stated that "there shall be one session of the General Assembly holden [*sic*] annually on the last Tuesday in May at Newport and an adjournment from the same shall be holden annually at Providence." But there still remained two capitals, because Newport had been one of America's largest cities in 1790, and the Newport Colony House was the oldest and most imposing of all statehouses. Each new legislature had assembled there on the first Monday in May since 1681. The Superior Court was also located at Newport. But Newport was stagnating,[8] and the economic growth of Providence as well as its political influence in the General Assembly soon reduced Newport's influence. The decision to build a new statehouse helped to convince everyone that a unique capital would be more practical. The Special Committee for a New State House was set up in 1873, composed of two senators and three representatives. They reported that "after full inquiry, the Committee became satisfied that there was great unanimity of opinion among the people in all sections of the state, to the effect that, if a new State House is to be erected, it should be in the City of Providence, and, with this opinion, the Commission unanimously concur" (*Report of the Special Committee* 1873, 9). As usual, some time elapsed between public acceptance and legislative action (Newport leaders needed time to accept the situation). It was not until May 1890 that the General Assembly accepted Providence's donation of $200,000 for the purchase of a site for a new statehouse. Providence still had to wait for the completion of the new statehouse in 1900 to see the state constitution amended (in its Article IX) to make it the sole capital. In 1900, Providence's population stood at 175,597, while Newport's was only 22,441. Providence was also easier to reach from all parts of the state than Newport, which was located on an island at the end of the eastern part of Narragansett Bay.

Readjustment

Readjustment occurred in seven states: Alaska, Colorado, Kentucky, Maryland, Montana, Oklahoma, and Oregon. Contrary to the westward movement in other states, capitals did not move far from their first locations. However, small mileage did not mean small changes. Indeed, a short trip often brought quite different political or economic conditions, or both. This was especially true in territories where population had long remained concentrated near a major industry (e.g., mines or agriculture).

Maryland was granted to Lord Baltimore by the king in 1632. The first capital, the port of St. Mary, was founded in 1634. In 1689, Maryland became a crown colony. A new governor arrived in 1694 and wanted the capital moved from St. Mary's city, because it was the seat of Calvert (the family name of the Lords Baltimore) and Catholic power. He chose Arundelton, which was to become Annapolis the next year, "not because a successful town was already in place or because the area had an unusually prosperous local economy or a deep natural harbor. His choice was a central location where the rural population was politically stable and non-Catholic and where there was a peninsula on rising ground with a reasonably healthy natural environment" (Baker 1986, 192). Annapolis, founded in 1648, was also a good port with a boatyard. As a result, St. Mary's was definitively abandoned, although it revived in the twentieth century as a historical and archaeological park.

While the smallest American state, Rhode Island, experienced a total of seven rotating capitals (five at a time), the largest state, Alaska, only had a readjustment between two small towns. When the population of a territory is sparse, small changes can greatly alter the economic balance and therefore the political one. Under Russia, the capital was situated in the southern part of Alaska, on the coast. The last one was Sitka, on Baranof Island. Created in 1799 under the name of Novo-Arkhangel'sk or New Arkangelsk (Paul's Harbor had already been created in 1792), it soon became the largest town of Russian America, owing to the establishment in 1804 of the headquarters of the Russian American Company by Baranov. After Baranov's retirement in 1818, the governors of Russian America used it as their capital (Naske and Slotnick 1987, 36).

When on October 18, 1867, the United States bought Alaska for $7.2 million, it remained at first under military authority, and Sitka naturally became the military headquarters. But the town's prosperity did not last long, owing to the absence of lighthouse, mail, and land titles, as well as to the paucity of local population (in 1880, there were only 430 white men in Alaska). In 1884, a civil government replaced the military one. Alaska became a district, with a civil governor but without a legislature. Sitka was designated the seat of government, although it never was officially named the capital (DeArmond 1995, 218). The court held its meetings alternately in Sitka and Wrangell. Sitka was also the home of the Alaskan Land Office. Gold fostered a population boomlet, so that in 1896, the year of the Klondike gold discoveries that led to the rapid creation of mining towns (Skagway, Nome, and Fairbanks), Alaskans began to ask for self-government. The government complied in 1900, selecting Juneau over Sitka as the new capital, because it was a larger town, near mines and near the center of population. The governor, a resident of Sitka for twenty-two years, waited until 1906 to move to Juneau. The new capital was still located in southeastern Alaska, only ninety miles east of Sitka. The transformation of Alaska into a territory in 1912 did not change anything, although in 1923 the most populated part of Alaska, the southeast, wanted to secede. A vote was organized, but because Sitka was not informed in time and could not vote, secession was refused. When Alaska became a state on January 3, 1959, Anchorage, built by the Alaska Engineering Commission, had become the state's leading city (44,237 inhabitants in 1960, due mostly to the presence of the army and to the railroad), but Juneau (6,797 inhabitants) remained the capital.

Apparently Erratic or Perpetually Mobile Capitals

For ten states (20%), the spatial pattern makes no immediate sense; their capitals seemed to move endlessly. The title of a book about California's capitals, *The Capital That Couldn't Stay Put*, aptly describes the apparent frenzy that seized legislators during the first years of the new state of California (Oxford 1995). California had seven successive capitals in four years and three months (from December 1849 to February 1854). The Mexican capital, Monterey, was first abandoned for San Jose, then moved to Vallejo, then (temporarily) to Sacramento, then back to Vallejo, then to Benicia, and then definitively to Sacramento. Californians dubbed it "the capital on wheels," echoing the suggestion of a newspaper editor during Alabama's 1846 capital relocation debate, who had proposed that "the capitol be constructed on wheels so every part of the state could enjoy having it for a time" (Rogers et al. 1994, 148).

Even before American settlement, the capital question had never been fully resolved. California was first a Spanish religious colony (organized around Jesuit missions, which were secularized in 1833). In 1825 it became a Mexican territory, under the name of Alta California. Los Angeles and Monterey vied to be the capital; the governor lived in the former, and the latter was home to the legislature and also the only legal port of entry. Los Angeles was founded as a pueblo in 1781 and was granted city status in 1835. In 1846 it had three thousand inhabitants. Monterey, which had fifteen hundred inhabitants in 1849, was described by Commodore Thomas ap Catesby Jones in 1842 as follows: "Most of the garrison were off at work in the fields, fort and guns were in their usual state of decay . . . and everything . . . was quiet, peaceful, and normally dilapidated" (quoted in Cleland 1959, 98). In 1845, an American consul settled in Monterey, for Alta California was attracting American pioneers. Already eight thousand Europeans had settled in 1846, the year of the Bear Flag Revolt and the short-lived California Republic. After a brief spell under military government and the payment to Mexico of $18.25 million, according to the Treaty of Guadalupe Hidalgo, California became a state directly in 1850, gold enabling it to bypass the territorial period. Population had increased to 264,435 as early as 1852, but it took some years to settle the governmental seat.

California legislators were partly responsible for the wandering capital. At the constitutional convention in Monterey, in September 1849, they had indeed provided a clear legal framework for the choice of the capital, but it was far from being stable. Article XI, "Miscellaneous Provisions," section 1, of the 1849 constitution stated, "The first session of the Legislature shall be held at the Pueblo of San Jose which place shall be the permanent seat of government, until removed by law; provided, however, that two thirds of all the members elected to each house of the Legislature shall concur in the passage of such law." Legislators obviously knew that they did not concur enough. The two-thirds rule had been added precisely because of the disagreement among members about the location of the capital.[9]

San Jose had been selected mostly for economic reasons. California's first town (neither a fort nor a mission), the Pueblo de San José de Guadalupe, was founded on November 29, 1777, to make the territory self-supporting through agriculture and handicrafts. During Spanish and Mexican rule, southern California was favored, but the gold rush swung the power toward north-central California. Los Angeles became therefore too southern to have any chance of becoming capital. San Jose, a mere village until the end of the 1840s (524 residents in 1831), began its boom during the gold rush, as it was one of the supply centers for the Sierra foothills mines. It grew so fast that in 1849 it was the logical choice for the state capital. The first California Legislature convened in San Jose on December 15, 1849. This session passed 146 acts and nineteen joint resolutions in four months, making it one of the most active ever. But accommodations were poor in San Jose, and the town was considered too Mexican and too isolated. Therefore, just after the admission of California to the Union on September 9, 1850, a new election was held to choose legislators, state officers, and the capital's location. On October 7, Vallejo was the winner, with 7,477 votes; San Jose received only 1,292, and Monterey, 399. Vallejo at that time only existed in the mind of a prominent legislator, California's wealthiest man, General Don Mariano Guadalupe Vallejo, a leader since Mexican times. He promised 156 acres of his land (out of the 146,000 he owned), and $370,000 for the buildings (from state house to botanical gardens and university) in the "truest center" of the state for commerce, population, and travel, in the great San Francisco Bay. He proposed to name it Eureka, but the legislators changed it to Vallejo to honor him. But when the legislature convened in San Jose in January 1851, many contenders claimed their advantages over the paper town. At last, Vallejo won in February, and in March commissioners selected sites for the capitol (on a hill overlooking the Bay and Napa Valley), the university, the Lunatic Asylum, the Deaf and Dumb Asylum, the Blind Asylum, the Orphan Asylum, and the Penitentiary![10] But work was slow, and the legislators arrived on January 5, 1852, to find an unfurnished capitol. A steamer had to be converted into a hotel to accommodate them. As this did not suffice, the legislature voted, on January 12, for a temporary removal to Sacramento, which took place on January 16, aboard the steamer *Empire*. But on March 7, 1852, Sacramento was flooded. An act was therefore passed on April 8 that designated Vallejo as the permanent capital. On January 3, 1853, the legislature regained a "town" that was still far from finished. Its main promoter, General Vallejo, even asked to be released from his promises. Benicia, which had already offered to become the capital, now donated a newly built, red-brick City Hall with Doric columns for the use of the legislature. Sacramento responded by offering its Court House, but another flood followed by a fire discouraged the legislators. Meanwhile, blatant bribes from the promoters of Benicia convinced them to make Benicia the new "permanent capital" on February 4, 1853. A few days later, all the legislators came to Benicia, and General Vallejo was at last released from his commitment. Although he had suffered great losses, he still owned the land on which Vallejo and Benicia were built (General

Vallejo had also donated the fifteen square miles on which Benicia stands). Benicia had been chosen by the navy, the army, and the Pacific Mail Steamship Company as their headquarters, since it offered a deepwater harbor at the head of San Francisco Bay. In 1853, around one thousand people lived there.

But although the act declaring Benicia permanent capital of California had been confirmed by a new act on May 18, Sacramento still wanted to be the capital. The town could count on the governor, John Bigler, a corrupt Sacramentan, and on the inadequacy of smallish Benicia; City Hall was perfect, but the other buildings and accommodations were not, and Sacramento had been brilliantly rebuilt after the 1852 flood and fire. It had thirteen thousand inhabitants and now boasted several hundred brick buildings and new levees for protection against the Sacramento River.[11] It could also rely on David C. Broderick, a very ambitious young man who had arrived from New York and had rapidly become rich. He wanted to become state senator and persuaded the legislature to advance the senatorial election to 1854, a year earlier than normal, by pledging to support the removal of the capital to Sacramento. This was done on February 24, 1854. On March 1, the legislators met in Sacramento.[12] The story was not yet finished, however, because the Supreme Court refused to come to Sacramento and even declared San Jose the legal capital, because the legislature had never respected the two-thirds majority required by the constitution. But, despite another great fire in Sacramento on July 13, 1854, and despite the building of a new capitol (or City Hall) in San Jose, an appeal before the Supreme Court convinced it to uphold Sacramento's claims. The replacement of a justice who had died by one of governor Bigler's men also helped. The new statehouse welcomed the legislators in January 1855.

California's story exemplifies the roughness and instability of state-building, as well as the passage from an old political and economic order (Mexican agriculture) to a new one (American fortune seekers). A populous state, it was full of speculators trying to exploit all the opportunities of the gold rush. That explains the location of all the main contenders for the capital; they were only a few miles apart (Benicia and Vallejo), and all of them were close to mining operations. The city that eventually won was the largest one and the only "real city" among the contenders. The speculators were not all new arrivals; they came from the East or from old Spanish-Mexican stock (General Vallejo). The wandering of the capital was also due to bribery among legislators. But in this case, bribery did not result in dreadful laws that would deprive citizens of their rights or money (except for the people who had invested in the failed capitals). The cities that could have stayed capitals may not have fared as well as did Sacramento, but, far from disappearing, most of them now belong to the great San Francisco metropolis and are still proud of their old capitol buildings and of their brief moments of fame.

A Quiet Twentieth Century after Decades of Struggle?

Although some state capitals were afraid of losing their status during the twentieth century, the only capital that actually lost it was Oklahoma's Guthrie, in 1910. Contrary to common knowledge, however, during the early twentieth century, several states considered moving their capitals (e.g., Georgia in the 1920s),[13] even some states with apparently strong capitals, such as California. Berkeley tried in 1907 to attract the seat of government, followed by San Jose and Monterey (both former capitals), but with little effect. The monumental 1874 capitol in Sacramento still hosts the legislature (Hansen 1967, 252). The same is true for Tallahassee, capital of Florida since 1824. In 1831 a commission was formed to choose a new capital, but failed to agree. In 1832 and 1843, removal bills likewise failed. In 1881, a bill removed the capital to Gainesville, but it was vetoed by the governor. In 1854 and 1900 referenda on the question were won by Tallahassee (Ellis and Rogers 1986, 26–27). The city's website (2003) reminded people of the fight to retain the capitol:

> Almost since being named as the Capital, Tallahasseeans have fought back various attempts to move the Capital to another City. After the turn of the century, business men have promoted hotels and lodging houses to insure that legislators had places to stay. In an effort to beautify the town, hundreds of dogwoods and oaks were planted along streets and in front yards and have become a symbol of Tallahassee. In the 60s, the town even organized "Springtime Tallahassee," an annual parade and celebration, in an effort to keep legislators from moving the Capital. With the dedication of the new Capital Complex in 1978, the threats of moving the Capitol were put to rest for the time.

Some states, without actually moving their capitals, maintain extensive offices in cities other than the capital (e.g., Washington State). In Nevada, some offices are located in Las Vegas and Reno, and in New Jersey, the governor's mansion is located in Princeton, whereas all the other offices are in Trenton. The influence of New Federalism is also evident. During the 1990s, several states talked of decentralizing some of their offices, allegedly to bring them closer to the people (Lemov 1993, 48). In Oklahoma, a study was commissioned to assess a possible removal of the Department of Commerce to Tulsa. This proved to be too expensive (we are no longer in the situation of 1850s California, when removal costs could be counted in thousands of dollars), and the department remained in Oklahoma City, although some state offices are today located in Tulsa. In Ohio, Pennsylvania, and Florida, several cities and some legislators also used democratic arguments, as well as technological ones (physical proximity is no longer necessary for good and coordinated work), or economic ones (a state has to bring employment to its more distressed regions) to try to capture some state offices, but none of these arguments yielded any tangible result. Are these arguments just the usual complaints about the capital by less happy competitors, or do they belong to a new way of looking at state politics? The first answer seems the best.

There is one exception to the current stability.[14] Alaska's capital, Juneau, is experiencing a very long "narrow escape." The movement began with the talks about statehood at the 1956 Constitutional Convention, during which the delegates from the Anchorage area argued for linking capital relocation with statehood. A compromise was reached, according to which the constitution designated Juneau as a transition capital that could be changed by amendment or popular initiative. The capital question resurfaced quickly, during the first Alaska legislature in 1959, and was raised again by five initiatives (Newton 1977, 165). The first ones, in 1960 and 1962, were defeated. After the discovery of petroleum, however, money was at hand to finance a relocation, and a new relocation initiative, launched in 1972, was adopted in August 1974, despite the traditional opposition of all major state newspapers. A nine-member Capital Site Selection Committee was appointed to nominate no more than three alternate sites. "Each site had to be located west of the 141° meridian or in the main body of the state, which is known as Western Alaska. Each site had to be located on 100 contiguous square miles of state-owned land (or land that the state could acquire at no cost). No site could be located within a radius of 30 miles from Anchorage or Fairbanks. Each site was to be accessible to the Alaska public by air, highway and rail communications" (Newton 1977, 166). Three sites, all situated on the Anchorage-Fairbanks railroad axis, were nominated in December 1975: Mt. Yenlo, Larson Lake, and Willow. The selection committee approved the choice of Willow, seventy miles from Anchorage at the November 1976 election, although the predicted cost was $2.6 billion until 1990. In 1977, the legislature therefore established a New Capital Site Planning Commission. But three years later, section 1, chapter 54, of *State Legislature of Alaska* repealed the chapters of Alaska's constitution linked to the capital relocation and abolished the Alaska Capital City Development Corporation. In 2012, the capital was still Juneau.

This story poses some interesting questions about the status of state capitals. Asked in 1977 by a survey what they thought of the question, Alaskans did not propose clear answers, torn between the wish to have a more accessible capital (Juneau cannot be reached by road or rail), and more important concerns: only "6.5 percent feel that 'moving the capital' is the most important problem facing the state" (Newton 1977, 169). Juneau was preferred by 43% of the respondents; Anchorage by 15%, and Willow by only 13%. Juneau was preferred because it had, according to a survey, "the most spectacular natural setting of any state capital." It is therefore not surprising that the electors rejected a bond issue in 1978 that allocated more than $900 million for moving the capital, thus preventing de facto the removal of the capitol, if not de jure (Naske and Slotnick 1987, 183). The *Alaska Statutes* state that "the purpose of AS 44.06.050–44.06.060 is to guarantee to the people their right to know and to approve in advance all costs of relocating the capital or the legislature; to insure that the people will have an opportunity to make an informed and objective decision on relocating the capital or the legislature with all pertinent data concerning the costs to the state; and to insure that the costs of relocating the capital or the legislature will not be incurred by the state without the approval of the electorate."

To explain the situation, Newton contended that, as the capital relocation movement was fundamentally divisive, Alaska's political leaders always opposed any relocation. Another argument is the secondary importance of the exact location of the seat of government; the third argument is that decentralization of state government and the improvement of communication facilities are making this process useless. Indeed, geographic centrality is no longer a problem. It is more a question of accessibility, and that is why frosty Alaska could still, in the second half of the twentieth century, discuss such a matter. In addition, capital relocation debates can revive geographic factionalism, because they reopen the door to parochial town boosting. But because Juneau is far from symbolizing the state (Anchorage or Prudhoe Bay would be better choices), it is easy to understand the persistence of debate upon the subject.

This initial approach to the process of selecting state capitals clearly underlines that numerous causes lie behind each geographical pattern. Selecting a capital was part of building a state and reflected the multiple factors at work in that process. However, the scientific status of the previous discussion is far from perfect, since it is based either on arguments put forward by stakeholders in the selection process or on their contemporary reinterpretation by social scientists. It is necessary to rank the causes and try to explain them through models or theories. The next two chapters examine these factors to determine how they need to be intertwined, paving the way for the construction of a global explanatory model.

4

In Search of Explanatory Models

Why was a new capital necessary? His [Thomas Jefferson's] stated reasons included Williamsburg's growing inconvenience to the western counties, its vulnerability to attack in the event of war, its unhealthy climate, and its poor situation for the conduct of trade and commerce. Equally important, though, was Jefferson's conviction that Williamsburg provided no fitting display of architectural magnificence.

(WENGER 1993, 77)

Oklahoma's governor in 1890, George Steele, was perhaps right when he stressed that "the public institutions should be located by competitive bidding among the communities that wanted them" (Forbes 1938, 14). But this would reduce the choice to a mere capitalistic deal, whereas many more considerations influenced the process of choice. Competitive bidding was never approved by a territory or a state as the means of selecting a capital, except perhaps for Frankfort and Bismarck: a capitol is not a factory. Likewise, Daniel Boorstin was oversimplifying reality when he argued that the wanderings of state capitals were further proof of the functional conception Americans had of their governments. According to him, when a capital ceased to be useful, or when another location offered more advantages, the capital was moved (Boorstin 1965).[1] If indeed economy and politics were (and still are) intertwined, the cases studied and the epigraph above emphasize that there was often no unique factor. To find a simple explanatory model seems therefore out of reach. A first approach was provided by Clyde E. Browning, who proposed three main causes for the selection of capitals:

1. Centrality, be it geographic or demographic
2. Distrust of large cities, which resulted in a small capital city

3. Compromise between two large cities in opposition to one very large city or between two different parts of the state.

He nevertheless added that this list was "neither exhaustive nor necessarily true for every state capital" (Browning 1970, 40). This chapter goes beyond this first study and scrutinizes each factor to show to what extent it works and to what extent it must be combined with one or several others to come closer to understanding the complex process of choice.

After the preceding chapter's diachronic narration of the geographic movements of capitals, this chapter draws on a diachronic perspective to examine the processes of choosing capitals within the larger framework of American territorial construction. I first synthesize the causes proposed by the sources I found (and interpreted) to further explore their validity. Chapter 5 provides an in-depth analysis of the major causes in order to provide a basis for proposing a global explanatory model.

Seven causal elements are successively analyzed and hierarchized. First, tables 4.1 to 4.3 encompass an initial ensemble of measurable causes. These causes are interrogated after the elaboration of a first model based on chronology. The first causes to be analyzed are perceptions (distrust of cities), because they are common explanations (see Browning 1970). This also provides the opportunity to discuss one of my initial questions, the relationships between Americans and their cities. Economic explanations are studied next, because they figure prominently in tables 4.1 to 4.3 and echo another of my initial questions (Why did capitals often remain small towns?). Finally, we examine politics, the most commonly mentioned cause, along with geography through the notion of centrality, because they are often linked. Other approaches were possible, but this one allows me at once to hierarchize probable causes and to open a discussion of the interpretations I am proposing.

Choosing a capital reveals all the factors of division in each state: geographical divisions (east against west or lowcountry against upcountry), economic divisions (agriculture versus trade and industry), and political divisions, as well as cultural, ethnic, and religious ones, all of which are naturally intertwined. A capital has to be a symbol of unity, but all the factions struggle to represent this unity. The well-known fight in the United States between individualism and community is emphasized and symbolized by this process.

Before beginning to look more closely at the processes at work, we should address two questions. First, was there a difference between the selected towns—did they already exist, at least in an embryonic way, or were they to be created ex nihilo and ad hoc? Second, was there a difference between the choice of the first capital and the subsequent choice(s)? Inquiry shows that although there were differences, similar patterns were at work, as shown by tables 4.1 to 4.3, which also underline the complexity of the explanatory quest. Those three tables present, for each state, the successive capitals and the reasons for their choice. Tables 4.1 to 4.3 do not claim to cover the full wealth of the processes at work, but only to provide the reader with an initial and overall assessment of the complexity involved in choosing the fifty capitals of the fifty states. Since there were often multiple causes, the tables provide the total number of capitals concerned (163) as well as the total number of stated causes (255).[2] Forty-six cities out of 163 were specifically founded to become capitals, an important figure amounting to more than one quarter of the total (28.2%).

The figures that follow present the same processes according to the main factors of change, first generally (fig. 4.1), and then divided into four periods (figs. 4.2 to 4.5). The label "entry point" in these figures refers to the establishment of a first presence in a new world, whether by Puritans from England, commercial companies, or later pioneers beyond the inland frontier. These tables and figures are based on all the sources I could find on the history of American states and state capitals: general and local historical books (e.g., Littrell 1806; Cabbage 1999); articles, pamphlets, and pageants; web resources (e.g., Conley and Campbell 2012) and archives when available, and their verification when possible (when there were several narratives at hand);[3] and on the interpretation of the sequence of events I was able to make with the help of general interpretative frameworks, mostly from the historical, geographical, and political points of view (see the references).

Nine factors were at work. Browning's three factors are indeed present: quest for centrality, rejection of large cities, and compromise resulting from sectionalism. However, they only amount to less than half the explanatory factors and do not cover the whole time-span of the selection process. Six more have been found: compromise resulting from factionalism (shifts between first Federalists and "state-rightists" and later between Democrats and Republicans), defense, point of entry (for newly founded colonies or territories), religion (a motive for migrating to the New World or founding a new colony after seceding from a too Puritan one), economy (be it boosterism or a town's role as supply center for one local staple), and

Table 4.1 Successive capitals (first to third) and the causes of their choice

State	First capital	Second capital	Third capital
Alabama	St. Stephens, 1817–18 Sectionalism, entry point	Huntsville, 1819–20 Temp., to wait for Cahawba	Cahawba, 1820–26 Sectionalism, centrality
Alaska	Paul's Harbor, 1792 Entry point	Sitka, 1799 Commercial and military	Juneau, 1900 Larger (economy)
Arizona	Prescott, 1864 Military, politics (anti-secession) and mines	Tucson, 1867–77 Politics and accessibility	Prescott 1877–89 Politics
Arkansas	Arkansas Post, 1819 Military, entry point (but remote; mosquitoes)	Little Rock, 1820 New town, centrality, speculation	—
California	Monterey and Los Angeles, until 1849 Military and commerce	San Jose, 1849–51 Supply center for mines (but poor accommodations)	Vallejo, 1851 Speculation; new town offered on prosperous S. F. Bay.
Colorado	Colorado City, 1861 Near mines	Golden, 1862–67 Mines and bribes	Denver, 1868 Better located, boosterism
Connecticut	Hartford 1636; New Haven 1638 Entry points, separate colonies (religion)	Two of the four capitals of the Confederation of the United Colonies, 1643–65	Hartford, 1665–1701 NH integrated with Connecticut
Delaware	New Castle, 1704 Transportation hub	Dover, 1777–1779 War and politics	1779–1781: Wilmington, Lewes, Dover, New Castle Politics (factions)
Florida	St. Augustine, 1565 Point of entry, military	St. Augustine and Pensacola, 1763–1824 Two provinces, military	Tallahassee, 1824 Compromise and economic centrality
Georgia	Savannah, 1733 Entry point	Augusta, 1779–1782 Demographic centrality	Savannah and Augusta, 1782–86 Politics
Hawaii	Honolulu, 1850 Royal capital	—	—
Idaho	Lewiston, 1863 Accessibility, mines	Boise City, 1864 Gold and demographic center	—
Illinois	Kaskaskia, 1809 Important French town Economy, politics, religion	Vandalia, 1820–37 Centrality	Springfield, 1837 Lincoln, bargain, and new centrality
Indiana	Vincennes, 1800–1813 French center, "Paris of the West"; economy, politics	Corydon, 1813–24 More central after boundary changes	Indianapolis, 1825 Exact center and compromise
Iowa	Burlington, 1838–41 Entry point (temp.)	Iowa City, 1841–57 New town, centrality	Des Moines, 1857 Further centrality
Kansas	Fort Leavenworth, 1854 (temp.)	Shawnee Mission, 1854–55 Accommodations (temp.)	Pawnee, 1855 Paper town, speculation
Kentucky	Lexington, 1792 Provisional	Frankfort, 1793 Boosterism	—
Louisiana	Mobile, 1717–22 Entry point	New Orleans, 1722–1830 Economic capital	Donaldsonville, 1830–31 Politics, anti–large city

Maine	Portland, 1820–31 Main port (politics, economy)	Augusta, 1832 Centrality	—
Maryland	St. Mary's, 1624 Colonial entry point	Annapolis, 1694 Religion, politics, health	—
Massachusetts	Newtown, 1630 Entry point, religion	Boston, 1630 Better location	Boston, 1692 Creation of Massachusetts
Michigan	Detroit, 1805–47 French colonial entry point	Lansing, 1847 Military, centrality, anti–large city	—
Minnesota	St. Paul, 1849 Entry point, designated by Organic Act	—	—
Mississippi	Natchez, 1798–1802 Economy	Washington, 1802–1820 Politics	Columbia, 1821–22 Temp. (design, 1817)
Missouri	St. Louis, 1805 French capital (politics, economy)	St. Charles, 1821–26 Temp., more central, anti–St. Louis	Jefferson City, 1826 Centrality, anti–St. Louis
Montana	Bannack City, 1864 Governor's choice, economy	Virginia City, 1865 Center of population (mines)	Helena, 1874 New demographic and political center
Nebraska	Omaha, 1855–67 Politics (governor's choice), speculation (fraud)	Lincoln, 1867 Sectionalism, centrality, economy (great salt fantasy)	—
Nevada	Carson City, 1861 Economy, centrality, speculation	—	—
New Hampshire	Portsmouth, 1679–1775 Colonial port, Anglicans	Exeter, 1776–84 War, politics (anti-British)	Wandering capital (8), 1784–1807 Politics
New Jersey	Elizabethtown, 1665 Port of entry	Burlington, 1677 Perth Amboy, 1686 Economy	Wandering capital, 1776–88 War
New Mexico	Santa Fe, 1610 Defense, accessibility, agriculture	El Paso, 1681–93 Temp. during Indian uprising	—
New York	New York, 1626 Port of entry, defense	Rotating (Kingston, Poughkeepsie et al.) War and politics, 1776–84	New York, 1784–88 Politics
North Carolina	Wandering capital, 1664–1792 Factionalism	Raleigh, 1792 Compromise, centrality	—
North Dakota	Yankton, 1861–82 Entry point, patronage (Lincoln)	Bismarck, 1883 (temp.; 1889 perm) Centrality, railroads and boosterism, compromise	—
Ohio	Marietta, 1788 First "Western city," (economy)	Cincinnati, 1790–1803 Centrality	Chillicothe, 1803–1810 Politics
Oklahoma	Guthrie, 1889–1910 Politics, railroads	Oklahoma City, 1910 Politics, economy	—
Oregon	Oregon City, 1849 Centrality, politics, economy	Corvallis, 1855 Repealed act	Salem, 1852 Politics
Pennsylvania	Philadelphia, 1682–1799 Colonial port, religion	Lancaster, 1799–1811 Fever, centrality, anti–large city	Harrisburg, 1812 Centrality, speculation

(*continued*)

Table 4.1 (*continued*)

State	First capital	Second capital	Third capital
Rhode Island	Providence, 1636 Religion Four rotating capitals Democracy	Providence, Newport 1854–1900, twin capitals Practical, economy	Providence, 1900 Practical, influence
South Carolina	Charleston, 1680 Port of entry	Columbia, 1790 Politics and centrality	—
South Dakota	Yankton, 1861–82 See North Dakota.	Bismarck, 1883–89 See North Dakota.	Pierre, 1889 Demographic centrality, railroads
Tennessee	"Rocky Mount" (mansion) 1790–92 First governor's choice	Knoxville, 1792–1813 Sectionalism	Nashville, 1813–16 Sectionalism, centrality
Texas	Los Adaes (1729–72) Politics	San Antonio, 1773–1824 Main town (economy, politics)	Saltillo, 1824–33 Politics
Utah	Salt Lake City, 1847 Religion, economy	Fillmore, 1851–56 Official capital, centrality	Salt Lake City, 1856 Back to centrality, religion
Vermont	Rotating (14 cities), 1791 Sectionalism	Montpelier, 1808 Compromise	—
Virginia	Jamestown, 1609 Port of entry	Williamsburg, 1698–1779 Health, centrality, politics, speculation	Richmond 1779 Security, health, centrality
Washington	Olympia, 1853 Port	—	—
West Virginia	Wheeling, 1863 Principal city, politics	Charleston, 1870–75 Politics, centrality	Wheeling, 1875–85 Accommodations, politics
Wisconsin	Belmont, 1836 Temp. (entry point, politics)	Burlington, 1837–38 Temp. (entry point, accommodations)	Madison, 1838 Speculation, centrality
Wyoming	Cheyenne 1868 Railroad	—	—

Note: "Entry point" refers to the establishment of a first presence in a new world, whether by Puritans from England, commercial companies, or later pioneers beyond the inland frontier.

provisional, or temporary, capitals (approved until a more suitable capital could be chosen). Each factor has to be exactly weighed, and their combination must also be more precisely studied, as well as their possible sequence in time. The idea of progressive stabilization—already emphasized in chapter 3 in the light of geography—permits the creation of an initial and still crude explanatory model, presented in table 4.4.

The first phase, called "entry point" in table 4.4, refers to the establishment of a first presence in a new world, whether by Puritans from England seeking the freedom of faith they had been denied at home, commercial companies seeking fortunes, or later pioneers searching for happiness and wealth beyond the inland frontier. Although religion and defense account for only 10.6% of the causes, they were important. Not surprisingly, their

Table 4.2 Successive capitals (fourth to sixth) and the causes of their choice

State	Fourth capital	Fifth capital	Sixth capital
Alabama	Tuscaloosa, 1826–46 Health, sectionalism	Montgomery, 1847 Centrality, compromise	—
Arizona	Phoenix, 1889 Compromise (centrality, politics, economy)	—	—
California	Sacramento, 1852 Economy (but floods)	Vallejo, Jan. 16, 1853 (temp.)	Benicia, 1853–54 Military, speculation
Connecticut	Hartford and New Haven, co-capitals, 1701–1873, Largest cities and hist.	Hartford, 1873, sole capital Practical, economy	—
Delaware	Dover, 1781 Politics, compromise	—	—
Georgia	Augusta, 1786–95 Economy, centrality	Louisville, 1795–1807 Centrality	Milledgeville, 1807–68 New town, politics
Kansas	Lecompton, 1856–61 (but Lawrence, free-state de facto capital, 1858–61)	Topeka, 1859 (temp.) 1861 Politics (antislavery)	—
Louisiana	New Orleans, 1831–49 Accommodations, "city life" (politics, economy)	Baton Rouge, 1849–64 Politics, anti–large city	New Orleans, 1864–82 Politics
Mississippi	Jackson, 1822 Centrality (and politics, against Natchez)	—	—
New Hampshire	Concord, 1807 (1814 perm.) Politics	—	—
New Jersey	Burlington and Perth Amboy, co-capitals, 1788–90 Politics	1790, Trenton Accessibility, politics	
New York	Albany, 1797 Politics, centrality		
Ohio	Zanesville, 1810–12 Politics	Chillicothe, 1812–15 Politics	Columbus, 1816 Centrality, compromise
Tennessee	Knoxville, 1817–19 Sectionalism	Murfreesboro, 1819–26 Sectionalism, centrality	Nashville, 1826 (1843 perm.) Centrality, economy
Texas	San Felipe de Austin, 1836 Founder, entry point	Harrisburg, 1836 Speculation	Houston, 1837–39 Speculation, war
West Virginia	Charleston, 1885 (referendum) Centrality	—	—

Note: "Entry point" refers to the establishment of a first presence in a new world, whether by Puritans from England, commercial companies, or later pioneers beyond the inland frontier.

Table 4.3 Successive capitals (seventh to ninth) and the causes of their choice

State	Seventh capital	Eighth capital	Ninth capital
California	Sacramento, 1854 Large supply center for mines; corrupt governor	—	—
Georgia	Atlanta, 1868 Military HQ, railroad	—	—
Louisiana	Baton Rouge, 1882 Politics	—	—
Texas	Austin, 1839–42 Economy and politics	Washington on the Brazos, 1842–45 Wartime capital	Austin, 1845 Back to normalcy

Note: "Entry point" refers to the establishment of a first presence in a new world, whether by Puritans from England, commercial companies, or later pioneers beyond the inland frontier.

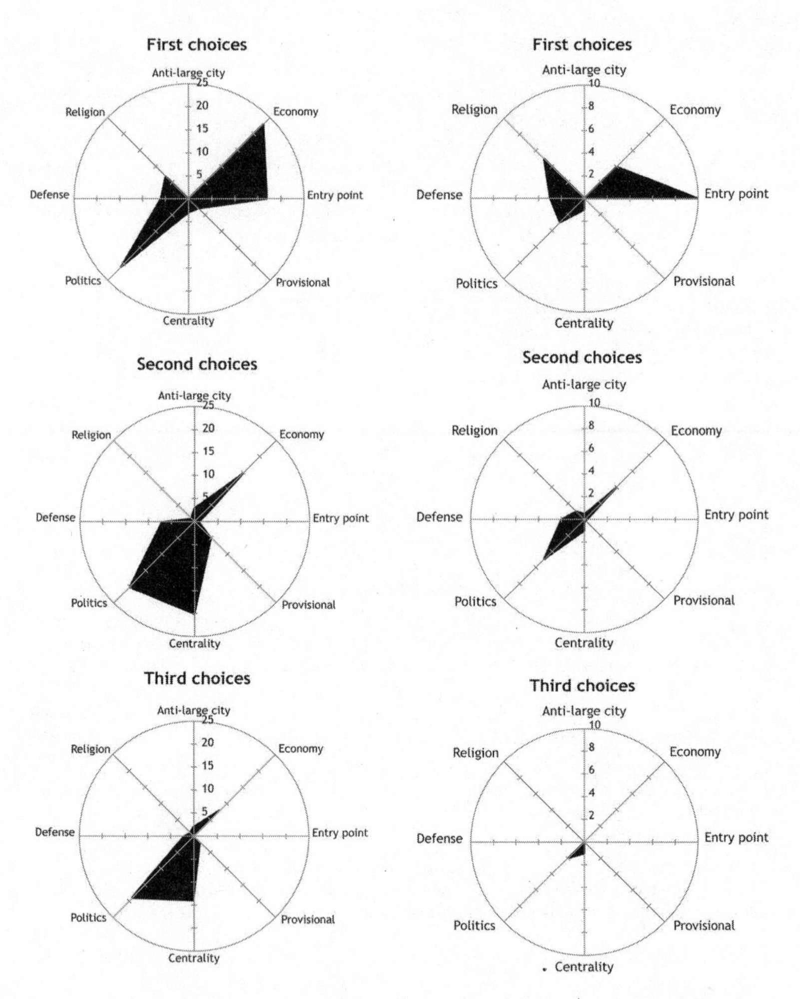

FIGURE 4.1 (LEFT) Major causes of the choice of capitals (in general). Credit: Christian Montès and M. L. Trémélo.

FIGURE 4.2 (RIGHT) Major causes of the choice of capitals before 1776. Credit: Christian Montès and M. L. Trémélo.

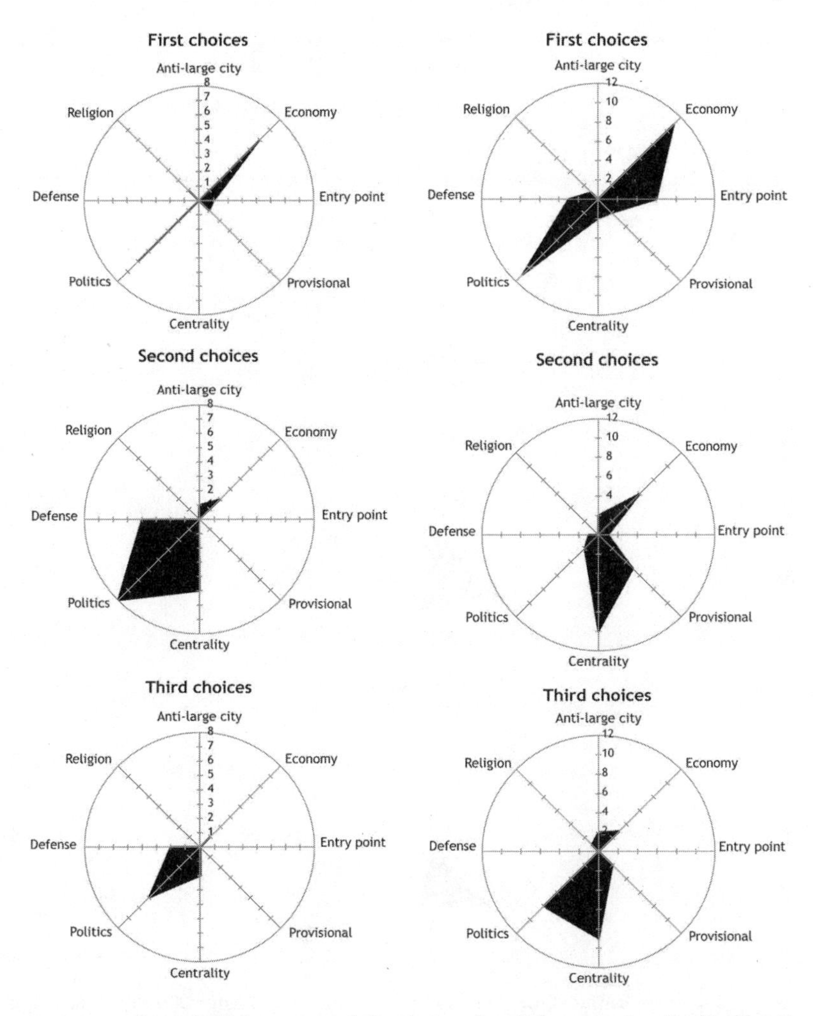

FIGURE 4.3 (LEFT) Major causes of the choice of capitals, 1777–1812. Credit: Christian Montès and M. L. Trémélo.

FIGURE 4.4 (RIGHT) Major causes of the choice of capitals, 1813–65. Credit: Christian Montès and M. L. Trémélo.

influence was strongest mainly during two periods: before 1776 and between 1813 and 1865. But why did the pilgrims land on the northeastern coast, the future seat of the megalopolis? The common thought at the time was that they had been sent to the most inhospitable part of the continent (Earle 1992, 59–87).[4] Speculative colonies were located on the southeastern coast, which was believed to be the richest one. Since the first European colonies were mostly sugar islands, and spices were in demand, the warmest parts of

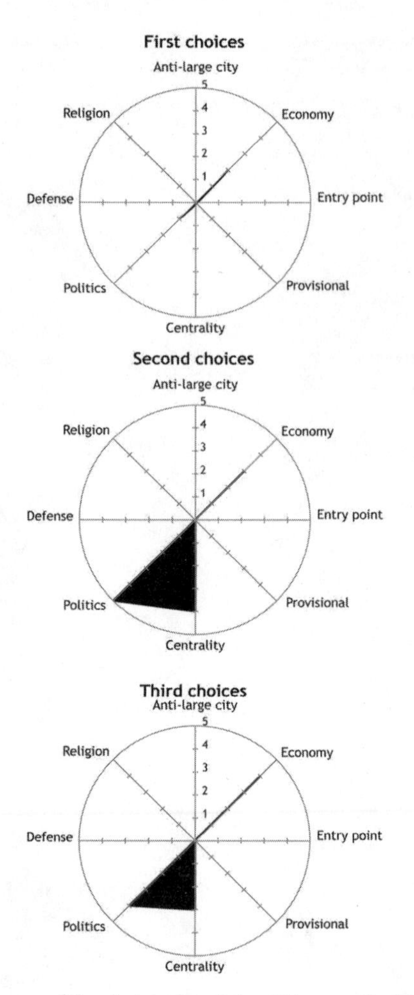

FIGURE 4.5 Major causes of the choice of capitals, 1866–1910. Credit: Christian Montès and M. L. Trémélo.

Table 4.4 An initial explanatory model

Phases	Name	Main events
First	Entry point	Colonial port, inland gateway; role of founder of colony, Congress, or of first governor
Second	Building territories	Successive capitals: health, lack of accommodations, sectionalism, westward movement
Third	Choice of permanent capital	Search for centrality and compromise between sections, after settlement was more advanced
Fourth	Doubt	Failed attempts to unsettle the capital, resulting from spatial inertia and political standstill
Fifth	Definite stability	Occurs in the twentieth century only, sometimes as late as in the 1950s

Note: "Entry point" refers to the establishment of a first presence in a new world, whether by Puritans from England, commercial companies, or later pioneers beyond the inland frontier.

the new colonies were thought to be the most hospitable, with the richest agricultural promises. The most refined rural and urban societies would arise in Charleston and Savannah. New York is an exception to this pattern because it was interesting from a strategic point of view, being located at the junction between the northern part of the British American empire, Canada, and its southern part, the thirteen colonies, accessible through the Hudson and the Mohawk valleys.

Salt Lake City was founded for similar reasons when many Mormons had to flee the ire of their fellow Americans and find a remote shelter. After the assassination in 1844 of the founder of their religion, Joseph Smith, and their expulsion from their first ideal city, Nauvoo, Illinois, they fled to the West in 1846, where they created a new "state" in 1848 in a quasi-desert, called Deseret, which means "honeybee." It was far from the main existing land routes; the Oregon Trail ran north of it, and other trails were to the south. Salt Lake was chosen as main settlement and capital because it appeared rich in these barren lands and had a strategic location. Economy was thus not forgotten. The initial impression of the pioneers was indeed very positive. The Salt Lake valley had "fertile soil, good vegetation, a generally favorable climate, plenty of timber in the mountains, abundant water in the mountain streams, and a place that promised golden opportunity for agricultural success" (Alexander and Allen 1984, 23).[5] In 1850, Utah became a territory with Salt Lake City as the official capital.

Defense was important in the West. The first western capitals were often either forts or founded near one. When new territories were created, treaties still granted the Indians part ownership of them, and forts were built to "protect" the pioneers. Fort Leavenworth became capital of Kansas in 1854; Fort Des Moines, of Iowa in 1857; Prescott, of Arizona in 1864; Sitka, of Alaska in 1867; and (in a different context) Honolulu (with the Pearl Harbor Navy base), of Hawaii in 1898. All are examples of the links between defense and capital choice.

The second phase belongs to the territorial construction of the United States of America. Since this period was highly unstable, numerous states had many capitals in rapid succession. First, independence induced changes in capitals because early capitals were often associated with the hated British rule, although Americans had known a certain degree of liberty during colonial times. Such was the case in Virginia. The new capital, Richmond, should express, according to its promoter, Thomas Jefferson, "the social and political order then taking shape" (Wenger 1993, 81). But there were also some very simple causes. The small size of local governments and the low cost of removal certainly played a role in some apparently "unnecessary" movements. When Michigan's legislators decided in 1847 to move the capital from Detroit to the new town of Lansing (then still called Michigan), only one wagon was necessary to move the state's records. In many other states, British occupation during the War of Independence or earlier fires had considerably reduced the amount of archives and records to move. For California, the cost of the removal from Monterey to San Jose (1849) was only $1,100; the subsequent move from Vallejo to Benicia (1853) cost $12,000 (Oxford 1995, 62). Second, during this early period, fires were common in capitols. The buildings could be replaced easily, however, since they were small wooden structures, often not built for that purpose at all (former city halls, courthouses, or even inns were used). The fact that most other houses were made of wood and the lack of proper firefighter corps allowed fires to spread and destroy vast areas of many towns.[6] Williamsburg's capitol burned in 1698, Montgomery's in 1849, Austin's and St. Paul's in 1881, Madison's in 1904, and Bismarck's in 1930. In Raleigh, the third big fire (in 1831, after those of 1816 and 1821) that destroyed the statehouse, with its famed marble statue of George Washington by Antonio Canova, was the last straw for the legislators, who proposed to move the capital "to anywhere else." For a single vote, the capital would have been removed to the confluence of the Cape Fear and Haw Rivers (Waug 1967, 8).[7] Securing the archives was sometimes the most important factor. Austin, Texas, almost lost its status as the capital only three years after it obtained it. Commercial trails were still unsafe and unreliable, native Americans were not always friendly, and Mexicans definitely were not. After the latter invaded San Antonio in 1842, the capital left Austin for three years. Austin's citizens sternly refused to allow the removal of the archives to the new capital, Washington on the Brazos, which "probably saved the embryonic city" (Wheeler 1968, 5).

Third, the tumultuous conduct of politics early on could deter some cities from retaining the capitol. Houston was the capital of independent Texas from 1837 to 1839. When President Lamar transferred the capital to Austin in 1839, Houston's commercial leaders, instead of crying foul, were relieved because, during the legislative season, "statesmen brawled with each other outrageously, often without cause, and partisanship flamed high. The removal of the capital thus permitted the town's promoters to center attention upon business" (WWP Texas 1940, 293). By 1841, they had created the Port of Houston, key to the town's future growth.

Brawling and partisanship were at the heart of the second, the third, and the fourth phases. But this model only tells part of the story, primarily because there were many exceptions to this sequence of events: eighteen states did not need three phases to select their permanent capitals, and sixteen states needed at least a fourth phase to attain stability. Second, and more important, the causes at work in a particular phase were not identical. In Georgia, for example, which was under military rule after the Civil War, the army's chief, General Rope, chose the location of the convention elected in 1867. He rejected the former capital, Milledgeville, on the official ground that its innkeepers were refusing to house the thirty-seven black delegates (Coleman 1991, 211). Such racist behavior was certainly an important factor in early Reconstruction times, but Atlanta was chosen mainly because the Yankees had made their military headquarters there, due to the city's primary importance as a railroad hub. The causes were therefore military, moral, political, and economic in varying degrees.

We must therefore go beyond this first chronologically based attempt and look more closely at the processes at work. They can be summarized under what I call the "BBC theme": boosterism, bribery, and compromise (the latter divided in two *B*s: bargain and barter). Underneath this "theme" lies the cultural aspect of the question. Since capitals embodied territorial and urban visions of the nascent United States, it seems logical to begin with culture. The "distrust of the cities that resulted in a small capital city" was the second factor of explanation put forward by Clyde E. Browning (1970). It is therefore necessary to explain why there are only 7 occurrences of such feelings out of the 255 causes found (a mere 2.7%).

Capitals and Puritanism; or, the Erroneous Religious Explanation

A TRIP TO COLUMBIA

Come, *Citizens*! take up your parts,
To *Columbia's* sweet Banks we'll repair;
Let's quit the gay scenes of the Town,
Pregnant with suits and despair.
Near the centre of State we'll convene,
Sagely devised by the *People*;
Adieu, to your pompous parade!
Adieu, to your Balls and your Steeple!
Our roads may be sandy and ruff,

Margin'd with fine lofty pine,
Directing our drover along,
With herds of noble fat swine.
Our amusement, is *Business of State,*
Centrically placed for our ease;
Without shrimps, or oysters in pies,
So *delegates* fare as you please.
Your fair may out-dress us we own,
An artless appearance we boast;
The feather which fills you with pride,
Is not *the Columbian toast.*
This movement of Records may be
American plodding or Dutch;
But the object well answers the end,
And happy the state that has such.
Avaunt, then, you Nabobs of pomp;
You fine powder'd Bucks of the city;
With curl irons fixed in each pate,
With Lasses sweet scented and pretty.

State Gazette of South Carolina, December 31, 1789 (quoted in Moore 1993, 49)

This satirical poem leads us to wonder if the choice of a capital results from the fight—very Jeffersonian or Puritan—between pure countryside and rotten cities. Are capitals the "Christian Spartas" of Samuel Adams? Raleigh's capitol, finished in 1794, was intentionally built to face east, toward Jerusalem (Waug 1967, 5). Salem was the name given to thirty-two American cities and towns besides Oregon's capital, more than for any other capital. The Puritan division of cities between Jerusalem and Babylon goes back to the Bible. In St. John's Book of Revelation, Paradise takes the form of a city, Jerusalem (Rev. 21:1–22:5), while men built Babylon (Rev. 17–18), a city that was destroyed because it deified humans and their works. Thus the New Jerusalem had nothing to do with the actual Jerusalem, which was just another example of mankind's errors. The New Jerusalem is "the house of God with men" (Rev. 21:3), where citizenship stems from communion with God and not from the link with the Beast and the paramount importance of economic, political, and social status. New Jerusalem is therefore on another level: it belongs to faith and not to historical reality. To flee Babylon is not actually to flee the city, but to live inside it in a new way, "being in the city but not of the city, knowing that the true city, the one from which one receives his identity, is elsewhere" (Cuvillier 2001, 8). This means that the wish to build a terrestrial New Jerusalem is a too literal and false interpretation of the Bible. It stems from the way new Americans understood their country. America was seen as a new beginning for humanity, as a New Jerusalem, in pastoral surroundings.[8] Such a view belongs to myth, because it envisions the Millennium, the thousand years of Christ's reign, beyond history (Marienstras 1991, 62, 76).[9] Although early colonizers tried hard to realize such a wish (think of the notorious Salem), it soon became clear to the majority that a saintly life was better achieved on a personal basis than by trying to impose sainthood on entire, unwilling communities. Likewise, chapter 2 showed that the model followed by capitals was a lay and not a religious one. Capitals reproduce democracy, not theocracy. They therefore aim at reviving the antique Agora, where Aristotle's Peripateticians wandered, discussing the world.

Focusing on Puritan ethics leads us also to forget that the old colonies considered ports (i.e., trade) as important from the beginning. William Penn was certainly a Quaker seeking religious freedom, but he likewise wanted to organize a profitable colony based on market towns. The Puritan ideal ("the city upon the hill" with a covenant) disappeared as early as during the eighteenth century in favor of towns founded on real-estate speculation (Lingeman 1980, 33, 53). When Jefferson expressed his ideas from Virginia, where cities remained rare, New England had already gone beyond the agrarian community modeled on those of England and followed urban and capitalistic logics. This was especially the case in capital cities. Eighteenth-century Americans "believed that the mere honor of hosting the capital would bring a place tremendous gains in population, trade, business, and revenue" (Zagarri 1987, 25).[10] Such processes proved even truer for the westward expansion: in the mountains, the frontier was urban, and farmers were fast becoming rural businessmen (Smith 1966, 211). Jefferson himself, at the end of his life, was aware that his ideas were utopian. The American "character" and society since the Jacksonian Era present social visions and practices that are far removed from Puritan ideas. Indeed, during the nineteenth century, several religious minorities once again pursued the same utopian purpose—the Mormons being the most well known—but mainstream America had long become more practical than zealous, materialism being "the transcendent American value according to most contemporaries" (Pessen 1985, 24). Pessen added that Americans were hypocrites who hid only prudishness behind their moral stances and went to church as much out of social conformity as out of faith (67–70).

Thus, we cannot accept the Jeffersonian model, which oversimplifies the reality. Only one capital bears a variation of the name of Jerusalem, Oregon's Salem, founded in 1840. It was christened by Methodist missionaries who made the native village of Chemeketa their headquarters in 1843 after fleeing the floods and unhealthy climate of Oregon's first "capital," Champoeg. But the choice of Salem by the legislature in 1850 owed less to divine intervention than to the spacious rooms the Oregon Institute could provide, the "most important and best used building west of the Rocky Mountains" (Minto 1902) and to the power of the state's Democrats. The website of the city's Chamber of Commerce nevertheless boasts in 2012 that it has "twice been selected [as] an All-American City [first in 1961] by the Citizen's Forum of Self-Government / National Municipal League."[11] But we cannot entirely discard the Jeffersonian model, for moral arguments were often put forward during discussions about the removal of state capitals.[12] When Virginia's burgesses in the spring of 1749 proposed a bill (which eventually failed) calling for the planning of a new capital city near Newcastle, morality was one of the strong arguments. "Because the Morals of the Youth of this Colony educated at the College are greatly depraved by the evil Examples they see from the Numbers that flock to this Place at the public Meetings, the Impressions that are receive'd at those Times being too strong for all the Care of the Masters to overcome. And we are persuaded that while the Seat of Government is continued, that Evil will increase, and our Prospect from Corruption of the morals of the rising Generation is a very melancholy Consideration" (quoted in Reps 1972, 187).

In reality, the main reason for separating the state capital and state university was not the protection of the youths' morality, but the spoils system, which rewarded the losers in the capital location fight with such prizes as the state penitentiary or university. Later in the eighteenth century, the same moral opposition was brought forward to move Pennsylvania's capital from Philadelphia, out of "a symbolic rejection of the dominance of the state by the nation's largest city.[13] Western farmers felt the City of Brotherly Love harbored an unholy gaggle of merchants, bankers, pettifogging lawyers, old Tories, and, worst of all, government officials of all varieties. Moreover, the theatrical performances which took place there clearly marked the city as decadent and immoral" (Sneddon 1971, 349). But even here, the main causes were geographical. The western part of Pennsylvania was growing and had sufficient market towns, such as Harrisburg, Carlisle, Easton, York, Lancaster, and Pittsburgh, to be able to get along without Philadelphia.

During the nineteenth century, the moral theme appeared in the cases of St. Louis, Baton Rouge, and Detroit. Is it a coincidence that Catholic French had founded all three capitals and that the moral stance was taken by tight-lipped, Protestant Anglos? When St. Louis lost the capital, the causes were more cultural than purely geographic, although it was located far from the center of Missouri. St. Louis had been the political giant since settlement began, but this situation widened the gap between rural Missouri, which suffered crises like the banking crisis of 1819, and its prime city. Farmers mistrusted the capital's establishment, and as soon as they gained enough power by apportionment, the capital was moved to St. Charles in 1820, and definitively to Jefferson City in 1826.[14] Mistrust remained high well into the twentieth century: "Even at the present time, St. Louis remains curiously independent of the rest of the State; nor has friction between it and the rural sections completely died out" (WWP Missouri 1959, 301). Note, however, that St. Louis had only fifty-six hundred inhabitants in 1821, up from twenty-six hundred in 1812. It had yet to become the metropolis of the West.

In Louisiana, due to an apportionment system that largely favored the countryside (and therefore the Protestants), the mood was against the big Catholic commercial city, New Orleans. In 1825, the legislature voted to move the capital out of New Orleans, which was supposed to have a bad influence on legislators because it was a "modern Sodom," according to a contemporary newspaper. Once a new statehouse was completed in 1830, legislators went to the new capital, Donaldsonville, but because of poor accommodations and a leaky roof, they soon returned to New Orleans (Davis 1959, 168). This did not settle the debate. Since agitation never ceased, the question arose once again at the 1845 convention. The capital's newspaper, *L'Abeille de la Nouvelle-Orléans: Journal Officiel de l'Etat* (the *New Orleans Bee*), which naturally favored the city, was outraged. Its arguments were based on the idea that the legislators were denying the people's right to choose in the name of a dubious Puritanism. It ended by adding that the people could revolt against such a decision and that "a Constitution based on hatred and injustice would not last."[15] Strangely enough, the article was followed by a review of the opera played the day before—Bellini's *I Puritani* (1835)! The arguments put forward by the New Orleans promoters against the Sodomite theory were sound, showing that the issue was at least as much political and rhetorical as purely moral.

The next year, the legislature examined the candidates. Amid serious debates, a humorous event occurred when one New Orleans representative took the Puritan logic to its ridiculous end. The *New Orleans Bee* was naturally happy to provide with an account of the day, February 27, 1846. A certain Mr. Lyons had proposed to his colleagues at the Chamber a new, unknown town called Fontania. The proposition was seconded by a Mr. Bullitt, who, despite knowing absolutely nothing about Fontania, praised that town, arguing that since the convention considered cities as sores, the larger the city, the larger the sore. So, to avoid any sore, the convention should avoid any city at all, and thus make the fictitious Fontania the capital! Fontania was naturally discarded. But the following day, the convention had to consider about thirty candidates, some of them only slightly more alive than Fontania. Among them were a former capital, Donaldsonville, as well as Jackson, Harrisonburg, Opelousas,[16] St. Martinsville, Marksville, and Baton Rouge. Favored by the senate and far enough from New Orleans not to be "contaminated" by it, Baton Rouge was selected a few days later as the new seat of government. The capitol therefore left a city of about 115,000 inhabitants in favor of a town of approximately 3,000.

In Michigan, when the 1847 deadline for selecting the new state capital neared, "the outcry against the 'Detroit influence' and demands for a 'healthier political climate' grew more strident," according to William Upton, member of the Chamber (Upton 1990, 390). Detroit (with about twenty thousand inhabitants) was indeed dethroned, but for other causes, as New Orleans had been the year before.

Such high morality was not common. As early as 1795, in small Columbia (350 inhabitants), South Carolina's new capital, the attempt of some legislators to ban horse racing within twenty miles of the statehouse remained unsuccessful (Moore 1993, 48). Likewise, in 1834, Tennessee's delegates debated whether the new capital should be located in Nashville or Murfreesboro, a city criticized as lacking sources of amusement (Bergeron, Ash, and Keith 1999, 94). Unlike Baton Rouge, Murfreesboro lost: capitals were usually chosen according to politics, very seldom according to morality. The same happened when the village of Argyle invited the New York state capitol to relocate there in 1863, arguing that "it is a very moral and religious place, and can be recommended as a location on account of presenting no temptations to legislators to depart from the paths of virtue" (quoted in Roseberry 1964, 15). This moral stance did not strike a chord among the legislators. It could even provoke laughter. During the 1857 debate upon the possible relocation of Vermont's capital, Senator Benton stated that "he wished to say that he had previously intimated his private wishes to have the capital located at Lunenburgh. But on second thought, he believed that his constituents were a good moral people, and thought it prudent not to place temptation in their way. [laughter]. He had therefore voted for White River Junction" (*The Capital of Vermont* 1857, 10). In early twentieth-century Oklahoma, Senator Russell proposed a new capital to resolve the angry battle between Guthrie and Oklahoma City. The senator had proposed in 1908 to build a new capital city, to be called New Jerusalem, near the geographic center of the state. An initiative petition of 22,891 signatures was submitted in 1910. Oklahoma City responded with urban embellishment: a landscape architect designed a system of parks and boulevards. New Jerusalem was first defeated in the House and then at a statewide election. "At Oklahoma City special trains were arranged to tour the state and canvass each town for votes. . . . Oklahoma City claimed the capital should be there because that city was the center of amusement, recreation, shipping, and the grain and meat-packing industries" (Forbes 1938, 31). Amusement and recreation apparently carried more weight than a stern New Jerusalem.

In fact politicians usually used morality as a mask for other interests. Consider the example of Colonel George, one of the promoters of Concord, capital of New Hampshire. When the capitol had to be enlarged in 1863, Manchester offered $500,000 for a new capitol to be located in Manchester, the state's largest city.[17] Colonel George spoke ardently against the proposal. "He argued that the General Court would be subjected to possible evil pressures if it were moved to Manchester, because its burgeoning textile mills corporations were financed by Massachusetts capital. He praised Concord citizens as "steady, sober, and industrious," compared with an "80 percent migratory" population in Manchester. George capped his presentation by stressing that Manchester mill hands lived in boardinghouses, "without roots," and were given to "passions and excitements and tumults which might be harmful to legislative deliberations." His opponent in Manchester, Judge Clark, was quick to see the bias in the argument and "suggested that Concordians had a lot of nerve to bespeak corporate influences when so many of its prominent citizens (including Governor Gilmore) were on the payrolls of the railroad" (Anderson 1981, 335).

Clark rightly added that Manchester's mill-owners had more than enough money to travel to and lobby in Concord, which was only fifteen miles distant. Concord won by 158 votes to 132 (Anderson 1981, 327).[18] This story shows that small towns no longer retained an image of purity and that state capitals were already considered lacking in morality. In 1859, Springfield, capital of Illinois since 1839, counted no less than twenty gambling houses for about 9,000 inhabitants—among them Abraham Lincoln—"some of which were fitted up as elaborately as the glittering establishments of St. Louis and New Orleans" (Angle 1950, 195), cities with, respectively, 160,773 and 168,675 inhabitants.

There are, as usual, exceptions. Salt Lake City, Utah's capital, under the religious and political influence of the Church of Jesus Christ of Latter-Day Saints, is a well-known example. But Salt Lake City was also the leading commercial and financial center of Utah from the beginning and resembled any other American city by the 1880s (Alexander and Allen 1984, 4, 11). A less-known example was Lincoln, Nebraska's capital. When the writers of the Works Progress Administration described it in 1939, they remarked that it was "once known as the 'Holy City,' because it had well over 100 churches, or one for every 700 people, [and] was avoided by criminals. The city is a religious and educational center rather than an industrial center" (WWP Nebraska 1939, 177). But Lincoln did not become the one "American Jerusalem" among state capitals; the 1939 text went on to say that "moral corruption" began in 1936 when the ban on Sunday movies was lifted.

Nowadays, the city is often seen as the only place where one can achieve wealth and happiness. "Although the cities relegate minorities to the ghettos, Texas is the New Canaan."[19] The style is still biblical, but the waters and wealth of Canaan replace the Great Temple of Jerusalem. The Bible was therefore not the main factor behind the choice of capitals. It nevertheless remains true that capitals are often small cities. It is now time to look at the small-town ideal, which draws more upon secular than religious themes. This enables us to analyze how Americans look at their cities.

Capitals and the Small Town Ideal

The country town is one of the great American institutions, perhaps the greatest, in the sense that it had . . . a greater part than any other in shaping public sentiment and giving character to American culture.

(VEBLEN 1923, CITED IN LINGEMAN 1980, 7)

Much has been written about the relationship between Americans and their cities that favors the small town as the ideal type of American community (Weil 1992, 9–29). A small capital would be the realization of such an ideal. Morton and Lucia White provided a famous seminal work about this question (White and White 1962) that were critical of large cities. Although they concentrated on a specific, small social group—the intellectuals—many other writers contended that most Americans were against large cities. In reality, the picture is less sharp, and the relationships more diversified and less Manichean, as is evident from diaries, newspapers, booster pamphlets, and travelers' stories.[20] The city is commonly seen from a moral perspective as a corrupted place, but ambiguity reigns, because it is also considered a progressive place that fosters social and economic mobility (Fries 1977, xv, 161). Americans tried to accommodate both visions in a nation that was more rural than urban before 1920. The country consequently held the political power without the help of malapportionment or gerrymandering, tricks that were later used by rural leaders (Rourke 1964).

There have indeed been some heated political arguments in the United States against large cities, similar to the findings of Ted Margadant for Revolutionary France during the process of choosing the *chefs lieux de département*. French citizens had recourse to a "rhetoric of denunciation" and to "egalitarian slogans and populist rhetoric that challenged the 'aristocracy' of large towns" (Margadant 1992, 442). Democracy was thus linked to the multiplication of places of power. Likewise, Tocqueville argued that the absence of a "great capital city" was "one of the first causes of the maintenance of republican institutions in America" (quoted in Bender 1987, 9). The other side of the coin was the creation of the myth of the idyllic small town, the prototype of which is described in Harriet Beecher Stowe's *Old Town Folks* (Höbling and Franzens 1990). But such rhetoric was not universal. Other views of the city existed simultaneously, such as the myth of urban pastoralism that described America as an urban-pastoral society, where nature and culture could be harmoniously combined, as was already the case for Washington, DC, where Pierre L'Enfant wanted to create a "compleat garden" (Machor 1987, 5–7).[21] Although strongly linked to the Bible, such a vision was also tied to the historical conditions of the colonization of America and the progressive theory of history.

Another conflicting view contended that cities had democratic advantages over the country. Among the arguments put forward by the *New Orleans Bee* during the 1845 Louisiana capital controversy was the idea that laws passed in cities were better than laws passed in small towns. According to the newspaper, cities provided better information and nurtured an enlightened citizenry and press; in the country, plotters would have an easy time, due to the lack of a proper public opinion with conflicting views. The editor also noted that smaller capitals like Albany (about forty thousand inhabitants at the time) and Harrisburg (with about seven thousand) were notorious for inequities that would never have been tolerated in New York or Philadelphia. He was quite right. Even popular literature, although filled with traditional values, was not as anti-urban as one might have thought: these books, "filled with numerous ambiguities that revealed both the lights and the shadows of the urban world, nevertheless broadcast the lure of city life" (Siegel 1981, 176).

The image of the city changed with the growth of municipal services and public works, to the extent that "by the close of the century, American city dwellers enjoyed the highest standards of public services in the world" (Schultz 1989, 215).[22] The idea of manifest destiny was inseparable from the movement of "civilization"—and the development of cities—toward the West, although the celebration of the western city kept a place for the cowboy image even after the 1890s (Hamer 1990, 231–32). At that time, "for three quarters of the nation, the small town was Mecca, Athens, Rome, and Babylon all rolled in one–on a much humbler scale, of course" (Lingeman 1980, 262). Indeed, the pre–Civil War belief that Western expansion would spare the United States the evils of European manufacturing towns had proved false, and Americans had accepted manufacture as well as the large cities that came with it. Moreover, American thinkers, compared to French, British, and German ones, were the most pro-urban of all from 1850 to the end of the 1930s because of the urban conditions of life and the "exuberant optimism of the American people" (Lees 1985, 311). Cities were seen as places of opportunity, and capitals were no exceptions.

It was only between the two world wars that popular literature reinvented the small-town myth, at a time when the "real" small towns were fast disappearing. William A. White provides a typical text from this period: "These western county-seat towns house their citizens more satisfactorily, give them more breathing space, provide [more] of the physical and spiritual blessings of life today for the average citizen than any other kind of human habitation. . . . These new thriving county trade centers of the West are the social safety valves which carry forward into the twentieth century all the energy which realized the vision of the nineteenth century pioneers" (White 1939, 26–27).

But some writers did not concur, such as Sinclair Lewis, who, in his novel *Main Street* (1920), ferociously criticized all the shortcomings of small towns. This did not prevent the small-town myth from being revived once again after 1945, especially in light of the lurking urban crisis, although an alternate version was offered, for example, in Grace Metallious's novel *Pey-*

ton Place, which became a popular television program in the 1960s. Still, the suburban ideal, the federal Main Street program (in effect since 1980), the new towns of Seaside (1980s), and the Disney Company's celebration of small-town life in Florida (1994) testify to the enduring favor enjoyed by the images of the picket-fence house and the shop around the corner in contemporary America (Agnew and Smith 2002, 281–87).[23] Richard Francaviglia concludes rightly that "Main Street is, in fact, one of our most important and persistent statements about America in time and space" (Francaviglia 1996, 192). He uses the word *statement* rather than *reality,* because the small town today owes much more to myth than to reality. The nineteenth-century small-town ideal was already a tentative re-creation, fostered by local elites fighting regional decline, of the New England "city upon a hill" envisioned by seventeenth-century people (Wood 1997). A 1998 book bemoans the "thrashing" of small towns by federal policies. According to its author, they "have become the repository of aging men and women who are living out their lives along shaded side streets, getting by on modest pensions, carefully monitored savings accounts, Social Security, and Medicare, . . . and 1990 census data confirm that those 11,897 towns of less than ten thousand people contain larger concentrations of the poor and elderly, on a percentage basis, than American cities" (Davies 1998, 192–93).

Capitals are therefore no more the concretization of the secular ideal of the small town than they were the concretization of its religious one. American identity is built around shared ideals far more than around one specific territory; hence the ideological expansionism of the nation (Agnew and Smith, 2002). This is perhaps partly explained by the fact that most state capitals were chosen before the Civil War, when local and state identities were strong, and when a national identity was still to come. Gary Nash provides a literary summary of the real forces at work in American cities as early as the eighteenth century: "The novus ordo sec[u]lorum that large numbers of city dwellers sought was not Adams's 'Christian Sparta' but an unfettered stage on which a drama composed by Adam Smith could be played out by actors who spoke not of the fair wage and the just price but of the laws of supply and demand and the greater benefits to the community of unbridled competition and consumer choice freed from the restraining hand of government" (Nash 1987, 23). The booster model therefore seems better for explaining the choice of a capital, for 23.1% of the causes presented in tables 4.1–4.3 were economic. If we include "entry point" (7.5%), which also has an economic sense, the figure climbs to 30.6%.

Capitals and the Booster Model

Boosterism—that uniquely American combination of faith in the future and vociferous promotion.

(ABBOTT, LEONARD, AND MCCOMB 1994, 71)

Capitals fully participated in the construction of urban America, in a context of intense competition among communities. To be the colonial, territorial, or state capital had (and still has) not only political importance, but also an economic significance: it meant that the town was supposed to survive, to expand, and not to dwindle into oblivion like so many beautiful paper towns that ended as ghost towns. This was well illustrated by the arguments that supporters of Fayetteville put forward after North Carolina's capitol in Raleigh burned in 1831. According to them, Raleigh was only a small village, a mere paper capital, "with no promise of ever becoming a real political or business center. The superior commercial advantages of Fayetteville, if combined with the political advantages of the state capital, would convert it into a real metropolis, something the state badly needed" (Powell 1989, 272).[24] Numerous towns therefore fought hard to be chosen as capitals. A paradoxical result of such competition was often to prevent the concentration of power in a single city, which partly explains the dissociation between political and economic state capitals.

The fight often began with the location of county seats. Indeed, most future territories were created as new counties in an already established state or territory. County seats meant judges, lawyers, officials, and taxes, but the main gain was more psychological than material—the aura of permanence—although becoming county seat did not always ensure survival. "A sort of negative determinacy principle was at work, under which a town that was designated did not necessarily survive, but on the other hand a town that lost the designation inevitably languished or disappeared–unless, of course, it had some other economic base" (Lingeman 1980, 159). Augusta and Des Moines are examples of surviving county seats. In 1846, commissioners had been appointed by the Iowa legislature to select a seat of the newly established Polk County. Speculative paper towns were created that based their future prosperity on winning the county seat. Tom Mitchell, of Fort Des Moines, used "legislative creativity" to crush his rivals. Conscious that the county would be lopsided with Des Moines as the county seat, he borrowed four townships from Warren County to "even up." Once "the

Fort" was selected, he returned the four townships to Warren County.[25] Mitchell then platted a new town near the fort. This clearly stresses the importance of centrality as well as imaginative town boosters.

Imagination was fundamental, because legislators were faced with a wide range of choices when deciding which town was to be the capital. Since all these towns were similar and "were perceived in this light by commentators at the time" (Hamer 1990, 10), each town had to present itself as the best candidate for present and future growth. This is called *boosterism* by historians, and would be called urban marketing or geomarketing today. Boosterism is usually described as a nineteenth-century phenomenon, but that does not mean that towns founded previously had nothing to do with economics. As noted above, the biblical ideal of the "city upon a hill" soon faded, even for Boston and Philadelphia, whose founders were strongly religious, and was replaced by capitalistic speculation.

Boosterism was especially strong in the United States. In other British colonies, such as New Zealand, Australia, and Canada, boosterism and uncertainty remained limited because most towns were founded by governments or by monopolistic companies, whereas in the United States this process was left mostly to private enterprise. This means that most towns were founded by people who wanted to attain prosperity through rising land values. Success depended on the skills of local entrepreneurs, called "town jobbers" by Lingeman (1980, 101), even more than on the town's environment.

A quarter of all state capitals were built from scratch. Five present ones were created by speculators who intended them to become capitals: Little Rock, Arkansas; Denver, Colorado; Boise City, Idaho; Madison, Wisconsin; and Cheyenne, Wyoming. Boosterism was essential for them. Economic factors were naturally at work, particularly in the West, where a capital was often one more virtual boomtown.[26] Funding came from eastern seaboard corporations, among which were railroad companies. Cheyenne owed much to the chief engineer of the Union Pacific, Grenville M. Dodge. He made Cheyenne possible by selecting a central route through the mountains for the transcontinental railroad, instead of the South Platte Valley route near Denver or the old Oregon Trail past Fort Laramie. District engineer James Evans had been caught in a dreadful snowstorm when exploring the Denver route in 1866. Dodge himself selected the site for future Cheyenne in 1867, at the foot of the Rocky Mountains, because it was to become the "division point between the company's two sections of the road—the plains division from the Missouri River to the mountains, and the mountain division to Salt Lake. To him, the central location seemed particularly well-suited for a future metropolis of the region" (Stelter 1967, 6). Land speculation was likewise essential for Denver. On November 17, 1857, William Larimer, a land speculator, bought the "city" of St. Charles Town, founded not long before by gold seekers, and organized the Denver City Company. In a June 29, 1859, letter to his family, he wrote, "We are bound to have a territory if not a State, and [its] capital will be Denver City with the state house near Will's and my claims" (quoted in Abbott, Leonard, and McComb 1994, 53). He was pretty confident, despite the fact that four competing cities had been platted in the immediate vicinity. Soon only three were left: Denver and Auraria, on opposite sides of Cherry Creek, and Highland, on the bank of the South Platte, opposite to both Denver and Auraria. In February 1859, Larimer was able to write, "I am Denver City!" In that year, Denver became the sole name of the three competing towns "in return for a barrel of whiskey to be shared by all" according to Denver's website. It is thus not surprising that Denver's first permanent structure was a saloon. To boost land sales, town jobbers like Larimer donated land for public use—up to one-third of the area of the new towns. Such donations were almost always their "only public-interest consideration" (Lingeman 1980, 158).

Besides these five "private" capitals, fourteen others have been created by legislatures: Annapolis, Maryland; Austin, Texas; Columbia, South Carolina; Columbus, Ohio; Tallahassee, Florida; Indianapolis, Indiana; Jackson, Mississippi; Jefferson City, Missouri; Lansing, Michigan; Lincoln, Nebraska; Raleigh, North Carolina; Richmond, Virginia; and Salt Lake City, Utah. Temporary capitals (e.g., Iowa City, Iowa, and Knoxville, Tennessee) followed the same path. Here the processes might seem closer to the colonial British model than to the American booster model, but such a view is misleading because legislatures were short of cash, as they are today, and had to rely on the sale of lots to finance construction of the town and the magnificent public monuments they wanted to erect. Moreover, since states wanted their capitals to flourish economically, they used booster methods to advertise and sell their towns. For instance, although Indianapolis's town-lots did not bring great returns for years in the then unhealthy new town, their sale fulfilled the state's expectations, yielding about $125,000, a sum sufficient to construct the court house, the "executive mansion," the clerk's office, the house and office of the state treasurer, the first statehouse, and the state prison at Jeffersonville. Long-term investors made excellent bargains. Lot number 7, on the northeastern corner of

Pennsylvania and Washington streets, which was sold in 1821 for a mere $300, was taxed in 1910 for $330,000 (Dunn 1910, 33).

When speculators were well connected, success was even greater. This was the case in the newly formed Dakota Territory in 1861. The ruling Democrats had decided on making tiny Sioux Falls the capital, but the Republican Abraham Lincoln became president, and Sioux Falls fell out of grace. The election gave ideas to a cousin of Mrs. Lincoln, Captain John B. S. Todd. After his resignation from the army (he had been stationed at Fort Pierre) he became sutler at Fort Randall and acquired real estate in Yankton,[27] a new, small settlement, where he persuaded the newly appointed territorial officers to meet (Potter 1932, 27). In Yankton's official hagiography, the outpost was described as follows: "An ideal location for a townsite. Look at the country, circled by hills, natural drainings, fine river front, rich soil" (Hanson 1916, 22). The choice of Yankton as the capital is recorded under the heading "Arrival of Governor Jayne, June 1861": "I shall make this town my official residence until the legislature locates the capital here or elsewhere." Governor Jayne was President Lincoln's doctor, and they were old friends (39).

Besides connections, newspapers were useful in supporting or opposing the claims of competing towns. When Lancaster, capital of Pennsylvania, was threatened with losing its status to more western cities like Harrisburg, supporters sent a letter to the *Lancaster Intelligence* on February 19, 1805, supposedly signed by Harrisburg citizens with names like Nicholas Frogpond, Samuel Tiggleginn, Timothy Shiverwell, and Simon Spindleshanks:

> If, however, Your Honors should not find the living here so comfortable as at Lancaster, there will be this advantage attending it, viz, by using a Spare Diet, your mind will be more fitted for the arduous duties of Legislation. . . . We have likewise been informed, that our Place is objected to, on account of the streets being sometimes inundated by water. We assure your Honors, that this seldom happens; never except in times of heavy rain; and should they, unfortunately, at any period during the sitting of the Legislature, be in that state, we hereby pledge ourselves to have canoes at all the public crossings in order to ferry over the Member free of all costs, the same to be paid out of the Corporation. (McInerney 1994, 26–27)

Some newspapers even initiated the campaign for removal of a capital. The *Alabama Journal*, in its morning edition for January 11, 1843, after congratulating the editor of the *East Alabamian* for an article on the same subject, stated:

> During the last few years, since it has become apparent that there must be a removal of the seat of Government, we have endeavored to impress our citizens with a sense of the great importance of this matter to their interests, and to induce them to arouse from their apathy, though we had deemed a public discussion of the matter at present premature. Since, however, the ball is opened, we are prepared to go into an examination of the claims of Montgomery, and at the proper time to show from statistics that will hardly be questioned, that from its central position, its facilities for direct communication by steamboat, railroad and stage with every section of the State, its salubrious and healthy location, the fertility and productiveness of the adjacent country, its prosperous and increasing trade and population, and its possession of all those advantages deemed requisite in the location of the seat of Government, there is no place in the interior that can for a moment hold comparison with it.

The enthusiastic editor had to wait until 1846 to see his dreams fulfilled, after sixteen ballots in the legislature (Williams 1979, 140).[28] There was fierce opposition, since other newspapers favored other cities, the most contentious of them being in Wetumpka, which was favored by east Alabama. On December 13, 1843, the editor of the *Alabama Journal* answered the easterners' claims:

> Some of our contemporaries at Wetumpka, La Fayette, Selma &c., taking for granted from the signs of the times, that the people will desire a removal of the seat of government at no distant day, have been recently agitating the subject with reference to early action. The Wetumpka papers are specially [*sic*] zealous in setting forth the pretensions of that place as the site for the contemplated removal. The "Times" says that at the proper time it will be prepared to show, in that respect, the superior claims of Wetumpka &c. Where may lie, or in what they may consist its superior claims over others, we have no hint, or knowledge, further than that afforded by a free imagination and liberal faith. In what single respect are its claims superior or equal to those of Montgomery? Is it nearer the centre? Or superior in health or salubrity of location? Are its lines of communication with the State such as to make it more accessible to the people? In which of these respects is it superior, or any way equal to Montgomery?

In Colorado, Golden City and Denver contended for prominence. Golden won the first contest in 1862 and became state capital, thanks to backing from the Boston Company that had founded it in June 1859. The president of the company was George West, a writer for the *Boston Transcript,* who tirelessly promoted Golden City as "the right location, where the canyons flattened out into the plains. Travelers and supplies bound for the diggins are forced to use the canyons as highways and our town is right at the crossroads" (quoted in Leonard and Noel 1990, 298).

Besides newspapers, pamphlets were also used to spread competitors' ideas.[29] "Verbal battles between rival communities were often as exciting to western residents as the bloodiest Indian battles or the well-publicized confrontations between outlaws and peace officers" (Stelter 1973, 189). In 1892, according to Montana's 1889 constitution, which had left the permanent location of the state capital up to a vote of the people, an election was scheduled to select Montana's permanent capital. The two remaining contenders were Helena, already the temporary capital, and Anaconda, a rich copper-mining town. The capital fight was also a struggle between two local barons. W. A. Clark was the champion of Helena, where he had investments, allies, and friends. His foe, Marcus Daly, was Anaconda's founder and chief promoter. Daly was certainly behind the publication of a funny pamphlet, *Helena's Social Supremacy,* which posed as a "supplement" to the Helena Capital Committee's *An Address to the People of Montana: Reasons Why Helena Should Be the Permanent Capital of the State.* It "lampooned Helena's claims to social grandeur, played on the Queen City's purported anti-labor stand, and implied that Helena was a greedy octopus eager to control the state's economy" (Newby 1987, 68). Following are some excerpts (68–72) that emphasize a strong anti-intellectualism and anti-urban feeling. To industrial Anaconda, a financial and political city implied uselessness and haughtiness, leading to plotting and dishonesty, that were misplaced in the rough and hard-working pioneer state of Montana. "As it is, all she asks of the people of Montana is to bear in mind that her society is doubtless the best society on earth. . . . There are people in the city of Helena that can trace their ancestry back two or three generations without discovering a single flaw in the form of a day's work or a night's sobriety." The text then provided statistics purportedly taken from an annex to the 1890 Census:

Patches on knees of trousers: Helena 0, Anaconda 253
Patches on conscience: Helena 1,691, Anaconda 8
Average number of children per family: Helena ½, Anaconda 5¾
Poodle dogs: Helena 774, Anaconda none
Skeletons in closets: Helena 1,343, Anaconda 16

Among the ten reasons why Anaconda was deemed "unfit to be the capital" were the following: "(1) Anaconda is lamentably lacking in tally-hoes, four-in-hands, drags, waxed floors, dress suits, Browning [the English poet] clubs, theosophical societies, ceramics, art coteries, eight-course dinners, ten-button gloves, skirt dancing and other social facilities. . . . (7) Neither Marcus Daly nor any other citizen of Anaconda has ever asked Helena's permission to live." And among the ten reasons why Helena "should be the capital" were these: "(1) Helena is the only aristocratic city in Montana. . . . (3) Helena's criminal classes are uniformly courteous and gentlemanly. . . . (7) By a special arrangement with Providence, Helena is exempt from disease and death. . . . (8) Helena enjoys the marked esteem and friendship of that most noble, generous, large-hearted, charitable, philanthropic corporation, the Northern Pacific railroad."

Educated Montanans certainly laughed heartily when reading this pamphlet, but the 1892 election failed to produce a majority. In the runoff election of November 1894, Helena won with 27,024 votes to 25,118 for Anaconda. Perhaps the railroad won over the copper mine. It was estimated that Daly spent more than $2.5 million and Clark at least $400,000 ($56 for each vote). Marcus Daly never overcame his defeat. W. A. Clark became a hero, but Helena's image was very bad in the country due to these fine pieces of "journalism." Journalist C. P. Connolly called Helena "a city hysterical with guilt and greed," adding that "the morning salutation with everyone was 'what's the price of a voter today?'" (Malone, Roeder, and Lang 1991, 219).

On the other side, a newspaper hostile to the state's political rulers could prove disastrous. One of the main reasons that Guthrie, Oklahoma, lost the capital was the violent attacks by the editor of the town's staunch Republican daily, the *State Capital*, against both the Democratic legislature and Governor Haskell. His wife "disregarded Guthrie's society women to spend her days with embroidery in the executive office" (Forbes 1938, 25). A wealthy Guthrian tried to buy the newspaper and make it Democratic and governor-friendly, but the price asked was too high. The newspaper was not the only reason, however: African-American officials in Logan County, of which Guthrie was the seat, had alienated the state, which was racist at the time.[30] There were also competing railroads: the Santa Fe for Guthrie, and the Saint Louis & Oklahoma for Oklahoma City, which had secured the construction of that railroad in 1894, when Guthrie had been too late in accepting the extension to it of the St. Louis & San Francisco Railroad.

Boosterism and Capital Buying

Since classic promotional means were not always successful, unsavory tactics were sometimes used. In Alabama, when Montgomery's (uninsured) capitol burned in 1849, the cause of the fire was never determined. One of the most likely suspects was the doorkeeper of the House of Representatives (equivalent to the chief of staff and police), who was strongly in favor of the former capital, Tuscaloosa (Blue 1963, 245). But evidence was lacking, and the capitol was rebuilt on the same spot. In Montana, the governmental seat had been moved in 1865 to Virginia City, a mining town and the center of population. Virginia City was losing population to Helena, however, and both cities struggled over the location of the capital. Virginia City retained it in an 1867 election, but its decline continued, and another election was held in 1869. Those ballots were destroyed by a very suspicious "accidental" fire in the office of the territorial secretary at Virginia City (Malone, Roeder, and Lang 1991, 109). Fortunately, such violence remained exceptional. Bribes, however, were commonplace: politicians, like almost all pioneers, had materialistic values (Pessen 1985, 113).[31] The best way to induce legislators to vote for a particular location was to give them land in the proposed city. In Arkansas, the removal of the capital from tiny Arkansas Post (two hundred inhabitants in unhealthy lowlands) might seem to have been dictated by geographical causes, for the influx of pioneers from Missouri called for a more central capital (Bolton 1998, 36; Moussalli 1997, 61–62). Arkansas Post had been chosen because it was the only settled part of the new territory, on its eastern border. The choice of the new capital by the legislature, in October 1820, had been well prepared for by William Russell, the land speculator who had just settled a new townsite up the Arkansas River, called Little Rock. Many important politicians, among whom was the territorial secretary, owned a part of the site before the vote. When government arrived in 1821, the land had already been divided between speculators and politicians, the general public having been largely excluded.

The Nebraska Territory was created in 1854. The first governor, Burt, exhausted by the long journey to Bellevue, died two days after he had taken the oath of office. Burt had not yet fixed the territorial capital, although he had apparently planned to convene the first session of the territorial legislature at Bellevue. He was replaced by his secretary, Thomas B. Cuming, an ambitious young man. He was linked to Iowan influence and especially to the Council Bluffs and Nebraska Ferry Company, the principal proprietor of Omaha City (Potts 1988, 173). The rivalry between Bellevue and Omaha was intense because the territorial capital would certainly be on the route of the future transcontinental railroad (Olson and Naugle 1997, 81–82). The census of 1854 counted 2,732 persons in Nebraska, two-thirds of whom lived south of the Platte River. Cuming therefore created four counties north of the Platte, with seven councilmen and fourteen representatives; the southern four counties had been granted only six councilmen and twelve representatives. He also included an unwilling Bellevue in the same county as Omaha, which he designated as the territorial capital on December 20. Despite hot arguments and meetings accusing Cuming of having blatantly accepted bribes, Omaha remained the capital during territorial times. Cuming had indeed accepted a 5% ownership of the capital city and was denounced as "the Governor of Council Bluffs and Omaha City Ferry Company" (Potts 1988, 174). The opponents of Omaha used bribes as well in their attempts to gain votes for the removal of the capitol, offering, according to the *Nebraskian,* sixteen shares of Douglas City per vote. Federal funds had already been given for the construction of a capitol, however, and the governor could veto the removal because it contradicted the Organic Act. This did not end the sectional controversy. Some South Platte residents even wanted to be annexed by Kansas. Neither Congress nor Kansas agreed to this annexation, however, and residents were eventually content with the election of their own delegate, owing to the introduction of party politics in 1858–59, which weakened sectionalism. This controversy came back with a vengeance during the debates upon the location of the permanent state capital thirteen years later.

In Oklahoma's first capital, Guthrie, legislators were continuously entertained by representatives of the towns that wanted to become the capital. They sometimes offered more than the price of a good meal: $32,500 had been used during the session to buy votes for the capital measures. For instance, in 1890 four members from Kingfisher asked for six or seven thousand dollars in cash to help Oklahoma City steal the capital from Guthrie. They never got the money because the relocation bill was vetoed by Governor Steele. In an angry response, the *Kingfisher Journal* published the (perhaps falsely attributed) *Psalms of George Steele,* which included the following verses: "Why art thou cast down Guthrie? And why art thou disquieted? Hope thou in me for I shall continue to veto capital bills (for town lots)" (quoted in Forbes 1938, 17).

The line between speculation and politics was still blurred in the early

American period, as shown in the case of James Duane Doty, a land speculator born in New York. In 1835, he created the new town of Madison, Wisconsin, with a new territorial capital in mind. With the help of town lots and buffalo robes given to sixteen legislators, the clerks of both houses, and the governor's son (the governor had indignantly refused), he succeeded in obtaining the capital.[32] His success owed much to his knowledge of politics: he had been clerk of the territorial court at Detroit, had served as a judge, and was elected to one term in the Michigan territorial legislature before turning to land speculation. His partner in organizing the Four Lakes Company, which owned the land of the Town of Madison, was Stevens T. Mason, Michigan's first state governor. Therefore, when Doty came to Belmont to promote "his" town, he already knew many legislators. Several speculators like Doty were also in Belmont, scheming in favor of other would-be capitals. They resorted to the same tricks, but Doty was "the most adept, and eventually, the most successful" (Cravens 1983, 113). He even managed later to become territorial governor of Wisconsin.

In a similar case, the man behind the lobbying for the capital in Phoenix, Arizona, Moses H. Sherman, was a streetcar and real estate developer. In September 1888, he told an associate, "You see, Phoenix wants the capital, and the whole crowd is working to this end. It may take a little money, and if it does, we fellows are ready to put up some" (quoted in Luckingham 1989, 37). It certainly cost Sherman "a little money" to attract the capital, but it also brought some money back to him. In return for selling ten acres of land to the government for only one dollar, as a site for the capitol, he was allowed to build a streetcar line to connect his property, located west of Phoenix, to the center of the town.[33]

Most of the time, however, bribery was not so blatant and took the form of "presents" made to the Assembly as a whole (which had often asked for them). Golden, Colorado's first real territorial capital (1862–67), became the seat of government with promises of free accommodations, alcohol, and firewood to legislators (Leonard and Noel 1990, 25). The rooms were actually rented at ten dollars a week, which was costly for the time. Most of the would-be capitals had to pay at least part of the cost of the statehouse or of accommodations for hungry and thirsty legislators. Illinois grew so much after 1865 that the capitol was too small, and more lodgings for the legislators were needed in the capital. To counter cities that wanted the capital, Springfield citizens built a new hotel, the Leland, and offered a seven-acre site and $200,000 for the construction of a new capitol—an offer that the General Assembly accepted in 1867 (Carrier 1993, 127). Building the new capitol took twenty years (1868–1888) and cost $4.5 million, which indicates that money was certainly not the entire issue: Springfielders paid only 4.44% of the total. Lincoln's legacy certainly helped.

The most moving testimony to the demands such efforts made on citizens' purses is Montpelier's. After the legislature had made Montpelier the permanent seat of government for Vermont on November 7, 1805, citizens and the town of Montpelier (with the help of a tax of four cents on the dollar in 1807) paid $9,081.67 toward the cost of the first wooden statehouse (*Records* 1877, 428). A public-minded citizen gave the land. When the building became too old in 1832, the legislature passed "An Act Authorizing the Building a State House at Montpelier" on November 8: "Provided, the inhabitants of Montpelier, or any individuals, shall, before the first day of January in the year of our Lord one thousand eight hundred and thirty-three, give good and sufficient security, to the treasurer of this State, to pay into the treasury of this State the sum of fifteen thousand dollars; one half of said sum to be paid in one year and the remainder in two years from the passage of this act" (*Records* 1877, 287).

This time, Montpelier citizens paid $18,000. Twenty-five years later, on January 6, 1857, the capitol burned. The total cost of rebuilding and furnishing it was higher, $148,396.63. On February 27, 1857, the legislature voted "An Act to Provide for Rebuilding the State House": "Provided, the inhabitants of Montpelier, or any individuals, shall, before rising of this Legislature, give good and sufficient security to the Treasurer of this State, to pay into the treasury of this State, a sum equal to the whole cost of the work, mentioned in the first half of this act [$40,000]; one half of said sum to be paid in one year and the remainder in two years from the passage of this act, or on the completion of the work" (*Records* 1877, 289).

Montpelier was still a small town of about thirty-four hundred inhabitants, but ninety-three donors pledged sums that ranged from $10 to $500, for a total of $6,025. These brave citizens were compared to martyrs by J. Thurston, who added some words to the list of the ninety-three signatures: "The army of martyrs: This fund was raised in 1857—40% was collected and used in presenting and enforcing the claim of Montpelier for the new State House at the Extra Session of the Legislature in February 1857." He went on as follows:

> In order to raise the sum of Ten Thousand Dollars to be placed at the disposal of Elisha P. Jarrett . . . to be used in their discretion in procuring plans, estimates and an appropriation for the repairs at the State House at

> Montpelier, we the undersigned hereby agree to pay James J. Thurston the sums annexed to our names respectingly. Provided that if a less sum than Ten Thousand Dollars shall be expended, then the sum to be paid shall be in proportion to each of the respective subscription. (Third Subscription 1857)[34]

In December 1858, they pledged the additional sum of $52,795 for the same purpose. The capitol was then rebuilt in the same style, but a quarter larger and with a higher dome.

Bribes could also take a sweet form. Springfield, Illinois, was founded in 1821 as the seat of newly created Sangamon County. The commissioners who were sent in 1824 to select a permanent county seat (a great help in securing the capital) apparently decided on Springfield largely with the help of a "royal feast" put on by Mrs. Iles, the wife of one of the town's three proprietors, Elijah Iles. He was the owner of Springfield's first store and also platted the town in 1824 (Angle 1950, 15). The choice of Nebraska's permanent capital had a similar gastronomical motivation in 1867. The three commissioners charged with that task—the governor, the secretary of state, and the state auditor—had to examine three newborn "towns": Ashland, Saline City (also called Yankee Hill), and Lancaster. Here is a lively account of their adventures.[35]

> Their stop at Ashland necessitated an overnight stay with Governor Butler, as senior member, quartered in a screened area on the first floor of a new building with the other commissioners and the balance of the party on the open second floor. The mosquito attacks during the night ensured at least two negative votes for Ashland. At Yankee Hill the commission was entertained by John Cadman who had just lost the county seat to Lancaster and was determined not to let another political site slip through his hands. Mrs. Cadman, her sister and all the ladies of Yankee Hill set a table so sumptuous as to singly sway the vote. Their unfortunate choice of ice cream for dessert may have been their downfall, however. How they managed to supply the dessert puzzled everyone and some felt it was tantamount to bribery. Lancaster had only the salt flats to recommend her and a lack of fresh water chief among her deterrents and only the most optimistic would have bet on Lancaster as the commission retired to the attic of W. T. Donovan's cabin in the afternoon of July 29. On the first ballot, Lancaster received two votes, Ashland one. The second ballot was unanimous and so [was] announced by Governor Butler to the small group waiting outside. (McKee and Duerschner 1976, 1–2)[36]

The choice of Lancaster stems naturally from causes more important than mosquitoes and ice cream, since it was a compromise between South Platte towns, agreed upon by Nebraska City, which had favored Yankee Hill, as well as by Ashland. The commissioners noted in their August 14, 1867, report that Lancaster was "at the center of circle of about one hundred and ten miles in diameter, along or near the circumference of which are the Kansas state line, directly south, and the important towns of Pawnee City, Nebraska City, Plattsmouth, Omaha, Fremont, and Columbus" (Sawyer 1916, 81). Thus the booster model is part of the process, but it fails to explain that process fully. Contemporary historians have challenged the model, because it doesn't entirely depart from the "hero" model; it focuses on individuals acting in relative isolation for the advancement of their private interests (Ferraro 1998, 243–44). New approaches dwell more on the social cohesiveness of urban society seen as a whole.[37] The fight for capital status is also a translation of the fight for permanence, partly contradicting the idea that Americans have no strong local attachments, especially in the West. For the majority of pioneers, moving was more a necessity than a way of life, and the possibility of settling permanently, for instance, in a capital that promised to endure, was engaging. We must therefore look at broader economic explanations, such as the gateway concept.

Capitals as Mercantile Gateways

Among the factors put forward by historians to explain the process of urban growth in America, the first was gaining status as a mercantile gateway: based on "accessibility, market potential, and location" (Johnston 1982, 101), a gateway city exerts commercial control over a hinterland.[38] For western cities, an 1878 description of Des Moines, Iowa, summarized quite well the growth pattern: "the result of speculation and manifest destiny."[39] But urban growth was not steady; most cities had to overcome hard blows. For instance, despite ten major fires between 1653 and 1760, eight smallpox epidemics between 1640 and 1730, military casualties equivalent to those of World War II during four periods from 1690 to 1780, and two hyperinflation periods between 1740 and 1780, Boston was able to recover and grow (Nash 1987, 133).

Capital cities, like resorts and college towns, had uneven economic activity, with highs and lows that followed the presence or absence of the legislators. To get through these phases, in an era when bureaucracy was still in its infancy, and civil servants were scarce, they absolutely had to diversify their economic basis if they wanted to become major cities and not merely "legislative campuses." The most important factor was the relation of the capital to the transportation system. Transportation—primarily the lack of it—was almost always the most pressing problem of new territories and states. Capitals that were also transportation hubs were far more likely to succeed economically than those that lacked easy access to transportation.

In the seventeenth century, sea trade was foremost. Most colonial capitals were founded on the eastern seaboard or became important ports after initial religious experiments (e.g., Boston, Providence, and Philadelphia). Only Hartford was in the interior, but it was on the Connecticut River. In the south, the Tidewater Frontier (1607–1675) was also important: the first capitals there were Jamestown on the James River and St. Mary's on the mouth of the Potomac (Billington 1974, 52). Further south, Charles Town and Savannah (1732) were coastal ports intended to benefit from trade with England. During what Donald Meinig calls the "Atlantic America" period (1492–1800), the main lines of transportation between London and the provincial capitals in America superseded by far the internal links between these capitals. Of course, there were relationships—the Revolution proves that a common spirit existed—but inland transportation was difficult and relied on trails and streams. Trails were first used by Native American tribes and then by fur traders. This explains why almost all of the first capitals in the newly settled West were founded as trading posts or forts by the French (e.g., Kaskaskia, Vincennes, and Detroit), the British (e.g., Fort Nashborough, later Nashville), or the Dutch (e.g., Fort Orange, later Albany), where traders "bartered with the Iroquois for peltry" (Billington 1974, 94).

As cultivable lands were opened by pioneers (cotton was especially valuable in the southeast), trade between England and the colonies became less important than inland trade, and trails no longer suited the modernizing economy. Roads were needed, but the young nation lacked a proper road network. Thus the first cities to be linked by a trail, a turnpike, or by the National Road became more important. Capitals were of primary importance in the developing road system. Chillicothe, for example, was a nexus of roads in the Ohio country around 1800 and was briefly able to rival Cincinnati. The major road was Zane's Trace, which was opened in 1796 and ran from Pittsburgh to Wheeling and then to Fort Gillad, through Zanesville, Lancaster, and Chillicothe. It remained the only inland road for many years and contributed greatly to the development of Ohio (Brown

1948, 230).[40] This situation owed nothing to chance, for federal leaders argued that all links connecting the new states participated in the "political system of confederate Governments" and stressed the need "to connect the seat of the Federal Government, by the shortest lines of communications, with the capitals and great cities of the several states" (quoted in Meinig 1993, 342). For capitals, transportation and politics were strongly linked.

The most famous of the inland roads, the National Road, was built with federal funds and went ultimately (by 1840) from Baltimore to St. Louis, through four capitals: Wheeling, Columbus, Indianapolis, and Vandalia. The Upper Road linked Washington to Natchez, first under the name of Federal Road, an old military track that linked Columbia to Milledgeville, then through Alabama, where it became the Upper Federal Road through Montgomery and St. Stephens. Among the other main routes to the west were the Boston-Albany-Utica road, the Nashville-Natchez road, which prospered from the cotton trade, the New Bern-Raleigh-Knoxville-Nashville road, and the Richmond-Louisville road, with a branch ending at Frankfort, passing through eight capitals. In the West, the Santa Fe Trail, which ran from Franklin, Missouri, to Santa Fe, thrived when trade was opened with the United States in the 1820s.

The panic of 1837 temporarily ended all federal transportation endeavors, including the Atlantic Seaboard turnpike, which was to reach Tallahassee. Besides the financial panic, another important factor in the decline of inland roads was technological change, which first favored steamboats and later railroads. Such changes were far less favorable to capitals, and roads, even when they were called federal, were still only mud tracks through forests that did not allow quick journeys. To thrive, therefore, cities had to be hubs on the waterway systems, like the major gateways to the west: Cincinnati, Indianapolis, Chicago, and St. Louis, three of which were capitals. Several capitals were heads of navigation: Augusta, Providence, Hartford, Trenton, Richmond, Columbia (SC), Montgomery, Jackson, and Des Moines. Among former capitals, Milledgeville was head of navigation on the Oconee branch of the Altamaha River, and St. Stephens, Alabama, had that role on the Tombigbee (Vance 1995, 17). Steamboats stimulated the growth of capitals that were located on the Fall Line or close to it: Richmond, Raleigh, Columbia, Augusta, Milledgeville, and Trenton (Coleman 1991, 156). Capitals seldom profited from the canal mania, with the major exception of Albany, which benefited from the opening of the Erie Canal to the Great Lakes.

The last examples of capitals as gateways were in the Miner's Frontier, during the 1850s and 1860s in the Mountains area, where capitals were often mining boomtowns or trading and supply centers near mining areas. Because of the delayed settlement pattern and the fact that cities preceded the period of the Miner's Frontier, roads retained great importance. Salt Lake City, founded by the Mormons in 1847 and the only major settlement between the Midwest and the Pacific coast, was a stop for forty-niners seeking gold in the 1850s. Mormons also developed the "Mormon Corridor," from Salt Lake to San Diego (Provo and the temporary capital of Fillmore are located on that trail). Likewise, Carson City became the capital of Nevada in 1861; Lewiston, of Idaho in 1863; and Prescott, of Arizona in 1864. They were all located near mines and served as rear bases for miners. Capitals therefore also followed the rhythm of the depletion of ores. In Montana, Bannack City had to cede the capitol to Virginia City after only one year. Nine years later, the capital was removed to Helena, which became one of the leading trading cities in the mountain region because it was conveniently located on the trade route between Fort Benton, Bannack, and Virginia City. This example underlines the fact that in the formative territorial period of the United States, gateways were numerous and were soon challenged by new competing towns, multiplying the economic centers of a region and sometimes inducing a separation between political and economic capitals. This was still "the shakeout and selective growth" period, with shifts of county and territorial seats, selective survival of towns, and the frequent exodus of promoters, people, and capital (Meinig 1993, 271). Centrality was therefore temporary and leads to the study of geographic factors in the process of selection of capitals.

The Ambivalence of Geography

At a local scale, geography figured in the arguments used for or against a candidate town, through comments upon climate and the diseases and epidemics it induced. Since medicine was not as developed as it is today, the idea of having to live in an unhealthy town (or one supposed to be so) could deter legislators. Yellow fever epidemics, frost, heat, swampy grounds, and so forth were scrutinized when examining a town. Yellow fever cost Natchez and Philadelphia the capital, and harsh winter weather was used against Prescott to move Arizona's capital to "healthy" Phoenix. During the battle between Helena and Anaconda for the capital (1889–1894), promoters of Helena were quick to disparage the unhealthiness of industrial

Anaconda, in agreement with the hygienists' discourse at the end of the nineteenth century (Lang 1987). Perceptions of cities were at least as important as physical or economic facts, giving to the term *geography* a strong cultural flavor.

At the colonial, territorial, and state levels, heterogeneity reigned. Size was not really a factor in the mobility of capitals: vast New Mexico had only one capital, whereas small Rhode Island had five wandering ones for two centuries. More important was centrality, an issue often present in the debates over the removal of a capital (accounting for 18.4% of the causes found).

The principle of centrality seems at first to have been as powerful in the United States as it had been in revolutionary France (Margadant 1992, 265). However, locating the center was difficult, because of the ambiguity of the concept of centrality, which was interpreted as referring not only to geography, but also to demography and economy. Legislatures mostly debated about population—decennial censuses had been held since 1790—and territory. Geographic centers were easy to find and permanent,[41] but settlement at the time did not cover whole states and rendered geographic centers not very convenient. Larger states whose population was still sparse tended to locate their capitals at their demographic or economic centers (e.g., Harrisburg, Albany, and Lansing), while smaller states preferred geographic centrality (e.g., Trenton and Hartford; see Zagarri 1987, 9). Demographic centrality was the reason put forward by Wisconsin's first governor, Henry Dodge, to explain his choice of the "instant town" of Belmont as the temporary capital. The new town was indeed in the middle of the lead-mining region, but some contemporaries and later historians have hinted at another possible reason: Belmont had been founded by John Atchison, a land speculator who knew Dodge well. Whatever the reason, Dodge intended to allow the territorial legislators to choose a permanent capital as soon as feasible (Cravens 1983, 107–9).

Some towns that were not centrally located remained capitals, such as Frankfort, Kentucky. When the state was created in 1792, there were two candidates, Frankfort, near the demographic center, and Danville, forty miles south and nearer to the geographic center. The commissioners chosen by the legislature selected Frankfort because it had offered more inducements than Danville (Littrell 1806). According to another nineteenth-century local historian, Frankfort's prospects for remaining as the capital were not bright, but not because of Louisville's big inducements to attract the capital. Louisville, besides being far from central, was the metropolis of the state. It had 100,753 inhabitants in 1870, while Frankfort had a mere 5,396, and thus was looked upon with great suspicion by rural Kentuckians. Because of the coming of the Ohio & Cumberland railroad, the author reckoned that the centrally located city of Lebanon was well placed to become Kentucky's future capital (Allen 1872, 65). He was wrong.

Only thirteen capitals (26%) are geographically central: Annapolis, Columbia, Columbus, Dover, Hartford, Indianapolis, Jackson, Jefferson City, Little Rock, Oklahoma City, Pierre, Springfield, and Trenton.[42] Six were created by legislatures to become capitals, and one, Little Rock, by a land speculator. Indiana exemplifies such processes, since centrality was at the heart of its choice of a permanent capital. The territorial legislature even asked Congress to donate a township of land for the capital in the center of the state. The legislature had deemed it "good policy that every State should have its seat of government as nearly central as the local situation of the country will permit" (quoted in Carmony 1998, 107). Centrality was explicitly linked to the creation of an efficient state transportation network, that is, to economic growth. In 1818, Governor Jonathan Jennings said in an address to the legislators that Indiana's future network of roads and canals should be focused on the new permanent capital, which would serve as the Indiana's transportation hub. The 1820 commission charged with investigating the future site had to look for fertile soil and navigable streams. The site selected was thus located on a navigable stream and on the planned extension of the National Road. This allowed Indianapolis to become a major road hub and the largest city in the state as early as 1840. This process contradicts Zagarri's general assumption that larger states preferred demographic to geographic centrality, since central and northern Indiana were still to be settled in 1820.

The same link between economy and politics was at work in independent Texas. General Mirabeau Buonaparte Lamar, its second president (1838–41), explicitly cited Washington when planning the capital of the new Texan nation, Austin. His choice of a new town was first deemed absurd, because it was beyond the already settled areas, inaccessible, and on lands controlled by Mexicans and Indians. But Lamar thought of the future, when Austin would be the political and economic center of a fully developed Texas, on the roads and the inland water route he envisioned. He already knew the area, having hunted there some years earlier and been enchanted by the landscape.

Most of the time, in their search for centrality, legislators were unable to agree upon a city among the existing ones and decided to create a new town for which they acquired the land, as had been done for Washington, DC. Columbus, Ohio, is an example. Legislators were deadlocked between Cincinnati and Cleveland. The neighboring states of Indiana and Illinois had already placed their capitals in central and compromise locations. For Florida, compromise stemmed from natural hazards. In July 1821, Florida became American after its purchase for $5 million from Spain in 1819. After one year under a provisional structure (Jackson being governor), a territorial government was set up in 1822, with two federal courts in Pensacola (founded 1686) and St. Augustine (founded 1565). This stemmed from the division of Florida in 1763 into two provinces (East Florida and West Florida) with both cities as the capitals. The first Legislative Council was held in Pensacola, since Congress had ordered that annual sessions of the council should alternate between St. Augustine and Pensacola. But delegates to the 1822 and 1823 sessions experienced dangers and delays, suffering two shipwrecks (one near each capital) and one bout of yellow fever at Pensacola. Governor Du Val thus sent two commissioners to find a site midway between the two existing capitals. They chose a location described by the first commissioner as "high, healthy, and well watered"[43] and by the second as "a delightful high rolling country, clothed with excellent oak, hickory and dogwood timber on a soil of chocolate colored loam."[44] The chosen place was also charged with history. Here Apalachee Indians had once had a meeting place, Hernando de Soto had camped, Spanish Franciscan missions had stood, and Creek migrants had established two towns in the 1700s, both of them burned by Jackson's army in 1818. Although the two local Seminole chiefs with whom the commissioners had met had no intention of leaving, Tallahassee became the permanent capital in March 1824 (Gannon 1996, 210). Settlers soon arrived, and the first log capitol was built. Situated in rich cotton and sugar country, Tallahassee became a commercial as well as political center, and had 1,500 inhabitants in 1835 (but still only 2,981 in 1900).

Centrality and Representation

Centrality was thus far more than a matter of geography and was part of the way early American legislators defined politics.[45] They "argued that centrality was the principle of equality expressed in geographical terms. Americans envisioned the republic as a circle in which the legislature's meeting place occupied the center. . . . In the margins of the bill to move

Virginia's capital, Jefferson jotted: 'Central . . . Heart-Sun-Ch[ur]ch-C[our]thouse.' Just as the heart sustained an individual's life, so the capital was the source of life in a republic" (Zagarri 1988, 1240). Tables 4.1–4.4 show that centrality very seldom stands alone as the explanation for the location of a capital but is linked to politics, that is, to exactly one-third of all factors. Centrality is also a way to look at the republic envisioned by Jefferson, where, from the political point of view, all American citizens (i.e., white males who were not poor) would be independent, and where, from the geographical point of view, the same equality would reign. No large city would impose its power on a countryside equally divided according to the township pattern at the local level and according to the territorial and state pattern at the regional level. Such geography would create a balanced territory, where governmental seats would be equal and modest.

The manifold meaning of centrality was present from the very beginnings of British America. The first New England colonies, albeit divided by religious opinions, decided to organize themselves around a common body. In 1643, by the Articles of Confederation of the United Colonies, the four New England colonies formed a confederacy. A meeting place had to be found for the annual reunion of the eight commissioners. Article VI stated that "the next meeting after the date of these presents which shall be accounted the second meeting, shall be at Boston in the Massachusetts, the third at Hartford, the fourth at New Haven, the fifth at Plimouth, the six and seventh at Boston, and then Hartford, New Haven and Plimouth, and so in course successively. If, in the mean time, some middle place be not found out, and agreed upon, which may be commodious for all the jurisdictions" (quoted in Osborn 1925, 189). No middle place was found, probably because local pride overruled convenience by far and because the confederacy was too loose to bother finding a common capital. When the confederacy was renewed in 1672 under the name United Colonies of New England, New Haven was dropped from the list, because it was now part of Connecticut.

Independence induced a surge in petitions for the removal of "old" capitals (Zagarri 1987, 17–21).[46] Although petitions were mostly based on the desire to minimize travel to the capital, far more was at stake than better roads. Centrality was considered a democratic right for multiple reasons: outlying areas were unable to send as many representatives to the capital as central ones,[47] because of the time and cost of the travel; information often reached these areas after the end of the legislative session; representatives did not understand their "outer" constituents; and corruption would arise.

Removal would change that situation, breaking the traditional dominance of the East by allowing easy contact between all parts of the state and equal access to the seat of government for all citizens.

In states with rotating capitals, the quest for centrality was conducted at another level. County centrality was at stake instead of state centrality. It was crucial in the case of Rhode Island, where county seats were also part-time capitals in the state's rotating system. Bristol County was founded in 1680, and Bristol had been its seat since 1685, but its old primacy was challenged by the town of Warren when it was decided in 1809 to construct a new county courthouse. Warren's arguments were geographical, based on its central location, whereas Bristol's were legal (a provision in its Grand Deed gave it the right to be the county seat), as well as economic and fiscal: since it had the majority of the county's inhabitants, it paid a greater share of taxes. Bristol finally won because it offered a larger site, agreed to finance construction of the new building and be reimbursed by the state, and was home to a family that was very influential in the state, the DeWolf family (Conley, Jones, and Woodward 1988, 57). Bristol was chosen because it was economically more central to the state than Warren, although the latter was located only four miles north of Bristol.

Debates about the location of capitals must be linked to the apportionment question, for which the stakes were sometimes higher. Hosting the capital did not always mean having political power, but moving a capital to a more central location was a way to approach the apportionment question. From 1776 to 1812, one way to equalize representation was to move the capital: out of the thirteen original states, only six reapportioned their legislature on a population basis, but eleven moved their capitals. The American concept of actual representation lies at the heart of this process.[48] Actual representation is opposed to British virtual representation, which is based on the idea that representatives speak for the general good of the community. According to actual representation, legislators "should represent the particular interests of those who elected them; the assembly was to be a microcosm of the larger society. Unlike virtual representation, actual representation was inherently localistic. . . . Furthermore, as inheritors of the Real Whig tradition, . . . Americans viewed their legislatures with suspicion. Only the people's vigilance could safeguard liberty" (Zagarri 1988, 1241).

In South Carolina, for instance, the opposition between the lowcountry (the oldest part of the state) and the backcountry (the settlers' part, whose residents complained of the haughtiness of the former) was intense. Although the more populated backcountry got the capital quite early (in 1786, in Columbia), the lowcountry retained its delegates because of its wealth and the mode of apportionment. "Charleston, with only about 11 percent of the population, elected 39 percent of the members of the House of Representatives and 35 percent of the senate. St. Stephen Parish, with a white population of 226, had three representatives and one senator; so did Edgefield (9,785) and Pendleton (8,731) Counties" (Edgar 1998, 257).

Although it was never more at the heart of capital removal, the importance of apportionment lasted well into the nineteenth century.[49] When Alabama was preparing for statehood in 1818, the capital question and apportionment were clearly linked in a complicated give-and-take game related to slavery. The powerful northern part of Alabama was in favor of white-only representation, which would give Madison County twice as many representatives as any other county. Southern Alabama, which included the cotton belt, wanted the "federal ratio," which counted three-fifths of the slaves, thus giving it more weight than the northern counties. A compromise amendment was passed that located the permanent capital at Cahawba, a new town in the south, near Selma. Until Cahawba was laid out and a capitol constructed, Huntsville, the first real town in the state but near its northern border, would be the temporary capital (Rogers et al. 1994, 66). Huntsville was the capital for two years (1819–20), but Cahawba's "permanence" lasted only six years. The capital was then moved Tuscaloosa (1826–1846), which in turn gave way to Montgomery, the current capital.

Texas also experienced the apportionment problem. Austin's victory at the 1845 convention was aided by the overrepresentation of the western counties that had been depopulated by the hostilities with Mexico. Another reason was that the Lamar faction, which wanted further expansion of Texas, was opposed to any capital located in the eastern part, meaning Houston. Sam Houston led the opposite faction, which favored quick annexation to the United States (Moussalli 1997, 73).

Representation was also important in Connecticut, along with the fact that the state was formed out of two distinct settlements, one on the coast (New Haven) and the other in the Connecticut Valley around Hartford, populated by dissenters from Massachusetts. From 1701, two capitals, New Haven and Hartford, held the sessions of the legislature. Toward the end of the nineteenth century, the situation began to seem too costly to the people and governor. But arguments over a satisfying solution lasted several years until 1873 (Osborn 1925, 41). The legislature had to choose between the two cities, both of which had aging and uncomfortable statehouses. The fight was uneven, because Hartford, being the main financial center, had more money. The city soon bought the site owned by Trinity College on Capitol Hill and appropriated $500,000 toward the cost of a new statehouse. It received in exchange the old one, worth about $350,000. New Haven fought bravely during the 1871 and 1872 legislatures but lost the battle. In 1873–75, a new statehouse was built in Hartford at a cost of $875,000. A popular vote was held in October 1873: 36,530 votes were cast in favor of Hartford and 31,338 against. New Haven was more concerned with representation than with the capital question. An article in its newspaper, the *Palladium* (October 9, 1873) stated, "Under our two capital system the inequalities of representation—monstrous as they are—were to some extent counterbalanced. With the ratification by the people of the Hartford amendment thus making that city the sole capital of the state the need for reform becomes even more urgent. . . . The southern portion of the State is not represented today and has not that influence in politics and legislation to which its great population, large wealth and rapid progress entitle it" (Osborn 1925, 45).

The issue of apportionment continued to be important well unto the twentieth century and could still influence the location of a capital, and even the limits of the state, as in Georgia during the 1920s. At the time, the state capitol was neglected, and some thought of building a new civic center in the suburbs of Atlanta. This idea renewed the fear that Macon, which had fought for the capital during the previous century, might at last see its claims recognized as a more central location in the state. Atlanta could not count on its power in the legislature because of Georgia's mode of apportionment. Divided into 160 counties, the state's legislature was huge (approximately two hundred representatives and fifty senators), while Atlanta's representation was very small. Some citizens even proposed to solve the problem by dividing Georgia into two states, "with Atlanta capital of the Northern industrial and commercial region and Macon capital of the southern agricultural sections; but there is no indication that there will be any action in this direction" (James 1925–26, 391–92).

Apportionment perhaps explains why some large cities did not fight to retain the capital, but fought hard to remain well represented in—and thus be able to control—the assemblies. For Philadelphia, losing the capital did not mean the loss of actual political power. In fact, in 1799, it sent six representatives to the House (less than one-tenth of the total) and was on the same footing as Lancaster, which was ten times less populated (41,220 vs. 4,292 in 1800). But it was still America's second largest city. New York, New Orleans, and St. Louis tell similar stories. They were content to be metropolises with an economic influence that extended far beyond their own state's limits.

This chapter has reviewed seven causal elements. Their study confirms that we cannot point to a single cause for the selection of a capital. It also highlights the uncertainty that surrounded the selection process: most capitals were chosen during a period in which "centers" (whether economic or demographic) constantly changed with the path of colonization, in which the political landscape remained unclear, in which speculation might induce either tremendous growth or complete disappearance, and perceptions (helped by aggressive boosterism) played a major part in urban successes or failures. That study also showed that selecting a capital was essentially a political process, in the sense that politics subsumed all the causes found: all the divisive forces at work in a state are involved, but they have to reach a compromise, embodied in one specific location, the future capital (or in successive locations, if the compromise did not last). The next chapter is therefore entirely devoted to the study of the balance of powers that the choice of a capital revealed.

5

Capital Choice and the Balance of Power

This chapter has two main sections. The first examines the major factor in the capital selection process emphasized in chapter 4: politics. The aim is still to build a graphic model that shows the relationships among the factors at work in that process. This model is discursive and based on natural reasoning. It also is inductive; it does not stem from any theory or hypothesis, but from facts, notions, or representations that I deemed adequate for getting beyond the juxtaposed narrative of individual destinies. The construction of the model dwells not only on the elements discussed in chapter 4, but also on the separate aspects of the political factor in more precise analyses. First, we look at the evolution of the political system as a framework for the selection process. That study emphasizes that the role of individuals was stronger than that of parties or political machines, which were not entirely in place at that time. It also deconstructs the myth of the founding hero, which is present in numerous sources. Once that framework is roughly complete, the analysis attempts to determine the relative influence of the three pillars of the American democracy—executive, legislative, and judicial—on the selection process. This analysis takes into account the effects of political instability on democracy, especially on the role directly given to "the people." It is then possible to propose a tentative model called the "model of complexity."

The second main section is based on five case studies, which provide significant variations on the model (Certeau 1988). Case studies illustrate and modulate the model by applying it to specific times and places.

Politics

When the governor of Tennessee, James Chamberlain Jones, addressed the senate and house of representatives on October 3, 1843, he was fully aware of the difficulties that could arise concerning the location of the capital. He stated precisely the powers that each of the components of the American democracy—executive, legislative, and judicial—had in the process, the legislature having the upper hand. The following analysis conveys exactly the same idea.

> It is scarcely necessary that I should call your attention to the necessity imposed on you of locating permanently the seat of Government within the first week of your session. This requirement of the constitution is absolute and imperative and will not, therefore, of course be neglected. The duty is purely legislative, with which the Executive cannot interfere. The only anxiety I feel upon the subject is, that the injunction of the constitution may be obeyed, and in your decision, what place so ever may be selected, I doubt not every good citizen will cheerfully acquiesce. (Tennessee General Assembly, *House Journal, 1843–44*, 27–28; quoted in Mahoney 1945, 100)

Unfortunately, tricks and invective too often replaced cheers. When capitals were selected during the territorial period, legislators were either inexperienced or much too experienced.

An Evolving Political System. During colonial and territorial times, when administrative bodies were new, almost everything had to be conquered and built. The choice of capitals took place in a pioneer and still agricultural world, much rougher than it would become later. At the end of the eighteenth century, towns were almost independent, with an often quasi-autarkic economy, in which links with Great Britain were still closer than those with other American towns. The nineteenth century saw the birth of a fragmented transportation network that altered urban relationships and fostered competition and rivalries. Consequently, the political framework was mostly locally based, since the direct political influence of Britain weakened progressively during the colonial period.

From the colonial era to, roughly, the eve of the Civil War, when a total 141 cities became capitals, and the majority of the nation's permanent state capitals had been designated, the political arena was more confused than it is today (the two-party system was set up only in the late 1850s). Before the Jacksonian era (1825–1845), looser coalitions and local feuding could render the political landscape quite blurred and ever changing. Edward Pessen described state politics as "a sordid business, preoccupied with the pursuit of office by any means, nepotism, and partisan legislation" (Pessen 1985, 199). Affiliation with Whigs or Democrats was based more on hatred and ambition than on ideas. This situation began to change during the 1820s, when the old "aristocratic system" based on "deference to social superiors was giving way to one based on white male suffrage and on insistence on social equality as the test of true republicanism. . . . It was the commercial concerns of an urban community that came to the fore, providing a first glimpse of the issues that would divide the parties of the second American party system" (Goodstein 1989, 94). Another factor was the shift from property-qualified electors to "universal" suffrage—restricted to white males—in almost all the existing states in the 1830s. This new political order was indeed not as easy to understand as one would hope, but it underlined the growing influence of national political issues.[1] All but a few state capitals were selected before the late nineteenth-century birth of the party machines that once again altered the way politics worked and would ultimately alter the way legislators dealt with the capital question. In Nebraska, for instance, "the introduction of party politics in the territory in 1858–59, which resulted in the election of a South Platte delegate, weakened old sectional and factional alignments" (Potts 1988, 180). The capital location debate was postponed, and the issue was handled in 1867 without the acrimonious and heated debates of the 1855–58 period.

An additional factor was that, unlike citizens of states, territorial inhabitants could not vote in presidential elections, in which candidates and voters take a clear party stand. Without parties, debates in the territories were mostly blurred and often based on personalities. Factionalism was at work and was a major factor in the process of selecting a capital. It was not always obvious, because it was often hidden behind "scientific" or "rational" arguments, such as the necessity of centrality, or it simply coincided with these arguments. Factionalism, at least for the thirteen colonies, was nothing new. Their political history had been complex and sometimes violent, and powerful factions arose that did not disappear after 1776 (Bailyn 1967, 95). Factionalism was reinforced by the fact that patriotism—that is, the sharing of common values—did not really exist before the end of the nineteenth century, when America began to enter the international scene.

When it had a geographic basis, factionalism became sectionalism. Pennsylvania illustrates that trend. When the Senate and then the House agreed in 1799 on moving the capital from a yellow-fever-stricken Philadelphia to Lancaster, the cartography of the vote was clear: the nays were confined to the five southeastern counties. This means that sectionalism and not affiliation with national parties was behind the choice (Sneddon 1971, 359). Eleven years later, the same sectionalism was responsible for the selection of Harrisburg as the permanent capital. The analysis of the 1805 voting on the issue of New Hampshire's permanent capital shows similar results. The two contenders were the "old" port of Portsmouth and the "new" inland town of Concord. The votes showed that the "Old Colony" voted against Concord, while the rest of the state voted for Concord—Federalists and Republicans alike (Turner 1983, 231). Further west, early nineteenth-century settlers also resented the eastern establishment and wanted to become their own masters. "Petitions sent over the mountains to protest rulings of a distant legislature were strangely reminiscent of protests that colonists of an earlier generation showered on Parliament" (Billington 1974, 211). Wanting to avoid the mistakes England had made with its American colonies and thus to avoid separatist movements, the Northwest Ordinance (1787) provided the same rights for the new western colonies that were to become states as for the old states.

Mark Twain's *Roughing It* (1872) gives a vivid account of the beginnings of new territories, the improvisation of early legislatures and the difficulties they experienced under tight federal financial reins, and the primary importance of town promoters, who did all they could to attenuate the hardships of political commencements and provide for the legislators' well-being. Twain lived in the Nevada Territory from 1862 to 1868, having accompanied his brother, who was then secretary of the Nevada Territory. Twain's account emphasizes that behind tricks, compromises, and bribes, behind new towns, there were actual men. Since parties did not yet closely control politics, there was more room for individuality and local diversity.[2] Local histories therefore often revere "the father of the town," from John Winthrop to Joseph Juneau, sometimes to the point of transforming into single-handed epics what had usually been a collective endeavor. The heroics involved in the selection or removal of a capital should therefore be addressed before proceeding to the "true" analysis of political processes.

The Myth of the Hero as Capital Founder. First, there was little space for women. Women could not vote, let alone be elected and participate in the actual process of choice.[3] During early colonial and pioneer days, women were far less numerous than men.[4] That is not to say that women lacked influence. In Illinois, Mrs. Iles secured the county seat for Springfield, and in Nebraska, Mrs. Cadman was foremost in the failed attempt to make Yankee Hill the new capital. The ladies of Pierre also helped make that little town the permanent capital of South Dakota. We have also seen that some capitals were given grand female names (Atlanta, Olympia) or the Christian names of the spouses of the leaders (Benicia). But that influence remained indirect, and men were at the forefront.

But they were not the average men. The founders were seldom plain Joneses; they were usually men of influence seeking higher positions in what was considered "virgin" land and society. The men we will consider had good situations or were well connected: an army general, a friend of President Lincoln. Positions had been granted since colonial times according to a patronage system that "had been tied to political parties at least since Jefferson's election in 1800" (Adrian 1999, 54). Capital founders were far from the classic image of the struggling pioneer, a family man seeking better times who was richer in hopes than in cash. Ray Billington justly wrote that for the West, "pioneering was a selective process" (Billington 1974, 649), but this was already the case in the East. Proprietors were close to the king and often belonged to the nobility or the gentry, and dissenters were people of education and means.[5]

When the capital was an already established place, a real town bourgeoisie had already replaced the original boosters. This "landed-money class, . . . if it was enterprising and kept up with the times, reformulated itself into a wealthy financial and business-owning class, with ties to the eastern financial establishment" (Lingeman 1980, 101). Most of these men belonged to the W.A.S.P. community. During the main period of capital choice, the migratory pattern was still centered on Great Britain and Germany. Some immigrants were Roman Catholics, but they settled mostly in states where the capital was already firmly established (such as New York or Massachusetts). When Irish people came to the United States, they were first considered distinct and hostile, and had little political power. Native Americans were "naturally" not enfranchised, and slaves could not vote. Black leaders were nevertheless sometimes courted in later capital fights. Booker T. Washington helped Charleston become the capital of West Virginia in the 1877 referendum by campaigning in its favor among members of the African-American community.[6] The black community of Guthrie also tried to retain Oklahoma's capital, in vain this time.

In the West, the old Spanish leaders did not fare well under the new American administration, as proved by the sad story of General Vallejo's failed attempts at creating a new capital for California (see the website of the Virtual Museum of the City of San Francisco). Mariano Vallejo (1808–1890) was born in a distinguished family of old Spanish stock and rose rapidly to California's leadership—always on the liberal side. He established the Sonoma pueblo in 1835 and in 1838 became commanding general of the territory. He was one of the richest Californians, with a 175,000-acre estate and a baronial castle near Sonoma's plaza, where he always helped American settlers. Today considered the "father of modern California," he was one of only nine *Californios* at the 1849 constitutional convention and subsequently became a state senator. His attempts at locating the capital in Vallejo failed, and he was treated as a foreigner by the new rulers and progressively deprived of almost all of his estate by scandalous—but all too frequent at the time—legal decisions. He spent thirty fruitless years trying to recover it. He died at his more modest last home, Lachryma Montis (prefabricated in New England!). As belated tributes, his house was listed as a state historical monument in 1933, and the Navy christened a Polaris submarine the *U.S.S. Mariano Vallejo* in 1966.

East and West were therefore linked, often from the outset, for "the purpose of every frontiersman was to build in the West a replica of the social order he had known in the East, but with a higher place on the scale reserved for himself" (Billington 1974, 651). When the Northwest Territory was created by the 1787 Ordinance, it quickly attracted settlers. General Nathaniel Massie was one of them. A civil war veteran, surveyor, and speculator in furs, salt, and land, he founded the city of Chillicothe in 1796. Massie was certainly a land speculator, but he seemed to have the public good in mind, for he asked only for a nominal fee when selling his land, thus transforming Chillicothe into one of the busiest towns in the Northwest Territory (Renick 1896, 19). Massie found allies in the Northwest territorial assembly, the leaders of which came from Virginia. The assembly met in Cincinnati for its first formal session on September 24, 1799. The first president of the House of Representatives was Edward Tiffin. Born in England, he became a physician, immigrated to Virginia, and moved to Chillicothe in 1797 with his brother-in-law, Thomas Worthington. "Both were men of property who sought a more open society and new economic opportunities in the Northwest. Both wished to escape the baleful influence of slavery; both were committed to the political philosophy of Thomas Jefferson" (Knepper 1989, 87). Owing to their combined influence, Chillicothe soon became the capital of the new territory that was carved out of the eastern part of the Northwest Territory. This example represents a common trend. Local legislatures were composed of "men of affairs and eminence," and cities "were governed by the propertied for the propertied" (Pessen 1985, 98). Around the mid-nineteenth century, many territorial and early state leaders were likewise men of position, although carpetbaggers, for instance, were common in the young Nebraska Territory.[7] Among them all, the practice of distributing useful bribes was common. These men were later called "speculators"; they certainly had profit in mind, but they also wanted to be remembered as heroes of the Western expansion of the United States and to link their names with those of ancient town builders such as Alexander the Great.

The man behind the selection of Sacramento as California's capital, David Colbreth Broderick (1820–1859) was one of these would-be heroes (see http://bioguide.congress.gov). His father had emigrated from Ireland to work as a stonecutter on the national Capitol. The family moved to New York City in 1823. After an unsuccessful campaign for election in 1846 to the Thirtieth Congress, Broderick moved to California in 1849, where he sought his fortune in gold. He was first elected to the state senate (1850–1851) and then to the United States Senate in 1857. Unfortunately, his rapid upward mobility ended prematurely on September 16, 1859, when he was killed in a duel with David S. Terry, chief justice of the California Supreme Court! The status of the family is further proved by two of Broderick's cousins. The first, Andrew Kennedy (1810–1847), was Indiana state senator from 1836 to 1840, before becoming a US representative for Indiana from 1841 to 1847, when he died from smallpox. The second cousin, Case Broderick (1839–1920), was Kansas state senator from 1880 to 1884, justice of Idaho's territorial Supreme Court from 1884 to 1888, and US representative from Kansas from 1891 to 1899.

The three men instrumental in creating Boise City and making it Idaho's permanent capital also hoped for future fame: a major, a farmer, and a journalist. The soldier selected the site of a fort, the farmer founded the town, and the journalist boasted of its qualities. In January 1863 Major Pinckney Lugenbeel was sent by the federal government to select a suitable site and build a military post, under pressure from the leaders in the West. The general site was selected by General Alvord, the new commander of Oregon, "forty miles east of the old fort up the Boise River, where wood, water, grass, and cultivable land can be found" (Chaffee 1963, 4). Selection of the exact site was left to Lugenbeel. He chose a transportation node in the Boise Valley that was near the mining districts, sheltered by a nearby mountain and with plentiful water and timber. As soon as the site for the fort had been selected, speculators came, eight of whom laid out a town nearby.

The most important speculator was Henry C. Riggs, who christened the town Boise City and was instrumental, as a member of the 1864 legislature, in making "his" town the territorial capital. Prospective businessmen were given land, trade routes were diverted from Idaho City to Boise in return for building lots, and letters were sent inviting legislators to visit the new town. The third man, who arrived in 1864, was James S. Reynolds, editor of Boise City's first newspaper, the *Idaho Tri-Weekly Statesman*, who lobbied hard to attract the capital. Success came in that very year, and Boise is still Idaho's capital.

The two men behind the removal of North Dakota's capital exemplify the alliances between "old" and "new" men, as well as the connections between East and West. The first man was a civilized eastern politician; the other was a western self-made-man. Both were considered corrupt. The first was the territorial governor, Nehemiah G. Ordway. Before becoming governor in 1880, he had been sergeant at arms for the House of Representatives, then a state senator in New Hampshire. Businessmen of that state who had invested in Dakota supported his nomination as governor. He soon was at the head of a "machine," and his patronage bought the friendship of newspapers—he placed may advertisements of public land sales. He found a useful ally in Alexander McKenzie, who had worked on Northern Pacific construction crews. He settled in Bismarck in 1873 and became sheriff of Burleigh County at the age of twenty-four in 1874, holding that office for twelve years. An imposing man, his easy manners won him power, and he was soon selected by the Northern Pacific, which was controlled by St. Paul interests, as its political agent. "McKenzie was to be the Republican national committeeman for North Dakota until 1908; he was to become 'Alexander the Great, Boss of North Dakota'; he was to die a millionaire, and, even though he had never held a state office, to be given a funeral in the state capitol at Bismarck" (Robinson 1966, 200). The removal of the territorial capital from Yankton to Bismarck was the first result of their alliance. When President Chester Arthur removed Ordway in June 1884 upon action by the Yankton group, the Northern Pacific immediately sent him to Washington as a lobbyist against statehood.

Juneau, Alaska, situated in a mining district opened during the winter of 1880–81, was the result of the work of forty gold miners, later known as the Ready Bullion Boys (DeArmond 1967). Three of them were typical of the enterprising spirits who were at the forefront of mining ventures: George E. Piltz (1845–1926), Richard T. Harris (1833–1907), and Joseph Juneau (1834?–99). At the time of their Gastineau Channel expedition, at the end of 1880, they already were seasoned gold seekers. A trained miner in his native Prussia, Piltz resumed his job after immigrating to America in 1867. In 1880, he raised money in Sitka and San Francisco to grubstake prospectors and sent a first party to Gastineau Channel, including Harris and Juneau, that came back empty-handed. Both men returned later that year and made Alaska's first major gold discovery. Piltz then came and served as chairman of the first miner's meeting and soon brought his wife and son. But when he left Juneau in 1887, he was broke. DeArmond attributes his failure to his "chronic inability to get along with other people." Harris also spent his life in frontier mining camps, after graduating from Girard College in Philadelphia. He stayed in Juneau, first as a well-off miner until 1885, when a judgment ruined him (he had made a critical mistake in locating and recording his claim), then as an employee, until ill health sent him to a sanitarium at Portland, Oregon, in 1905, where he died two years later. Joseph Juneau, born in Quebec, grew up in Wisconsin but went to California in 1849 at the age of fifteen to become a miner. After his 1880 expedition, he did not stay in Juneau; he went to the Yukon in 1894 and the Klondike in 1897 but without much luck, for he was running a restaurant in Dawson at the time of his death in 1899.

This story of the trio who discovered gold in Alaska and founded the future capital of the territory is not a happy one. All three left Juneau poorer than when they arrived. Piltz, instrumental in the creation of the town, seems forgotten, while Harris and Juneau are remembered locally. Despite his shortcomings, a street in Juneau bears Harris's name, and Juneau's remains, according to his last wish, were brought back to Juneau (with the help of a subscription) and buried in Evergreen Cemetery on Sunday, August 16, 1903.

Juneau's story makes it clear that those who founded capitals were not always as powerful as the sources generally state. Accounts of the period are often written in a traditional or official way, mainly by white men. There is a natural tendency to reconstruct the processes of choice, to create a mythical history of the state, in order to enhance the grand schemes presiding over its construction. And myths need heroes. Local narratives thus often dwell on the history of the great men (or the great man) whose genius triumphed over adversity to win the capital. This invites the idea of the helping hand of the Divine Providence in choosing a capital. Gaining the status of a capital would be a local and urban variation of Manifest Destiny. When reading the city pageants written for the first or second century of capitals (e.g., Hanson 1916; Langdon 1916), one is well aware of this tendency to overstress the single-handed action of one providential man.

Minnesota provides an excellent example of such processes. In 1857, when Minnesota was to achieve statehood, a debate was initiated over the removal of the capital from St. Paul, which was considered too Democratic and northern for the Republicans, and too powerful for many representatives from other large towns. The new location put forward by the "enemies" of St. Paul was St. Peter, a paper town that was to become the capital of a more rural (and Republican) state and benefit some representatives who were financially engaged in the project. The traditional history of the way St. Paul succeeded in retaining the capital goes as follows (according to a summary of the numerous versions of the story). The hero was a former judge called Joseph Rolette Jr., the mixed-blood son of Old Joe Rolette. Known for his pranks and love of the good life, he was nicknamed Jolly Joe. A councilman from Pembina, he held a strategic position in the House, being chairman of the committee on engrossed bills. He disappeared with the engrossed removal bill for one week (in fact 123 hours), only reappearing in the council chambers at the closing of the session. Another local hero, the council's sergeant at arms, conducted the search. Since he also supported St. Paul, his search systematically avoided the right place, a hotel room located right in the middle of St. Paul, where Rolette was being lavishly entertained by the locals. After his "reapparition," he was given a parade by the population, and according to an early historian of Minnesota, "Rolette became St. Paul's mascot, and there was no tribute of devotion its citizens were not willing to lay at his feet as an evidence of their gratitude" (Dean 1908, 15).

Such a "miraculous" story could not go without attempts to undo the myth. A contemporary history of Minnesota does this. According to William Lass, "his prank was actually all in vain, for during his absence another copy of the capital removal bill was passed on to Governor Gorman, who signed it" (Lass 1998, 122). More recently, a closer look has been taken at the main character of the story, who appears less "jolly" than in the myth (White 1999). First, the "half-breed," as he was portrayed in the late nineteenth century, was in fact the son of a well-off and well-educated trader from French Canada, with mostly French and British blood running through his veins. After receiving a private education in New York, he worked for fur traders in Pembina County. As for the "stealing" of the bill, there was no stealth, because, as he was chairman of the committee on engrossed bills, the bill was legally his. His duty was to certify the text and send it to the governor for signature. Rolette only avoided doing it.

Rolette's part in the history, whatever it was in reality, testifies to the end of the rough territorial system and its progressive replacement by a new one. His adventure could be used by everyone, and its jollity could absolve "influential and ordinary people from any accountability for a reckless, illegal, action that deprived a democratic majority in both legislative houses to the right to pass a bill that the governor was ready to sign" (White 1999, 41).

Minnesota thus also provides a current reassessment of history for historians eager to avoid the methodological pitfalls of former hagiographies. This "new history" nevertheless allows the process of capital choice to retain its symbolic meaning, which was and remains important, along with the people who "made" the state or at least have been at the forefront of the collective building of the state. In that sense Joe Rolette, although he died a pauper in 1871, remains a beloved figure in the collective mind of Minnesotans, without harming the state's ideal or record of progressive politics.

There are naturally many other elements at hand to undermine the "hero" thesis. The influence of president and Congress still remained strong; party politics reduced the influence of individuals; the eastern establishment and capital were behind the way the West was ruled; lastly, the weight of the numerous wars—especially the Civil War—cannot be discounted. Several high-ranking officers later became governors with the support of the powerful friends they had made in Washington during that time (Jackson and Grant had been generals before becoming presidents). When President Jackson nominated in 1836 the first governor of the new territory of Wisconsin, he chose Henry Dodge. Dodge had been a brigadier general during the War of 1812 and a colonel with the Iowa County militia during the Blackhawk War. Jackson commissioned him to lead a contingent of the United States Dragoons to protect settlers from Indian attacks. His association with President Jackson and the popularity he gained from his military "protection" of settlers ensured him the governorship (Cravens 1983, 105).

The governors of the Colorado Territory illustrate this trend very well (see the website of the Colorado State Archives). The first governor, William Gilpin (1813–94), was a scion of the "best" eastern establishment. He was born into an immensely wealthy Delaware Quaker family. After graduating from the University of Pennsylvania, he became a member of Fremont's party in 1843. A true believer in Manifest Destiny, he was also an important land speculator—in Oregon, Colorado, and New Mexico—and was one of the first to see Denver as a future railroad hub. He was governor of Colorado for only one year (1861–62). He was fired after having heavily indebted the territory to pay for the First Colorado Regiment of Volunteers that he created (although the regiment fought bravely during the Civil War). The second governor (1862–65), John Evans (1814–87), was born in Ohio. He founded the Illinois Medical Society (he was a doctor) and was one of the founders of Northwestern University. Through investments in the Chicago and Fort Wayne Railroad and the Chicago and Evanston Railroad, he attained wealth and political influence. As a founder of the Illinois Republican Party in 1852, he became a friend of Abraham Lincoln, who later nominated him as governor of Colorado. He was dismissed after the Sand Creek Massacre on November 29, 1864, and his political career was ruined. But his local influence did not cease, for he was behind several railroad investments, among them the all-important Denver-Cheyenne branch line. John Long Routt (1826–1907) was the territory's last governor and the state's first one, from 1875 to 1879 (Colorado became a state in 1876). The first governor born in the "West" (Kentucky), he was a carpenter and sheriff in Illinois. He rose to the rank of colonel during the Civil War, which allowed him to become a US marshal in Illinois and then governor of Colorado. He used his position to augment his own fortune, for which he was dismissed, but he had become a millionaire. He remained in the state, serving as director of the First National Bank, Mayor of Denver (1883–85), and governor again from 1891 to 1893, amidst political gangs and legislative violence. Although Western constitutions gave more power to the legislature than to governors, the initial territorial years (twenty in Colorado's case) were hardly democratic.

If we are to look beyond the "hero," then who had the upper hand in the process of capital selection? After the end of British rule, Congress let the new states write their own constitutions, allowing them to select their seats of government. The process has been mostly indirect, therefore, with the legislature and governor being of primary importance in the decision, not the people. Referenda were generally held to ease a legislative deadlock (e.g., in West Virginia) or were emptied of their sense by the multiplication of votes (e.g., in Oregon). No democratic vision ruled there. The constituents could nevertheless refuse the decisions taken by the legislators and go before the courts, be they local or national—another illustration of the balance of power in the American democracy.

King, Congress, and the Choice of a Capital. It is important to assess the degree of liberty the colonies, territories, and states had in choosing their own capitals. The legal framework for the processes of capital selection varied greatly. During colonial times, there were three options. If the colony was proprietary, the proprietor (or board of proprietors) was free to locate the capital where his representative stayed and where the burghers could have a meeting-place. If a colony was founded by a charter trading company, such as Virginia for the London Company in 1607, the capital was where the American headquarters of the company stood. For royal colonies, the king and Parliament could designate the capital, usually in the town where the governor stayed and the local assembly of freemen met. Among the many complaints against the king enumerated in 1776 was that "he summoned legislative assemblies in unusual and inconvenient places, remote from the public registers, with the only view to obtain, by exhaustion, their adhesion to his measures" (quoted in Brege and Crouzatier 1991, 106). The royal colony of North Carolina was freer in that respect, at least for its first forty years, during which it had no regular capital; the seat of government was wherever the governor chose to live (Powell 1989, 146).

After the War of Independence, there were two general situations. For the newly independent states, the capital was where the state's general assembly met. For the newly acquired or conquered territories, the role of the federal government was restricted to the territorial period of each state. The founding text was the Northwest Ordinance of 1787. Three stages of governmental development were set up from discovery to statehood.

- First stage: Congress appointed a governor, a secretary, and three judges, without any legislature.
- Second stage: when the population of "free male inhabitants of full age" reached five thousand,[8] a general assembly was set up, consisting of the governor, legislative council, and a house of representatives. "The governor and council were both the creatures of Congress; only the house was popularly elected. Furthermore, as a check against popular enthusiasms, the governor had an absolute veto over acts of the territorial legislature. He could also convene or dissolve the general assembly as he saw fit. The general assembly was authorized to elect one delegate to the federal House of Representatives who could introduce and debate legislation, but could not vote 'during this temporary government'" (Knepper 1989, 60–61).
- Third stage: the Northwest Territory was to be divided into three to five states. When its population reached sixty thousand free inhabitants, a

state would be admitted into the Congress of the United States. The new state would then form a permanent constitution and a republican state government.

This process (with later modifications) allowed the United States to expand in an already known land in an orderly manner under already prepared administrations and governments. Congress had first to draw the boundaries of new territories and states. After 1789, states were "to an important degree arbitrarily defined pieces of thinly settled national territory shaped in congressional committees and debates" (Meinig 1993, 446).[9] They were arbitrarily defined because, since a new territory was considered as a tabula rasa, geometry was able to rule, without any reference to already existing societies.[10] This process of delimitation exercised a major influence on the choice of the capital, by defining where the geographic centrality would be and by putting together similar or very different areas that would later nurture sectionalism or harmony. This was the case for Ohio, Indiana, and Illinois, the longitudinal division of which set them "in the path of three streams of migration that would fill them with diversity and tensions" (Meinig 1993, 447).

Once the boundaries of a new territory had been defined, selecting the capital was a major decision that testified to the territory's power. Congress was thus only able to designate the temporary capital when drawing the act creating a new territory. The congressional act of March 3, 1817, "establishing the Alabama Territory," contained therefore the following provision: "That the town of St. Stephens shall be the seat of government for said Alabama territory, until it shall be otherwise ordered by the legislature thereof." As the most important town of the new territory, St. Stephens was thought to be the best possible choice (Owen 1949, 186).[11] Congress could also set the rules for the designation of the permanent capital—generally by popular vote—and set aside a certain sum of money toward the cost of building the permanent capitol. It could also let the territorial governor select the meeting site for the first assembly; for example, the Nebraska Act of 1854 stipulated that "thereafter the legislature would determine the permanent seat of government" (Potts 1988, 173), as it did in Alabama.

If a permanent capital had been selected by the territory, Congress could also forbid any further change. Kansas illustrates this process (Adams 1903–4, 342). In 1858, the legislature passed an act for the removal of the territorial capital from Lecompton to Minneola, a paper town (located east of Centropolis) in which many of the members of the legislature had interests. Railroad companies were even chartered to conduct railroads to center there. But the governor vetoed the act, stating that the legislature had no power to make the removal. Since the legislature passed over the veto, the question of the validity of the act was submitted to the attorney general of the United States, J. S. Black. On November 20, 1858, he declared it to be in violation of the Organic Act, arguing that in 1855 the territorial legislature had already fixed the permanent seat of government in the town of Lecompton and obtained the $50,000 appropriation to build the capitol. He added, "Making a permanent location certainly did not mean a designation of a place merely for the purpose of getting the money, and then making another change. . . . Such a removal, if carried out, would defeat the manifest intention of Congress, violate the spirit of the act, and be a fraud upon the United States." The legislature complied and Lecompton remained the territorial capital.

Congress, however, was not the body most competent to choose capitals, being too far from state realities. State legislatures or constitutional conventions were closer to these realities. In Oklahoma, Congress had imposed Guthrie when the legislators had chosen Oklahoma City. Although the two cities were quite close, so that changing the capital was of little geographic importance, Oklahomans deeply resented the decision taken by Congress (Moussalli 1997, 74–75). It is therefore no surprise that Oklahoma City is today the capital.

The Legislature at the Heart of the Process of Choice. The contemporary growth of state powers has been justly emphasized, but this does not mean that early governments were powerless. On the contrary, since during territorial times, the government had been the only political institution and had important economic and social roles. State power was the more important because the federal government was small. When Jefferson was president, his "team" only amounted to six men, and the Civil Service was only created in 1883 (Fohlen 1992, 110). The decisions of Supreme Court Judge Marshall had certainly strengthened federal powers during the Jacksonian era, but state governments, although small, were powerful. They exercised their power in the delimitation of new counties, patronage over newspapers, licenses for ferries, contracts for public works, and appointments that held potential power—militia officers, sheriffs, and local judges (Cayton 1996, 229). Wisconsin was typical.

Most of the advantages of territorial status were reserved for those with political yearnings or an entrepreneurial temper who had schemes to promote that would benefit by some legislative action. Of course, there were services that a territorial government needed to perform for a growing population, and territorial status did serve to attract new settlers. But an examination of the index to the first legislative session of Wisconsin [1836] reveals that one-fourth of the entries are concerned with the incorporation of banks, railroads, toll roads, mining companies, and other such ventures. There were no general incorporation laws, and therefore each stock company had to get a special charter through a legislative act. Territorial politics were primarily concerned with office seeking, the dispensation of privileges, and appeals for federal largess to help develop the territory. This was a natural reflection of a sparse population, always short of capital, confronting a combination of opportunities and obstacles beyond its resources (Nesbit 1989, 125).

One reason for such conduct, which we would view today with scorn or reprehension, was that most new legislators lacked experience and even knowledge of the whole territory. Indeed, when the new legislators assembled in Wisconsin's temporary capital, Belmont, on October 25, 1836, none had any previous legislative experience, and almost all had arrived less than ten years before. They were far easier to manipulate than today's seasoned politicians (Smith 1973, 252). This was indeed what happened, in Wisconsin and in other brand new American territories and states, in the process of choosing the permanent capital. But to generalize such comments would be preposterous. During Michigan's early years (1805–1810), easterners appointed by Congress to govern the state were dubbed "small-minded" by historians, who compared them unfavorably to the "rough, untutored pioneers of the West" (Dunbar and May 1995, 117).

When Congress did not provide for it, the states had every liberty to choose their capital in any way they wanted. Table 5.1 provides some of the states' constitutional requirements. Stephanie Moussalli rightly points out that such specifications were examples of superlegislation, and cites Lawrence Friedman: "Ordinarily, behind each instance of superlegislation there lurks some concrete story, some concrete factional dispute . . . between interest groups" (Moussalli 1997, 60). To enlarge upon table 5.1, most states either gave the assembly the power to select the capital or gave such power directly to the people.[12] The latter was more often the case in the western states' constitutions, because they usually adopted the more democratic provisions of the eastern constitutions from which they drew their inspiration. They therefore preferred to give more power to the legislature than to the governor, and limited terms of office to prevent the formation of a political elite (Billington 1974, 652).

Table 5.1 Provisions for seats of government in six frontier constitutions

Constitution	Provision
Louisiana (1812)	New Orleans until legislature moves it.
Mississippi (1817)	Natchez at first session. Then legislature may move it.
Alabama (1819)	Huntsville at first session; Cahawba afterwards, until end of the 1825 session, during which legislature chooses permanent site. Final site can never be changed; governor's agreement not required; no appropriations for statehouse other than in Cahawba before 1825.
Arkansas (1836)	Little Rock at first session. Then legislature may move it.
Florida (1839, implemented in 1845)	Tallahassee at first session and the following five years. Then legislature may move it until ten years after first session, when permanent seat must be selected.
Texas (1845)	Austin until 1850; then at site chosen by popular election in March 1850; if there is no majority, run-off election between top two choices in October 1850. This site unmovable until 1870.

Source: Moussalli 1997, 60.

But constitutions can be changed. Thus, even if a capital seemed permanently chosen, a new constitution could provide for its removal, as occurred in Alabama, whose 1901 constitution contained the following section (No. 78): "No act of the legislature changing the seat of government of the state shall become law until the same shall have been submitted to the qualified electors of the state at a general election, and approval by a majority of such electors voting on the same; and such act shall specify the proposed new location" (Owen 1949, 186).[13] This provision must have drawn upon Alabama's earlier history. In fact, on January 21, 1846, the supposedly permanent capital of Tuscaloosa had already ceded the capital to Montgomery because the middle and eastern parts of the state had grown. The legislature had ratified during its 1845–46 session an amendment to strike out Section 29 of Article III of the constitution, which permanently located the seat of government (207).

Even if Congress, in its Enabling Act, had selected a capital, the state could change it. The Oklahoma Enabling Act had required that the capital remain at Guthrie at least until 1913. In 1910, however, the Democratic governor, Haskell, decided to transfer the state capital to Oklahoma City, away from Guthrie's "Republican Nest." On June 11, 1910, an election was held on the matter that did not mention the date of the removal. Oklahoma City obtained 96,261 votes; Guthrie, 31,301; and Shawnee, 8,382. During the night, Haskell had the state steal "stolen" from Guthrie and declared the Lee Huckins Hotel in Oklahoma City the new capitol. Guthrie brought the matter before the state supreme court. The court, in *Coyle vs. Oklahoma* (1911), upheld Haskell's action. In the same year Guthrie went on to appeal to the United States Supreme Court, in *Coyle vs. Smith*. The answer of the Court is very interesting: it wrote that a state was free to determine the location of its capital city, despite an enabling act mandate (Gibson 1981, 206).

Since most states left the matter in the hands of the legislators, it was easier for town boosters to lobby for their city. Many cities vied for the high honor of being capital, and factionalism and sectionalism were at their highest during these periods. "Temporariness" could feed itself with legislatures' factions, for factions led legislatures to select only a temporary seat of government. The official reason was that this enabled the legislature to form an ad hoc committee to find the most suitable permanent capital (e.g., Detroit and St. Stephens). Temporary capitals were not always changed (e.g., Olympia and Nashville), but the sword of Damocles threatened all of them because the selection of the permanent capital revived factions and sections. Maine illustrates such a pattern (North 1981). The first committee selected the town of Hallowell—part of which would become Augusta—on a purely financial basis: gathering the legislators at Hallowell was far less expensive than having them go to Portland, the temporary capital. The legislators should have been pleased with this defense of the public good. On the contrary, the report of the commissioners only opened a period of political bargaining that led to the appointment of many successive committees directed to do the same thing as the first one. Likewise, although the Mississippi legislature had chosen the new site of Jackson in 1821 as the "permanent" capital, other towns continued to seek it. The General Assembly discussed the matter for the next decade, but sectionalism prevented any agreement, although the majority was in favor of a removal. Vicksburg, Port Gibson, Clinton, and Westville in Simpson County were successively defeated. The constitution of 1832 did not settle the matter; it extended Jackson's capital status temporarily to 1850. The legislature was then to designate the permanent seat of government (McLemore 1973, 256). Jackson is still Mississippi's capital.

The process of selection therefore often ended in barter and compromise. Every American knows that Washington was a compromise location. During the Constitution's ratification process in 1788, many contenders arose in Delaware and Pennsylvania, boasting local facilities, defensibility, and easy access to the West. "Madison promised the necessary southern votes to assume the state debts if Hamilton could first prevent New England from undermining as it had in the past an existing deal between Pennsylvania and the South that called for a Potomac capital after a ten-year temporary residence at Philadelphia" (Bowling 1988, 178).

The Power of Compromise

The University at Tuscaloosa, the Penitentiary at Wetumpka, and the Capitol at Montgomery, would, under all the circumstances of the case, be but a fair and appropriate apportionment.

(*ALABAMA JOURNAL*, JANUARY 11, 1843)

At the end of a capital fight, there was only one winner, but the other contenders were not necessarily losers. They usually received consolation prizes: for example, universities were granted to Grand Forks in North Dakota, Saint Anthony in Minnesota, and Moscow in Idaho. The same pattern was at work when the capital moved: Golden, Colorado, gained the Mining School. The insane asylum or the state prison were sometimes preferred because they were larger grants. In Minnesota, St. Paul was selected as the capital in 1849 and confirmed in 1858 when statehood was granted. Among the other contenders, Stillwater, being larger and more influential, wanted and got the prison in 1851. St. Anthony's citizens did not want the prison and received the university. St. Peter got the insane asylum in 1866 (Abler, Adams, and Borchert 1956, 260). In South Dakota, after Pierre's victory in the 1889 capital fight, Sioux Falls had the next choice of state institutions and chose the penitentiary, while Vermillion had to take what was left, the university (for which appropriations were lower).

Such spoils were therefore not granted at random. Since most state constitutions were not as strict as Oregon's, which located all the state's public institutions in the governmental seat, the distribution of state spoils by barter in the legislature was often a way to reach majorities on the capital question. This was a common trait of the period.[14] Cities that did not want or had abandoned the idea of becoming the state capital had usually sold their votes to the remaining candidates in return for spoils. In Oklahoma, the Organic Act of 1890 had designated Guthrie as the territorial capital, and the first legislature met there on August 29, 1890. Instead of focusing on proper legislative work, legislators argued over the future location of the capital. The main contenders were Guthrie, with a population of 5,884; Oklahoma City, with 5,086; and Kingfisher, with 1,234. Oklahoma City won the first vote, and Kingfisher won the second, but both bills were vetoed by Governor Steele on the grounds that they were premature because the territory expected further land additions. Other cities used this delay to secure useful spoils: Stillwater (625 inhabitants) got the Territorial Agricultural and Mechanical College; Edmond got the Territorial Normal School, and Norman (764 inhabitants) got the Territorial University (Fischer 1975, 4).[15]

Springfield is one of the best examples of compromise. Indeed, it benefited from a master player, the young Abraham Lincoln (Angle 1950, 56–57). Since 1820, the capital of Illinois had been located in the new town of Vandalia. In the 1830s, the northern and pioneer part of the state was trying to attract the capital, while Vandalia remained a mere village, with slack sale of town lots. Legislators complained about the lack of adequate accommodations and Vandalia's unhealthiness. Lincoln linked the capital question with that of the internal improvement system, and was therefore able to trade Sangamon County's nine votes on the transportation question for votes on the capital question. He also succeeded—on February 25, 1837—in having the law locating the capital by popular vote repealed in favor of a majority vote by the legislature, after the internal improvement system had been voted upon (a very undemocratic agreement). On February 28, 1837, the first ballot gave Springfield 35 votes, more than twice as many as Vandalia, which came in second. On the fourth ballot, Springfield received 73 votes; whereas Vandalia received only 16; Jacksonville, 11; Peoria, 8; Alton, 6; Illiopolis, 3; and five other towns, 1. The causes for Springfield's success seem quite complicated (Burtschi 1954, 55). The town of Alton had won an 1834 referendum on the capital question (although no action followed), with 8,157 votes, compared to 7,730 for Vandalia, and 7,035 for Springfield (7,075 according to Angle, 1950, 56), and dreamed of becoming the permanent capital. In addition, in 1835–36, there was talk of extending the National Road either to Alton (north), or St. Louis (south). Vandalia preferred St. Louis, a position that deeply angered Alton, which changed its votes in favor of Springfield to take revenge on Vandalia for the National Road question. So, despite having built three capitol buildings (in 1820, 1824, and 1836), Vandalia definitively lost the capital and remained a small town (7,042 inhabitants in 2010).

But greed was not universal. In South Carolina, the losers, instead of substantial assets like a prison or a university, contented themselves with a more symbolic present. The commissioners of the new capital, Raleigh, "in deference to some of those towns which lost out in the Capital's location, axially radiated four of these streets out of Union Square—pointing them toward the direction of said cities, Fayetteville to the south, Halifax pointing north, New Bern to the east, and Hillsborough to the west. These thoroughfares, too, were to be ninety-nine feet wide, thirty-three feet wider than any of the rest" (Waug 1967, 4). Other losers proudly rejected the spoils they were offered. Port Townsend in Washington State, lost the 1860–61 fight for the permanent capital. The representative of the state's second oldest city had obtained the penitentiary, but the citizens of the town refused it with haughtiness. The town's journal proclaimed, "Our Councilman has been to the Legislature, and bought for his dear constituents, in exchange for the Capital, such a fitting institution, and so expressive of his appreciation of their remarkable character—a Penitentiary!" (quoted in Beardsley 1941, 270).[16] Legislative bargaining and bartering did not always settle capital controversies.

A Surfeit of Litigation. When citizens were not pleased with the result, they resorted to the third element of the American democracy, the judiciary. During early territorial times, however, the judicial system was in its infancy, and instability affected legal matters also, limiting the power of the judiciary. First, the examples of Minnesota and Idaho show that similar causes could result in opposite rulings. When Minnesota was gaining statehood in 1857, southern Minnesota boosters thought that a removal of the capital to St. Peter, in the Minnesota River Valley, would prompt the construction of a railroad and give the future state an elongated form in which St. Peter would be demographically central. They could count not only on the Republicans, who were opposed to St. Paul, but also on some rural Democrats who resented St. Paul's importance and on the legislators who were members of the St. Peter Company. In February 1857, when statehood appeared imminent, advocates for St. Peter succeeded in passing a capital removal bill in the territorial legislature. St. Paulites, alarmed over the threatened loss of the capital, took action. The change of governor and the lack of action by territorial officials left St. Peter's new frame capitol empty. The official capital's supporters therefore went to federal district court. The ruling by Judge Nelson in July 1857 was clear: the territorial legislature had exhausted its power to locate the capital when it had named St. Paul, and the removal act had no validity. St. Paul remained the capital (Lass 1998, 121–22). On the other hand, two residents of Lewiston, Idaho, challenged in court the vote by the second territorial legislature to relocate the territorial capital from Lewiston to Boise in December 1864. This was clear sectionalism, northern against southern Idaho. But incompetent handling of the case resulted in the upholding of the relocation by the Idaho territorial Supreme Court (McKee 1992). The Court's decision, on June 14, 1866, rested on two separate arguments. The first held that the removal act was valid because the session of the legislature had been regularly and lawfully assembled. The second was based on legal procedure and held that "as the case originated in a Probate Court, it should have been dismissed by the District Court on the appeal taken" (Bird 1945, 345).

Second, Dakota—a rich site for capital choice procedures—exemplifies the bypassing of the judiciary by corrupt town boosters (Potter 1932, 28; see also Dalton 1944). In 1883, the temporary capital, Yankton, finally had to yield to some more centrally located city.[17] A master schemer, Alexander McKenzie, was the primary agent of the move. He worked for the Northern Pacific Railroad, which wanted the new capital on its lines. The Northern Pacific drew its power from the federal land grant of 10.7 million acres, which meant that it controlled 24% of the state. McKenzie also aimed to make large real estate profits from the 160-acre tract he had bought with an associate (Ordway) on the northern edge of Bismarck. He succeeded in getting the legislature to create a small commission, which would be easier to control than a popular vote on the question. McKenzie was naturally a member of the commission. He cleverly structured a bill that could only have the outcome that he and the railroad wanted. The removal bill, passed on March 8, 1883, provided that every town contending for the capital had to offer at least $100,000 and at least 160 acres of land. Only a town with railroad money could pay such a huge sum; Dakota's towns at the time were tiny settlements (Yankton and Fargo were the largest in 1880, with 3,431 and 2,693 inhabitants). Although the bill met with almost unanimous disapproval, eleven towns bid for the capital: Aberdeen, Pierre, Bismarck (1,758 inhabitants in 1880), Mitchell (320), Redfield, Ordway, Canton, Frankfort, Huron (164), Odessa (now Devils Lake), and Steele. The story of the workings of the commission is so unusual that I quote it in full from Merle Potter's "The North Dakota Capital Fight."

> But before the commissioners could start their swing around the circle there was an ominous obstacle that had to be passed. An annoying provision of the capital removal bill required the commissioners to meet and organize in Yankton, and in Yankton they had sound reasons for believing they would encounter trouble. That town was not to be deprived of the capital without making a forceful show of resistance, and it based its objection to the commission on the ground that it was without authority to locate a capital for Dakota territory. It contended that the governor and the legislature alone had such authority. Therefore Yankton, the commission learned, was preparing to serve writs on the commission when it arrived for its organization meeting that would have delayed the work and placed the removal plan in a merry tangle.
>
> A discouraging prospect, but one that only spurred the ingenuity of McKenzie. He arranged for the commissioners to assemble at Sioux City,

> Iowa, where a special train was chartered and started for Yankton. On the way the secretary was instructed to prepare all the necessary papers for organization—the motions, the orders, and the other required formalities. The special proceeded on its way and when it reached the outskirts of Yankton, the commissioners found that friends had placed flags at the limits of the town and as the train sped through, a commissioner made the motions, they were passed unanimously, and the army of deputy sheriffs and court officers stood on the depot platform, armed with embarrassing legal papers but unable to serve them. Before the train had passed out of the western borders of Yankton, the commission had been legally organized. It was this triumphant strategy that gave rise to the expression, "The Capital on Wheels." . . . Everywhere "The Capital on Wheels" was received with all the ceremony and enthusiasm accorded today to transatlantic flyers or delegations of foreign notables. . . . There were bands and banquets, parades and pageantry, and, if numerous authorities can be credited, more wine and other liquors were produced on such occasions as seemed prudent. . . . Bismarck was the last town visited before the crucial ballots were taken, and among the arguments advanced for locating the capital there, were that it was on a great trans-continental thoroughfare, that it was "at the east end of one of the most gigantic iron bridges now in existence—in the vast commercial transportation flowing eastward from China, Japan, and the Pacific seaboard." Moreover, it was on a navigable river, was surrounded by a rich agricultural community, had a good climate, was near cheap fuel, had good building material close by, and possessed a host of other assets. (Potter 1932, 31–33)

The vote was taken in Fargo, and on the ninth poll Bismarck won, without much comment. But the legal battle was not over, and the *quo warranto* case was finally argued before Judge Edgerton, who declared the commission and all its work illegal. The matter was then brought before the Supreme Court of the territory, which overruled Judge Edgerton's decision. The territorial legislature of 1885 tried to take matters in hand and moved the capital from Bismarck to Pierre. But Governor Gilbert A. Pierce vetoed the move. This tragicomedy had another effect: resentment in the southern part of the Dakota Territory after Bismarck was chosen hastened plans for division and statehood (1889). Preparation for statehood resulted in another brief skirmish over the capital of North Dakota. This time, Bismarck's supporters succeeded in including the capital's location in the constitution instead of bringing the matter before the people, which greatly angered the residents of Grand Forks.[18] In 1890, Grand Forks was the second largest city in North Dakota, with 4,979 inhabitants, fewer than the 5,664 of Fargo but more than the 2,296 of Jamestown and the 2,186 of Bismarck.

Fortunately, justice was sometimes fairly dispensed. In Washington State, the 1860 "capital affair" is "still regarded as one of the outstanding legal battles in Washington history" (Beardsley 1941, 278–84; see chapter 3). It shows the limits that citizens can bear in legislative bargaining and bartering. The legal question was threefold: first, when the legislature passed the relocation bill that removed the capitol from Olympia to Vancouver in December 1860, the date of passage and the enacting clause were omitted, perhaps because the bill was hastily passed or because of tampering by the public printer (who was also the editor of the *Pioneer and Democrat*) or by some other Olympia citizens. Second, the acting governor refused to permit the removal of the territorial library to Vancouver. Third, out of a total of 2,315 votes cast in the referendum on the capital question held on July 8, 1861, Vancouver got only 639 votes, and Olympia received 1,239. The Supreme Court convened at Olympia in December 1861 to hear the case.

"The Court held that the legislature had exceeded its powers in declaring that the seat of government should be and remain at Vancouver; that the relocation act had been made contingent upon the decision of the people, as expressed by their vote in the next general election; and that an act without an enacting clause and without a date was void. . . . The decision had declared void the act relocating the territorial capital at Vancouver, but could not of itself fix the seat of government permanently at Olympia. There it would remain only so long as the people desired" (cited in Beardsley 1941, 285). All the previous examples show that it was common for legislators to take the capital matter in their hands and to distrust the people who should normally have been consulted in that important matter.

Political Instability, Democracy, and the People. Table 5.2 shows that two-thirds of current capitals have been chosen by legislatures or constitutional conventions. The capitals selected by Congress or governors are mainly cities that were chosen as the first capitals of colonies or of territories and were able to retain their status. The people were consulted only ten times, for one capital in five.

When the people were consulted, it was often the result of political instability—legislative deadlock between factions and sections (Alabama and West Virginia), democratic reaction against corrupt legislators (Texas), scheming (North and South Dakota), popular "revolt" (Iowa and Washington

Table 5.2 Responsibility for selection of current state capitals

Legislature	Congress
Phoenix	Juneau
Little Rock	Atlanta (army, 1868)
Sacramento	Honolulu
Denver	Salem (1855)*
Dover	
Tallahassee	Governor
Boise City	Annapolis
Springfield	Boston
Indianapolis	St. Paul*
Topeka	Carson City
Frankfort	Santa Fe
Baton Rouge	Oklahoma City*
Augusta	Austin (President Lamar)
Lansing	Cheyenne
Jackson	
Jefferson City	People
St. Paul*	Montgomery
Helena*	Hartford
Lincoln*	Des Moines
Concord	Helena*
Trenton	Lincoln*
Albany	Oklahoma City*
Raleigh	Salem (1864)*
Bismarck	Pierre (1889, 1890, 1904)
Columbus	Olympia (de facto)
Harrisburg	Charleston
Providence	
Columbia	Springfield : repealed
Nashville	Topeka: never implemented
Salt Lake City	St. Paul: never implemented
Montpelier	Bismarck: never implemented
Richmond	
Madison	

Note: Capitals with an asterisk followed two processes.

State)—or just to confirm the governor's or legislature's choice (Oregon, Montana, and Oklahoma) rather than from any political philosophy regarding "power to the people."[19] In four cases, legislatures even repealed or ignored requirements to leave the matter in the hands of the people.

In West Virginia the people were called upon through a referendum when legislators proved unable to agree. When West Virginia was carved out of Virginia as a separate state in 1863 after fifty years of political fighting, the selection of a capital might not have seemed problematic. But sectionalism and its political counterpart, factionalism, persisted, albeit on a smaller scale. The capital thus first wandered between Wheeling and Charleston (by way of steamboat and barges for the state archives), before settling definitely in Charleston in 1885. Wheeling, in the north of the state, profited from its economic prominence; it was the western terminus of the National Road, a thriving river port, and a manufacturing center. Wheeling benefited also from being Republican, while radical Reconstructionists dominated the legislature. In 1869, their power having receded, southern Charleston became the capital. But legislators soon complained that the town was too small (3,162 inhabitants in 1870, compared to 19,280 for Wheeling) and not accessible enough. In 1875, after a disastrous fire almost destroyed Charleston, Wheeling recovered the capital and offered a capitol building. Although the governor had not signed the bill, the move took place (Rice and Brown 1993, 163–64).

To definitely settle the matter, the legislature decided in February 1877 to hold a popular referendum on the capital question in August, between Charleston, Clarkesburg, and Martinsburg (both located about midway between Wheeling and Charleston). Representatives of each city were sent to all parts of the state to draw votes. At first, Charleston's representatives were unsuccessful, but they found an original way to address crowds. With the help of John Lowlow, a famous clown from Utica, NY, the representatives traveled with the John Robinson's Circus for a weak and were allotted five minutes to speak for their cause in the middle of each performance, which attracted as many as five thousand people (WWP West Virginia 1941, 184).[20] At the referendum, Charleston received 41,243 votes, Clarksburg 29,442, and Martinsburg 8,045. The transfer took place in 1885. But the major reason was certainly that, although Wheeling had been the largest city before 1861, it was not central enough to remain as the capital and had therefore withdrawn from the referendum.[21]

The people were also called upon when the process was deemed too undemocratic. In Texas, the selection process was altered because bribes had been too blatant. The choice of the permanent and definitive capital in 1845 was not left to the legislature, because the selection of Houston in 1837 had "opened the door to more fraud and corruption than anything else which has ever come before the people," in the words of the convention president, Thomas Rusk. The people chose the capital by popular referendum.

North Dakota is another example. On Sunday, December 28, 1930, the state capitol in Bismarck, erected in territorial days, burned to ground. The legislative decision to hold a referendum on a constitutional amendment giving North Dakota's citizens the choice between Jamestown and Bismarck was apparently not the result of democracy but of "piracy."[22] The battle to win the votes of the state's citizens was so fierce that "a newspaper openly charged that the plot to move the capital was 'one of the most wickedly conceived acts of out-and-out piracy' in the entire history of North Dakota" (Potter 1932, 25). Bismarck was nevertheless able to crush the plot and won overwhelmingly on March 15, 1932.

Events in Iowa showed that the people could refuse the legislature's choice. The capital, a new town founded in 1839 under the name of Iowa City, was no longer near the population center of the state because settlement was moving rapidly toward its western boundary. The legislature thus voted in 1847 to move the capital to Monroe City. Iowa City protested loudly, in spite of the state university it was granted in consolation. It was fortunately not alone in protesting; public opinion in general refused to accept Monroe City as the capital. The capital therefore remained at Iowa City (2,262 inhabitants in 1850) until Des Moines, only 27 miles west of Monroe, was selected ten years later (WWP Iowa 1959, 267). The reason for the public outcry is not recorded but was perhaps due to the settlement pattern, which was not dense enough to allow a western capital to be accepted, as it would be just ten years later, when Des Moines, which had been designated as the capital by the Constitution of 1857, was approved by a popular vote.

The Model of Complexity: Presentation and Case Studies

Chapter 4 set out the main causes behind the choice of state capitals, and this chapter has taken a closer look at the political components of the process. We may now propose a new model based upon those findings. The first step is to review how the current capitals have been chosen, using tables 4.1–4.4 of chapter 4. For the fifty capitals, eighty-nine causes have been found. The pattern is broadly the same, with one exception, the role of centrality, which has been far more important for current capitals than for former ones (31 compared to 16; see table 5.3). But centrality has several

Table 5.3 The pattern of selection of current capitals compared to all successive ones

	Defense	Entry point	Religion	Economy	Politics	Centrality/ accessibility	Anti–large city
Current capitals	4	2	2	21	27	31	2
Current capitals (%)	4.5	2.25	2.25	23.6	30.3	34.8	2.25
Former capitals (%)	9.0	11.0	4.5	24.5	37.4	10.3	3.2
All capitals (%)	7.4	7.8	3.7	24.2	34.8	19.2	2.9

Note: "Entry point" refers to the establishment of a first presence in a new world, whether by Puritans from England, commercial companies, or later pioneers beyond the inland frontier.

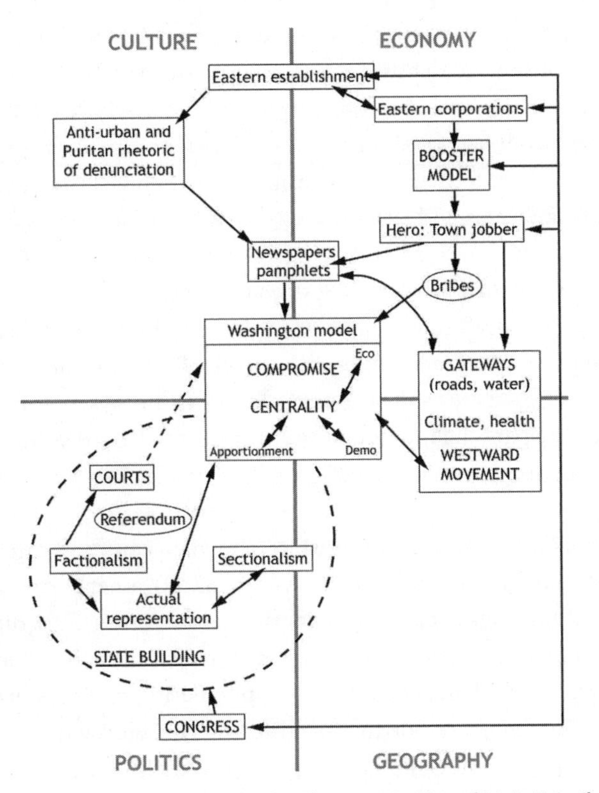

FIGURE 5.1 The model of complexity. Credit: Christian Montès and M. L. Trémélo.

meanings and either belongs mainly to the political sphere (half of the time, politics were joined with centrality in the causes stated) or is linked to the economic one (twelve times).

Figure 5.1 proposes a graphic image of the explanatory pattern of capital selection. The model emphasizes the links between the various factors that were influential during the process. The model is framed according to four broad themes: culture, economy, politics and geography. Browning's causes (centrality, distrust of cities, compromise) are present but at quite different levels. The two themes that the model highlights—at least graphically—are politics and the economy. However, the arrows that indicate relationships tend to show that cultural and economic factors influenced the process without leading it. At the heart of the process lies centrality, intertwined with compromise. The model shows that compromise between two large cities or two parts of the state was less important than Browning contended; it was more strongly linked to the apportionment system or to the power exerted by large corporations.

This model is not intended to force a particular interpretation upon readers, because it evidently simplifies processes that were not instantaneous and that also varied through time and space. However, its heuristic value is important because it provides readers with a basis for reflection on the explanatory pattern of the selection process. To test the model in relation to specific times and spaces, we analyze five case studies below: New Hampshire (1776–1814), Michigan (1805–47), Oregon (1843–64), Arizona (1863–89), and South Dakota (1889–1904). Although they do not exhaust the variety of processes at work, they stress the delicate causal balance needed to reach an agreement on the choice of a new capital, most of the time after fierce debates and false starts (in the form of one or more temporary capitals). Spanning more than a century, they also stress the continuity of the explanatory pattern.

New Hampshire. In New Hampshire, the main causes of capital change seem to be geographic, linked to the westward movement. The first capital was located on the seaside: Portsmouth was the major port directed toward England. In 1776, Puritan Exeter, nine miles inland, became the capital.[23] People from Massachusetts began populating new western regions, but the settlements were badly linked to the coastal towns, prompting a period of capital wandering. The capital was finally settled in 1814 in an inland town, Concord (Turner 1983; Heffernan and Stecker 1996, 73–78). Behind the obvious geographic causes, there were also political ones. Exeter became the first "American" capital because anti-English meetings had been held there. The second move owed much to the opposition between Federalists (Exeter) and Republicans (inland farmers, such as those living around Concord). Another way to read the opposition is to say that a new political class and newly settled parts of the state were challenging the "old order" of what was called the Old Colony, comprising all early coastal or subcoastal settlements, Exeter being one of them. In 1784, a new bridge—the Piscataqua Bridge—symbolized the rise of inland areas and the decline of coastal influence. Citizens of Exeter, known as the "Band of New Hampshire," nevertheless plotted to retain their power, and Republicans won in the 1804 and 1805 elections. At the same time, talks about a permanent capital began to yield some results. Concord became the temporary seat of government, and the eventual selection of Concord as the capital in 1814 is an example of Stephanie Moussalli's superlegislation. Officially in order to avoid long debates, the 1814 New Hampshire legislature devised an elegant trick to choose between three competing cities—Hopkinton, Salisbury, and Concord—that were all situated in the same inland area. Legislators decided that the capitol was to be a granite building and furnished free of charge by the chosen town. The ad hoc committee, with one member from each town, found that it had to give the capital to Concord, since it was the only town that had granite, on Rattlesnake Hill in West Concord (Anderson 1981, 327). This trick was testimony to the preeminence of the Concord group in the legislature. Hopkinton and Salisbury apparently received nothing in exchange, and they remain small towns.

Michigan. William W. Upton, a member of Michigan's House of Representatives during the period of choice (1847), wrote an account of the large number of arguments, involving very diverse logic, that towns put forward to attract the capital.

> It is almost incredible but there was no exception, the people of each embryo city could see abundant reasons why that city ought to be, and for hoping that it would be, selected as the capital of the state; some because they were central (and it was wonderful how many were in the center of population, or of the territory or of business, either present or prospective); others because of water power or steam boats or prospective railroad conveniences, or because they were lovely rural situations far from disturbing and corrupting influences of commerce and trade; others for sanitary reasons, and still others because they were beautifully situated on one of the three Great Lakes. (Upton 1990, 395)

Our analysis has to go beyond the obvious (centrality, for instance) to discover the other forces at work, be they political, economic, or something else. Today's Michigan was first created as the County of Wayne in 1800 in the Northwest Territory, the county seat of which was Detroit (Dunbar and May, 1995; Kestenbaum 1981 and 1986). The city was founded as a fort in 1701, by the French (Antoine de la Mothe Cadillac) and had become British from 1760 to 1796, when its population reached five hundred. Since new territories were soon carved out of the gigantic Northwest Territory, Wayne County was ruled by three successive capitals: Cincinnati, Chillicothe, and Vincennes. The latter was considered too far away by Detroit, which asked for a separate territory. The Territory of Ohio was created on June 30, 1805, and Detroit became its capital, as well as the arrival point of most pioneers who traveled by the Great Lakes from Buffalo. Ohio's population grew quickly from 8,765 in 1820 to 212,267 in 1840. Although Detroit's relative importance fell, it had 9,000 inhabitants in 1837 (there were 23,400 people in Wayne County). In 1823, Michigan became a second-stage territory. As its population grew, its center of gravity moved, and Detroit seemed less and less central. Moreover, the capital was deemed too insecure: the argument with Canada over the Oregon question was critical in 1847, the year set by the constitution of 1835 for the selection of a permanent capital site. That delay had been obtained by Detroit, the real "metropolis" of the state. But in 1847, three-fifths of the state's population lived in the southern two tiers of counties outside Wayne County, and Detroit's legislative influence had been accordingly reduced. Detroit nevertheless engaged in a fight with the rest of the state. There were thirteen proposed sites for the capitol building, including Jackson, Ann Arbor, Marshall (which still has a "Capitol Hill"), Battle Creek, Albion, and even the villages of Dexter, Byron, and Lyons.

As usual, all contenders sent people to the capital, Detroit, to lobby for their cause. William W. Upton wrote that "its adult population was fairly doubled in numbers by the lobbyists who assembled to aid in locating the capital in the respective cities intent on securing its location for themselves" (Upton 1990, 392). The House voted on all sites, but they were all defeated—Detroit by a large margin (18–43), and Marshall by a narrow one (29–32). The session then became very confused, with a majority of votes going first to the village of Lyons, then to Jackson, and then to Marshall. All the votes were later reversed.[24] Eventually, a place in the wilderness won, proposed by a representative of Ingham County. A big landowner in that county, James Seymour, had offered twenty acres and new buildings as commodious as those used in Detroit. He also offered to pay for the move and produced bonds to guarantee his good faith. Since the township offered only a mill, four small log houses, and a heavily timbered neighborhood, the proposal was first received "amid merriment," but it then met the approval of the legislators. The new capital was first to be called Aloda; it was then christened Michigan, and has been called Lansing since 1848.

The causes of Lansing's success are complex: Detroit was too close to foreign territory, Ann Arbor already had the university, Jackson had the penitentiary, and all, including Marshall, were considered too far south. But, according to Kestenbaum (1986), the causes were as much political as geographic. Faced with rival communities that deadlocked the debates (if they could not win, they acted to prevent any rival from winning), legislators could only agree upon a new town. The heated arguments obscured the real consensus that existed about removing the capital from Detroit and about locating it near the center of the state. In agreeing on these points, the legislature had a genuine prospective vision, hoping the new capital would enhance the settlement of the interior of the state as Columbus had done in Ohio. Lansing is indeed centrally located.[25] Centrality has here all the meanings stressed in the previous chapter: political, geographic, and economic.

Oregon. In Oregon, the process of capital choice was mainly political. Indeed, all contenders belonged to the same geographic unit, the Willamette Valley, the main area of settlement from the beginning. It is the largest valley in a mountainous state and is still Oregon's spinal column. In spite of such unity, the selection of Oregon's capital involved almost all the political processes. The location of the capital was settled only after numerous battles over legislation, legal issues, and referenda. The first settlers of Willamette Settlement, who appointed a civic committee on May 2, 1843, at Champoeg, set up the first territorial organization. The following year the first legislature of the provisional government met at Oregon City, then called Willamette Falls. Established in 1829 as a lumber mill near Willamette Falls, at the end of the Oregon Trail, it was the first city incorporated west of the Mississippi. In December 1845 the provisional government declared Oregon City the official seat of government "until otherwise directed." The next step was the establishment by Congress, on August 14, 1848, of the Territory of Oregon. Section 15 of that act of Congress provided that "the legislative assembly of the Territory of Oregon should hold its first session at such time and place in said Territory as the Governor thereof shall appoint and direct; and at said first session, or as soon thereafter as they shall deem expedient, the legislative assembly shall proceed to locate and establish the seat of government for said Territory at such place as they may deem eligible, which place, however, shall thereafter be subject to be changed by said legislative assembly" (Winslow 1908, 173–74).

The new governor, Lane, chose Oregon City as the first meeting place of the territorial legislature on July 16, 1849. During the second session, the legislature located the capital at Salem, the penitentiary at Portland, and the university at Corvallis. The governor, who resented not having been consulted on these important matters, declared the act unconstitutional, because it dealt with more than one specific object, a procedure forbidden by the Enabling Act. The Territorial Supreme Court sustained the governor's contention on the same grounds. The matter remained unsettled, for one member of the council and four members of the house of representatives kept meeting in Oregon City. They addressed a joint memorial to Congress that validated the removal act in 1852 (Beardsley 1941, 242).

Only three years later, the legislature relocated the capital in Corvallis, although the statehouse at Salem—chosen in 1851—was almost completed. The official reason was that Corvallis was at the head of navigation on the Willamette River and was nearest to the demographic center of the territory.[26] Opposition came this time from the United States Treasury Department, which refused to pay either the cost of a new building in Corvallis or the wages and mileage allowance of the legislators, officers, and clerks, because money had been appropriated by Congress for Salem and no other place. The legislature complied on December 12, 1855, passing an act relocating the capital in Salem. But the fight went on, and citizens were called in this time, for the legislature decided to hold a referendum on the question at the next regular election. The winner would need to receive a majority of all the votes cast. If this was not the case, the two cities having the highest number of votes should be voted upon at a special election. The referendum was held in June 1856, but there was no winner. Salem received 2,049 votes; Portland, 1,154; Corvallis, 1,998; and Eugene, 2,316. At the second election, the people refused to vote on the question, being tired of the matter, according to the *Salem Statesman*. Rather than exhaustion, the reason could be true democratic sense, since the citizens perhaps felt they were being used by the legislators for unsavory reasons. But Oregon still lacked a permanent capital. The legislators also seemed to be tired of the matter, because the bills that were introduced on the question during the eighth and tenth sessions were lost.

When Oregon became a state, the 1857 constitution required the question to be submitted to the people, in its Article 14, titled "Seat of Government":

> Section 1. The Legislative Assembly shall not have power to establish a permanent seat of government for this State. But at the first regular session after the adoption of this Constitution, the Legislative Assembly shall provide by law for the submission to the electors of this State at the next general election thereafter, the matter of the selection of a place for a permanent seat of government; and no place shall ever be the seat of government under such law, which shall not receive a majority of all the votes cast on the matter of such election.

Sec. 2. No tax shall be levied, or money of the State expended, or debt contracted for the erection of a State House prior to the year eight hundred and sixty-five.
Sec. 3. The seat of government, when established, as provided in section one, shall not be removed for the term of twenty years from the time of such establishment; nor in any other manner than as provided in the first section of this article; Provided, that all the public institutions of the State hereafter provided for by the Legislative Assembly, shall be located at the seat of government. (*Journal of the Constitutional Convention* 1882, 122)

The first referendum was held in 1862, but, as in 1856, no candidate received a majority, because there were so many. At last, at the 1864 election, Salem received 6,108 votes; Portland, 3,864; Eugene, 1,588; and all other places, 577. With a tiny majority of 57 votes, Salem was declared "the permanent seat of government" (Winslow 1908, 175). Countless sessions and four referenda had therefore been necessary to reach an agreement. Salem, which had been the capital for a decade, was already sufficiently ingrained in the hearts of the people of Oregon to appear as the "natural" choice, despite being a village of fewer than 1,000 inhabitants, while Portland already had 2,874 inhabitants in 1860 (8,293 in 1870), and Eugene had far fewer (861 in 1870).

Arizona. Our fourth example is Arizona (see Ehrlich 1981; Luckingham 1989; Mawn 1977). Arizona's history goes back to Spanish times,[27] but the capital story began in the middle of the nineteenth century, when Arizona was still part of New Mexico. After the discovery of silver and copper, Arizona wanted to secede, and a convention was held in Tucson, its oldest city, the Presidio San Agustín del Tucson having been settled in 1776. In 1863, Arizona became a territory, and the tortuous story of its capitals began. The first capital was a new town built in 1864, located in north central Arizona, near the Walker Mining District, where influential men had interests. Protected by Fort Whipple, it was christened Prescott. If the first choice was due to military and economic causes,[28] the second one stemmed from geopolitical ones. In 1867, the new governor, McCormick, wanted the state to have land on the Gulf of California (i.e., Mexico). He therefore moved the capital from Prescott to historic Tucson in the southern part of the territory. When New Mexico proved unable to seize land from Mexico, however, the capital came back to Prescott in 1877, for political reasons, Tucson having been deemed too close to Mexican interests by the new legislative majority. However, almost as soon as the transfer had been accomplished, the legislature began to quarrel over the permanent location of the capital. Besides the two usual contenders, Prescott and Tucson, a newcomer arose, Phoenix. All had assets and some hindrances. Prescott was a booming mining and military center, but it was also considered inconvenient because of its harsh winter, remoteness, and lack of railroad service. Tucson benefited from a better climate; it was the center of Pima County, which had about half of the territory's wealth and population (7,007 in 1880 for Tucson); and thanks to the 1880 reapportionment, it had more seats in the legislature than any other county. The Southern Pacific Railroad had arrived in the city the same year, owing to the discovery of Tombstone's silver mines. Tucson could thus boast easy access to most parts of the territory. The third contender, Phoenix, based its claims on its central demographic location in a rich agricultural valley. With the creation of Maricopa County in February 1871, Phoenix had become the county seat, despite resistance from Prescott, which already feared the rise of a potential rival. Phoenix was also the "service station" of central Arizona, surrounded by newly founded cities: Tempe, Mesa, Alhambra, Glendale, and Peoria. In 1887, the Maricopa and Phoenix Railroad linked the city to a transcontinental railroad. Phoenix also had a large new city hall able to temporarily host the territorial institutions, as well as better hotels and restaurants than its rivals.

The 1879–89 period was therefore a time of intense bargains. To retain the capital, Prescott, which directly controlled only a minority of seats, had to build political alliances to defeat the removal bills. It did it with promises. It promised Tombstone the seat of the newly created Cochise County in 1881. In 1885, during the "Thieving Thirteenth" legislature, it was the turn of the new territorial asylum (given to Phoenix), the prison (Yuma retained it), and the normal school (to Tempe). A first contender disappeared when Tucson seemed to content itself with the territorial university. But despite all these presents and the opening of a rail link in 1887, Prescott was unable to retain the capital. In 1889, the southern counties joined the anti-Prescott forces, realizing "that for too long the northern part of the territory had 'shrewdly played one of the Southern counties against another'" (Ehrlich 1981, 239). The fight was soon over: on January 24, 1889, both legislative bodies voted for the removal.

Why did Phoenix win? Prescott's answer was quite blunt. It accused Phoenix—rightly it seems—of having offered southern delegates about $10,000. Phoenix had also promised spoils to several counties and $50,000 for Tucson university.[29] Many of these promises were later broken. The contemporary press concurred with Prescott, considering the capital question a "corrupting element in territorial politics" and adding that the "sooner it is eliminated from Arizona politics the better it will be for the general interest of the territory."[30] Another reason could be that, unlike the other big southwestern cities, Phoenix had always been run by the social group that had the power, the Anglo-Saxons. Although bribes and some racism may have played a part in the choice, they certainly only hastened an inevitable process. Compromise in the case of Phoenix was multifold—geographic compromise first, between Prescott in the north and Tucson in the south. Tucson's weather was milder in the winter than Prescott's, and Prescott's summer was cooler than Tucson's. Political compromise also figured: the Prescott and Tucson factions having annihilated themselves, a third party settled the matter. Finally, there was economic compromise: Phoenix was situated in the most rapidly growing part of Arizona.

South Dakota. Our final example is South Dakota, where centrality was intertwined with the railroads' ambitions to explain the confirmation of Pierre as the permanent capital. Legislators had provided for the choice by the people of a temporary capital in 1889 and then of a permanent one, one year later. Pierre was selected the first time—the railroad having donated the capitol grounds—and wanted to retain it the next year. For that purpose, it resorted to every means available, even unusual ones. The main contenders were Pierre and Huron; the first boasted future centrality—with the opening of the Great Sioux Reservation—and railroads; the other, actual centrality. To boost their candidacy, their populations gave a helping hand. Women sewed ribbon badges boasting the merits of their town. Some of the badges became souvenir pillows: "Watertown for Capital. Presented by Ladies' Capital Club," or "Pierre for the Capital. O! My, we are in the center," or the alliterative "Peerless Pierre." Meanwhile, the leading men bonded the county, the city, and the school district to finance the campaign. Huron obtained in this way $237,000, and Pierre $462,000.[31] To gain more votes, Pierre sent forty volunteers to all of the state's counties (Schuler 1989, 35–36). Pierre beat Huron by more than seven thousand votes. But in 1904, Mitchell succeeded in calling for a new election for the capital. This time, it once more became a contest between railroads (or between Corn Palace and opera house?).

Then, beginning in June 1904, the Chicago & Northwestern Railway, whose line ended at Pierre, began giving free train passes to anyone who wished to visit that city, and the Chicago, Milwaukee & St. Paul Railway did the same for Mitchell, which was situated on its line. The people's response was immense. According to historian Doane Robinson, "South Dakota simply suspended business and went out for a grand sixty day's picnic." That fall, crowds flocked to the Mitchell Corn Palace to hear John Philip Sousa's band perform, and more than one thousand people a day visited Pierre, a city of about three thousand. Pierre entertained its visitors with moonlight ferryboat excursions on the Missouri River, picnics in the park, band concerts, nightly programs at the opera house, visits to the capitol, and daytime boat trips downstream to Farm Island or upstream to Scotty Philip's buffalo pasture. When the votes were counted, Pierre won with 58,617, compared to Mitchell's 41,155 (Schuler 1989, 39–40). A grand, million-dollar capitol was then decided upon to replace the old wooden one in Pierre.

The five case studies show that the model proposes a useful framework for understanding the process of selecting state capitals. They also show that to understand fully the complexity of the selection process, close analysis of each case is essential, in order to weigh each cause and scrutinize the relationships among them. The concept of centrality is indeed at the heart of the process, with its political, geographic, and economic meanings. For a full understanding of the process, however, one must return to the findings of chapter 2 and consider state capitals as places of memory. Choosing a capital is not only linked to the evident process of creating a state, but also to the process of building a state consciousness. The selection of a capital brings to light all the divisive issues in a state, but at the same time, it brings everyone to think about unity: one single capital in a space—the state—that is viewed as a valid entity. It would be too simplistic to say that choosing a state capital was the factor that consolidated state consciousness, but it helped. Not necessarily at once, but in the long term, all citizens were to accept the fact that the capital, however difficult the selection process had been, was to endure. This is clearly what happened in Idaho. Here are the concluding words of a 1938 article on the state capital fight: "The only long remaining results of the capital dispute were the repeated efforts of north Idaho to attain separate statehood. During the past two decades even this form of antagonism has passed away" (Chaffee 1938, 267). Other factors consolidated the movement, such as the Civil War, which divided the United States but also helped to unite divided states around a common purpose (with the exception of Virginia, which was divided by the creation of West Virginia in 1863).

The search for the most suitable location for a capital thus brought together various forces in early America. It revealed the way citizens defined democracy, territory, and their relationships with the urban world that was beginning to arise and dominate America. But locating the capital was not a way to translate the "democratic wish" of the citizens, because "the people" seldom chose directly. Most of the time, Congress, governors, or the legislature made the choice alone (or with the helping hand of lobbyists and town boosters), sometimes despite legislative requirements that called for popular referenda on the matter.

The last three chapters have tried to explain the process of capital selection as a revealing moment of crisis. But securing the capital did not necessarily mean becoming a large, prosperous, and revered city, as was initially planned. We must now investigate the position of capitals in the American urban system and their part in territorial structuring, through the study of their evolution.

6

Evolution of State Capitals to the 1950s

The "Purgatory Years"

As the following quotation clearly states, any measure of the "success" of state capitals must be carefully taken.

> What are the implications of emphasizing growth—in size, population, and economic development—as the measure of a city's success? As almost all urban biographies attest, growth has been the chief objective of urban leaders throughout the American past, and that objective has fueled intercity rivalries that in turn have accelerated the pace of urban development. When the growth imperative was hitched to industrialization after 1850 (or 1870 or 1880, depending on the city and region under examination), the result was cities humming with energy and excitement but also burdened with overcrowding, pollution, disease, poverty, crime, slums, corruption, violence, fiscal crises, and other afflictions that provoked a wide range of reform movements. (Chudacoff 1981, 120)

Success can be perceived through political and social indicators, as well as through demographic and economic ones. Recall that the American Constitution lists among the aims of the new country the "pursuit of happiness," and not of wealth.[1] Economic development must nevertheless be studied, for it was "the chief objective of urban leaders," to use Chudacoff's words,[2] although it is difficult to determine precisely the share of the economy of these towns that is due to their capital status.

In state and urban history studies, state capitals do not emerge as vibrant and well-considered cities during the nineteenth century and the first half of the twentieth. Quite the contrary, in fact, for state capitals are generally a travesty of the Jeffersonian vision: they remained mainly small, allegedly far from big towns' vices, but they harbored moral rottenness, factionalism, and economic apathy. The situation was actually more complex, and political life in state capitals underwent a cyclical evolution more than a march toward mediocrity. As for their social fabric, capitals evolved somehow differently from other cities. New capitals were often new towns or small villages, and they experienced almost as much difficulty as other pioneer settlements. As seats of government, however, they soon attracted a more sophisticated (or less rustic) society that differentiated them from their neighbors or competitors. This sometimes earned capitals the reputation of putting on airs. Their demographic and economic fate has been more straightforward and quite modest. That explains the title of this chapter: the usual denigration of capitals has to be tempered. After looking at their difficult debuts and formative years, we examine their evolution during the period of American industrialization and urbanization. In 1889, manufacture replaced agriculture as the dominant industry in the United States, and rural population declined from 72% in 1880 to 49% in 1920.[3]

Becoming the Capital: A Difficult Start

It was frequently observed that townspeople . . . appeared preoccupied with the future to the extent that they did not notice the present or at least were remarkably tolerant of its uncomfortable and unattractive aspects. Faith in the certain development of a great urban future was something which boosters worked very hard to foster.

(HAMER 1990, 16)

However well-platted a capital had been, and however beautiful the capitol was, the beginnings of capital cities usually proved as hard as for the rest of nascent urban America (Bridenbaugh 1966). First, news traveled slowly. After the designation of their village as the capital of the new territory of Minnesota on March 3, 1849, the citizens of St. Paul had to wait for four weeks until the news arrived via steamboat during a thunderstorm. Back then, the journey from an existing city to a new one still in its infancy was not without fears and difficulties. Indiana is a typical case. In November 1824, state records and property were moved by wagon from the former capital, Corydon, to Indianapolis, the new one. Ten days were necessary to make the 125-mile journey (Carmony 1998, 113). Among the other difficulties encountered was finding the town. When Texans chose a capital after their own War of Independence, they decided upon a brand new town,

Houston. Legislators made the journey in April 1837, but the settlement was so small that "the boat carrying the first officials missed it altogether and had to go back and search the shore more carefully for signs of habitation" (Moussalli 1997, 68). Most of the time a few wagons or a single boat were enough to carry archives and legislators. There were indeed few archives to transport, fires having often diminished their quantity.

Once a capital was chosen, economic growth was the main concern. Cities needed time to achieve stability in their status as capitals as well as their status as centers of economic activity. The process of choice, whether conducted for commercial reasons or for the common good, was hard and not always successful. How could a legislator, a merchant, or a speculator be sure that the site selected would develop into the metropolis he dreamed of? Such difficulties were the most acute in new territories, where towns grew like mushrooms and dropped like flies.[4] The booster model created many new towns that remained paper dreams or languishing villages rather than thriving as metropolises. Even in the East, being selected as capital did not necessarily bring a town the expected boom. Sixteen years after Columbia had been selected as South Carolina's new capital (1786), a visitor noted that, except during meetings of the assembly, "the town derives no particular advantage from being the seat of government."[5] After twenty years, in a township of two square miles, only one-third of the streets had been opened, and only one hundred houses had been erected. Since all town lots had not been sold, the specification that houses should be built on the lots within three years from day of purchase was abandoned (Green 1932, 149–61).

If capital status induced immediate growth, it was limited, in accordance with the modest economic pattern of the times. "A year ago, this was but forest. . . . Now there are more than a hundred houses, two hundred inhabitants. . . . Is this not magic?" marveled Prince Murat in 1825, one year after the creation of Tallahassee as Florida's capital (Ellis and Rogers 1986, 34).[6] A similar magic (on a smaller scale) worked for Minnesota's capital. Three weeks after the news of the selection had reached St. Paul, seventy buildings had been added to the seventy-two already standing, doubling the town's size. By 1854, St. Paul was incorporated as a city. But rapid growth did not always occur, for it was difficult to separate the wheat from the tares, and hasty speculation could often go awry. When Missouri became a state, its first general assembly met in St. Louis in September 1820 and appointed a five-man commission to select a site for the state capital (WWP Missouri 1959, 226–27). Article 11 of the constitution provided that the new capital be located "on the bank of the Missouri river, and within forty miles of the mouth of the river Osage." In May 1821, the commissioners first chose the only village they could find within the limits prescribed by the legislature: Côte Sans Dessein, on the northern bank of the Missouri River at the confluence of the Osage. But the legislature could not agree with greedy local speculators, and the commissioners were instructed to look further. They found a rudimentary settlement several miles upstream from Côte Sans Dessein, the City of Jefferson, and designated it as the capital on New Year's Eve. Five years later, the City of Jefferson became the permanent capital, but St. Charles served as the temporary seat of government during the time needed to transform the permanent capital into a real city. Even then, Jefferson City's survival was threatened by its location in a wilderness. The city was platted in 1822, but had only two resident families in 1823, along with a pub, a foundry, and a mission. When legislators arrived in 1826, thirty-one families lived in the capital, which had a general store, a gristmill, a distillery, tanneries, the Rising Sun Hotel, and a newspaper, the *Jeffersonian*, founded one year earlier. Since speculators from other towns were still asking for a different capital, Jefferson City had trouble growing, and lots sold slowly. The general assembly agreed in 1832 to build a state penitentiary to strengthen the town's position as the capital. The prison was completed in 1836. The burning of the capitol in 1837, instead of instigating calls for removal as usual, further strengthened the capital's position, for a new statehouse was soon built on the same site. The arrival of German immigrants induced further development, and Jefferson City was incorporated as a city in 1839, although pigs still wandered through its streets. In 1840, there were 1,174 inhabitants, including 262 slaves. The city was saved but was far from booming; by 1850 it had only 2,301 inhabitants, compared to St. Louis's 77,850.

Even when the speculators themselves had proposed the site and built the city, fortune was not always at the end of the road. In Columbus, Ohio, a private company proposed the following deal to the legislature in 1812: a site to build the new capital, ten acres (offered "in fee simple") for the public buildings, and another ten-acre lot for the penitentiary, because the speculators knew that penitentiaries were handsomely funded and would bring people to the town. The company would contribute up to $50,000 for the construction of the penitentiary and the statehouse (Studer 1873, 14–15). The four members of the company succeeded in fulfilling their contract, but two of them, Alexander McLaughlin and James Johnston, had failed in business in 1820. Since land sales did not thrive, they tried dropping the prices, which only worsened the situation by depressing land prices even more. A third partner, John Kerr, had his estate dissipated after his death in 1823. Fortunately for Columbus's prosperity, the fourth associate, Starling, lived a prosperous life, for himself as well as for the city (Condon 1977, 14).

The process of choosing Nebraska's permanent capital provides another example of economic miscalculation. The commissioners were influenced by "the great salt fantasy" to locate the seat of state government at Lancaster (Lincoln) in 1867. Lancaster was situated in a salt basin that was deemed very profitable by the first surveyor of the town site and numerous other "experts." Their reports described a basin of three hundred square miles able to support a thousand wells, each worth half a million dollars per year. John H. Ames published a pamphlet of thirty pages on the subject that concluded, "It is certain, then, that unless the old maxim, 'figures won't lie,' can be successfully controverted, that the people of Lincoln have a valuable interest in the salt basin, vested and indefeasible, except by some unusual providential dispensation."[7] The Nebraska Statehood Act therefore included a provision mandating that all the salt springs (twelve at most) would be granted to the state (McKee and Duerschner 1976, 89–90). But instead of the millions envisioned, the only royalties ever received amounted to $59.93, in 1871! The reason was not "providential dispensation" but the percentage of salt, which was far too small to be commercially profitable. The town had to find other ways to grow (the state university was founded in 1869). Fortunately for Lincoln, the initial sale of town lots produced $53,000, which exceeded the $50,000 the legislature had requested as a minimum to fund the building of the capitol and thus definitively settled the capital's location (Lincoln Chamber of Commerce 1923, 6–7).

Capitals nevertheless provided a major asset: an aura of permanence. Being a capital was not only a matter of pride, but also a matter of economic growth. Being only a temporary capital, however, often did not bring all the expected benefits, because the situation destroyed most of the "capital effect." Thus, temporary capitals sometimes had difficulties experiencing real growth, especially when they were built for that purpose, as was Vandalia, the capital of Illinois from 1819 to 1839 (Stroble 1992). In his autobiography, Frederick G. Hollman, a German immigrant, recalled the harsh early period of the town. Twenty persons died of illness in 1824–25, and he called the young German colony a "disastrous failure. . . . During the sessions of the legislature which occurred but once in two years business was pretty good, but all the other times Vandalia was a most dull and miserable village. Village property became almost worthless" (Burtschi 1954, 25). Although this view seems pessimistic, the population only grew from 700 in 1820 to about 800 in 1832, despite Vandalia's situation on the National Road (which had not yet reached St. Louis, seventy miles west) and on the Kaskaskia River, which was navigable for steamboats for six months of the year. When Vandalia lost the capital in 1837 (to take effect in 1839), its population stood at 850, compared to 600 for the former capital, Kaskaskia; 2,000 for Springfield, the new capital; and 4,000 for Chicago, the largest town in the state. It was difficult for Vandalia to compete against St. Louis, the door to the expanding Mississippi trade, which grew from 3,000 inhabitants in 1818 to 16,469 in 1840 (Burtschi 1963). But, beyond economics, capitals quickly became important social centers.

Capitals as Social Centers

THE SOT-WEED FACTOR
Up to Annapolis I went
A City Situate on a Plain
Where scarce a House will keep our Rain.
The Buildings framed with Cypress rare,
Resembles much our Southwark Fair:
But Stranger here will scarcely meet
With Market-place, Exchange or Street;
And if the Truth I may report,
'Tis not so large as Tottenham Court.

Ebenezer Cook, "The Sot-Weed Factor"[8]

As this satirical poem shows, a wilderness way of life was quite common for new or paper capitals. In South Carolina, low-country legislators were horrified when they arrived in the new capital, Columbia, in January 1790. Accustomed to the urban charms of Charleston, they found that Columbia "was little more than a clearing in a pine barren wilderness" (Moore 1993, 48). Besides taverns, there was only a jockey club, founded in 1788. The situation displeased the legislators, who had to sleep three in a bed, and the opposition between upcountry and lowcountry was revived. A committee was appointed, and a compromise embodied in the 1790 constitution. Columbia ("which shall remain the seat of government, until otherwise determined by the concurrence of two-thirds of both branches of the whole representation" according to the constitution), was to host all the meetings of the General Assembly. Charleston and Columbia were to host dual state offices.

After these initial inconveniences, the capital quickly gathered a great part of the state's elite, as had occurred since the colonial period. American colonial capitals tried to vie with their English counterparts and were sometimes successful. Williamsburg, incorporated in 1722, was the best example of the sophistication of social life during colonial times. During legislative sessions, the planter society converged in town "to occupy 'town houses,' comfortable rooms at inns or taverns, or to lodge with friends. Sycophants and adventurers swelled the throng. English visitors testified that balls, races, fairs, and other entertainment composed a 'season' not greatly inferior to London's in amusement and elegance" (WWP Virginia 1940, 315). This was partly the result of the tidewater territorial system, which was first based on tobacco. Small trade areas, with a radius of 12–14 miles, and relatively small tobacco production had fostered only small towns based on trade. The only exceptions were the capitals, Williamsburg and Annapolis, "which attracted luxury craftsmen who catered to the resident British officials and social elite that visited the capital during the political season" (Earle 1992, 104). Williamsburg, with only two hundred houses in 1759, hosted up to six thousand persons during court times (129).

Independence did not greatly alter the pattern (Wade 1959). In fact, until the 1850s, when transportation networks were only in their infancy and were often disrupted during winters, people found it difficult to stay anywhere but in the capital. They used taverns, boarding houses, and, when available, hotels. Some of these political hotels have become famous as "informal capitols": the Cornhusker in Lincoln, the Jay Hawke in Topeka (both bear their states' nicknames), and the Exchange in Montgomery (Goodsell 2001, 172).

When Springfield became capital of Illinois in 1839, it attracted many people and became the heart of the state because, even if state administration was not large, it was concentrated in a single town. "(In fact, the constant presence of members in Springfield was taken as such a matter of course that that body sat all day on both Saturdays and Mondays and contented itself with a single day's vacation at Christmas!). For a lawyer with cases in the higher courts not to remain in the capital for the entire term was hardly less feasible" (Angle 1950, 83–85). Wives and daughters also came, giving Springfield's social life an aura that no other city in the state could equal. But that golden period did not last long, since competitors arose for high offices: Chicago, which had thirty thousand inhabitants in 1850, compared to Springfield's forty-five hundred, also had United States courts, and after 1847 the Supreme Court also met at Mount Vernon and Ottawa. People came to the capital for shorter periods, and Springfield had to share social centrality in Illinois (Wallace 1983). The following excerpts present the evolution of the image of Springfield, which grew from an inconspicuous village to a real city in less than two decades and to a "fine town" ten years later. After only fifty years, Springfield was compared to an English shire-town and had become a respectable political center by 1950.

SPRINGFIELD, ILLINOIS, AS SEEN BY TRAVELERS

> [Patrick Shirreff (1833)]. Springfield is an irregular village of wooden houses, containing about 1,200 souls. It is three miles from Sangamon River, which is only navigable for small boats at the melting of the snow in spring. There are good stores of all description in the village.

> [Sir James Caird, agricultural expert (1858)]. This is a fine town, with good streets and shops, and the neighbourhood is diversified by timber. It is like all other places in this part of the country, surrounded by the wide prairie. The view from the top of the State house very much resembles that of the plain of Lombardy as seen from the Duomo of Milan, except that there is nowhere a boundary of mountains. [But he also noted "fever and ague."]

> [Sir John Leng, Scottish journalist (1876)]. Very much resembles an English county town like Doncaster, except that the bells of the tramway cars may generally be heard tingling in the streets. All the principal avenues are planted with trees that shade the sidewalks, which are planked, instead of paved or graveled. The best houses are in brick, but the majority are wooden-framed, neatly porticoed and balconied. . . . The State House—[is] the finest building of the kind in the United States.

> [A. J. Liebling, journalist at the *New Yorker* (1950)]. The city, which bears much the same relationship to Chicago that Albany does to New York, has a population of almost a hundred thousand, as against only nine thousand when Lincoln [who lived there from 1834 to 1860] left for Washington. It also has more politicians and lobbyists than it had in Lincoln's day, because

> the state has increased five times as much to lobby for. It has a fairly active industrial life, several good-looking department stores, and a generous quota of pubs. (Angle 1968, 132, 319, 403, 512)

The preceding quotations stress the rapidity with which new towns in the wilderness became "normal" towns. This was also true for the West. Contemporary historians have underlined that the "wild west" owes at least as much to literary and Hollywood mythologies as to reality. The beginnings of a new settlement could indeed be rather dangerous, before the vigilantes and then the true lawmakers came in, but that period only lasted one or two years. In a letter written in 1862, Mrs. Emily R. Meredith wrote about Bannack City, which was still in Idaho but was soon to be a leading city in Montana. "There are times when it is really unsafe to go through the main street on the other side of the creek, the bullets whizz around so, and no one thinks of punishing a man for shooting another" (Phillips 1937, 5). She was quite right; the years 1862–64 were a period of anarchy and bandits (Malone, Roeder, and Lang 1991). But the image of the wild west soon faded. After that short initial period, most of the contemporary descriptions of these boomtowns were full of comparisons with older Midwestern, seaboard, or even European towns (Doncaster in the case of Springfield). Houston will serve again as an example. Although legislators were living in tents and log cabins in April 1837, and records had to be stored in private houses, the population had already reached twelve hundred people in December of the same year (Moussalli 1997, 69). Denver, Colorado, experienced even quicker growth, along with its sister city, Auraria. They hosted most of the migrants of 1858 as well as the forty thousand gold prospectors of 1859, of whom "only" four or five thousand wintered. This growth stemmed from the excellent location of these cities near the junction of several important routes into the mining regions. At the end of 1860, a traveler wrote, "Everything here now, in the double capital, looks about as it does in any established city of the Western States" (Abbott, Leonard, and McComb 1994, 68). Prefabricated buildings helped to give these new towns an Eastern air. When a French mining engineer visited Cheyenne a year after its founding, he marveled at the sight of hundreds of prefabricated houses arriving from Chicago: "Do you want a palace, a cottage, a city or county home; do you want it in Doric, Tuscan, or Corinthian; of one or two stories, and attic, Mansard gables? Here you are! At our service!" (quoted in Reps 1981, 89).[9] At the end of the nineteenth century, electric lights came to new cities like Guthrie, Oklahoma, where the traveler and sociologist

Paul de Rousiers noticed that "there is no need for them whatever, but the settler, far away in his homestead, sees the brilliant point across the immensity of the prairie, and becomes confident of the future of Oklahoma" (Hamer 1990, 188).

Besides the growth in population, the opening of stores, and the arrival of professionals, culture also developed. At the beginning of the nineteenth century, only a few years after it was founded in 1796, Ohio's first capital, Chillicothe, displayed the rich social life of the aristocratic young Virginians who had settled there.[10] "Dances and dinner parties were frequent. The ladies and gentlemen who lived in the big stone houses were charming hosts and hostesses. Their freed negro servants knew the art of cooking and catering in the manner of the Virginian plantation. It was an exclusive society and in the opinion of the legislators from the north and the west, a snobbish society" (WWP Ohio 1938, 25).

Chillicothe was at the time nicknamed "the Athens of the West." Culture also flourished in Ohio's new capital, Columbus. The Columbus Literary Society was founded in 1816, only four years after the platting of the town. The Masonic Lodge was established the next year. The Juvenile Debating Society and the Columbus Mechanical Society began to meet in 1818, and the Haendel Society appeared soon after. In 1835, out of 977 buildings, only 41 log cabins remained (Cole 2001, 23, 27). A German visitor during the 1840s, Maria Ecker Wolf, wrote, "There is much in Columbus that reminds me of home. The cottages built close together along narrow streets, surrounded by gardens and grape arbors, are not so different from those of our neighborhood in Mannheim. The shops are owned by German-speaking people, the newspaper is printed in German, and everywhere we go, we hear our language spoken. It is almost as if, after three months of traveling, we have come home again" (quoted in Cole 2001, 208).

Like scores of American cities with cultural leanings, Nashville was nicknamed "the Athens of the South."[11] New western towns followed a similar path. Since the colonial period, their importance had been disproportionate to their size, because they reigned over large territories without real competitors. "Culturally, they became the matrix for well-developed social institutions, such as newspapers, theaters, schools, and churches" (Stelter 1973, 189).

But becoming a town did not always mean growing into a city. When a capital was quite new, early life remained more bucolic than elated. The social life of these boom-capitals sometimes lagged behind the image they proposed. Almond Gunnison, who visited Helena, Montana, in the early

1880s, wrote, "While the business streets have showy solidity, one has but to go to the rear to see that behind these ambitious fronts there is a conglomeration of shacks and shanties, the architecture being 'Queen Anne in front and Crazy Jane behind'" (1990, 188).[12] Even with a well-developed social life, however, capitals often lagged behind other cities in demographic and economic terms.

Demography: Waning Towns

The nineteenth century was the period of real urban growth for the United States, more than for the rest of the Western World: 1800 saw the first American city attain fifty thousand inhabitants; the quarter-million mark was reached in 1840, and the million mark in 1880 (see table 6.1). Adna F. Weber found an apt and often quoted comparison when he wrote that the United States was the "land of mushroom cities" (1899, 20).

Following the general trend, capitals experienced an absolute rise in population during the nineteenth century and the first half of the twentieth century, but most of them lost ground compared to their rivals in the state. Only twenty-five capitals belonged at some point among the one hundred most populated American cities during the 1790–1950 period. During that period, nineteen of these twenty-five capitals faded, and only seven capi-

Table 6.1 Number of US cities in each of five categories according to size, 1790–1930

Year	Towns (2,500–10,000)	Small cities (10,000–50,000)	Medium-size cities (50,000–250,000)	Large cities (250,000–1 million)	Great cities (more than 1 million)
1790	19	5	—	—	—
1800	27	5	1	—	—
1810	35	9	2	—	—
1820	48	10	3	—	—
1830	67	19	4	—	—
1840	94	32	4	1	—
1850	174	52	9	1	—
1860	299	77	13	3	—
1870	495	143	18	7	—
1880	716	188	27	7	1
1890	994	296	47	8	3
1910	1,665	488	90	16	3
1920	1,970	608	119	22	3
1930	2,183	791	154	32	5

Source: Elazar 1987, 150.

Table 6.2 Evolution of the rank of state capitals among the hundred largest US cities, 1790–1950

Capital	Best rank and date	Evolution to 1950
Montgomery	75th in 1850	Disappeared afterwards
Little Rock	100th in 1900	Disappeared; 100th in 1940; disappeared again
Sacramento	60th in 1860	90th in 1880; reappeared only in 1930 (96th)
Hartford	23rd in 1790	54th in 1950
Springfield	84th in 1870	100th in 1880, then disappeared
Topeka	98th in 1890	Disappeared afterwards
Augusta (Maine)	77th in 1840	82nd in 1850; disappeared afterwards
Boston	3rd in 1790	10th in 1950
St. Paul	23rd in 1900	35th in 1950
Lincoln	52nd in 1900	91st in 1910; disappeared afterwards
Concord	77th in 1850	86th in 1860; disappeared afterwards
Trenton	42nd in 1810	50th in 1890; fell to 80th in 1950
Albany	9th (1830,1840)	69th in 1950
Raleigh	59th in 1820	Disappeared afterwards
Harrisburg	55th in 1820	Disappeared after 1930
Providence	9th in 1790	43rd in 1950
Columbia	78th in 1830	95th in 1840; disappeared afterwards
Nashville	38th in 1890	Was 50th in 1830; fell to 56th in 1950
Richmond	12th (1810,1820)	46th in 1960
Capital	**First time ranked**	**Rank in 1950**
Denver	50th in 1890	24th
Atlanta	99th in 1860	33rd
Indianapolis	87th in 1850	23rd
Des Moines	80th in 1880	53rd
Columbus	70th in 1840	28th
Oklahoma City	87th in 1910	45th
Salt Lake City	93rd in 1880	52nd

Source: US Bureau of the Census, 1998.

tals that already ranked among the top one hundred for population in the nineteenth century underwent any growth during the twentieth century (see table 6.2).

Among the other capitals, three appeared among the top one hundred in 1950, with between 107,000 and 132,000 inhabitants (Austin at 73rd rank, Baton Rouge at 82nd, and Phoenix at 99th), but twenty-one remained small in states where either urbanization was under way (Miami, Springfield, Portland, and Seattle) or where, even if there were no metropolises, the gap between the capital and the largest city had greatly widened, indicating their developmental delay (Juneau and Anchorage, Dover and Wilmington, Frankfort and Louisville, Jefferson City and St. Louis, Carson City and Reno, Santa Fe and Albuquerque, Pierre and Sioux Falls, Montpelier and Burlington).[13] For example, in 1850 the population of Dover was 2,000, and Wilmington's was 13,979; in 1950, the figures were, respectively, 6,223 and 110,356.

Capitals as Economic Laggards: A First Approach

Capitals experienced difficulties in broadening their economic bases. Government employees always made up a high percentage of their workforce. Concerning the links between capital status and economic development, the rare studies of capital cities generally take one of two positions. The first group aims to prove the position that the economic development of capitals owed little to the presence of the seat of government. "It has been sometimes said by carping and envious persons that there was no vitality in Columbus, save what it obtained through State patronage. But such assertions have been, within the last few years, proven to be not only false, but utterly groundless" (Studer 1873, 104). The author referred to the coming of railroads and the growth of manufacturing. When discussing early Indianapolis in 1910, Jacob Dunn compared the capital with a college town (Dunn 1910, 81). "Capitol campus" was also the expression used by the mayor of Olympia in 2003 (March 2, 2003, pers. comm.). However, these statements testify as much to the way local historians wanted to present their cities as to any in-depth analysis of the influence of capital status. State capitals still had to fend off attacks from competing cities; critics argued that capitals owed everything to their status and nothing to the enterprising spirit of their leaders. One contemporary assessment of capital status appears in a study of the capitals of the thirteen original states, which were chosen between 1776 and 1812 (Zagarri 1987). The author presents two arguments. First, the egalitarianism of American society is said to have inhibited the emergence of grand capitals.[14] Second, capitals were located "in places that had demonstrated no prior need for a large urban center. As modern central place theory shows, towns develop throughout the hinterland in accordance with the importance of the functions they provide to the surrounding areas" (32). Americans having overestimated the attraction of government, capitals did not thrive. Since they lacked the real hinterland they needed to arise, central places were unable to appear, at least before the railroads came (Conzen 1975).[15] But for early Cheyenne, the presence of the capitol and the railroad did not ensure growth for the struggling new capital; its economic successes were due to the enterprising leadership of local businessmen and not to government, as was the case in most American cities (Stelter 1973, 193).[16]

The second position acknowledges the importance of capital status. "In government the capital had its origin; government has made its fiery history, and government remains the force propelling the city forward in prosperity and eminence" (WWP South Carolina 1941, 219). Such were the closing words of the sketch the Works Progress Administration published on Columbia, South Carolina. This statement stemmed from the fact that Columbia was not only the capital but also a big federal military center and the regional headquarters of the Farm Credit Administration. A contemporary geographic study of Pennsylvania also stressed the economic influence of capital status. Harrisburg was portrayed as a multifunctional city: a capital; a railroad center with extensive classification yards; and an active trading center with some manufacturing. Its capital function was seen in the state buildings around the capitol, the large number of good hotels (for representatives, lobbyists, and conventions), and the number of retailers patronized by state officials and employees. "The net result of this is a city that is much larger and more active than it would have been had Harrisburg not been selected as the State Capital" (Murphy and Murphy 1937, 266). The reasons for such capital advantages were brought to light by Daniel J. Elazar, whose analysis of Springfield, Illinois, examined the complex interaction between a politically conscious city and its economic status (Elazar 1970, 85). Becoming capital in 1837 brought Springfield more sophistication and cosmopolitanism than the average city of its size. Although Chicago and St. Louis were too close to allow it to become an economic metropolis, Springfield experienced a population boom in the 1880–1930 period, based on agriculture, agricultural manufacture, and insurance.[17] During the twentieth century, expansion of government and the deep involvement of the local community in political matters on every level transformed Springfield into a busy center of decision. Being a social capital brought economic benefits because the political and social circles largely overlapped.

Both positions are in fact rather close, as neither pretends that the presence of state government was the only or even the main engine of growth.[18] Whatever was the exact influence of capital status, it is clear that capital status could only improve a situation that was already reasonably good, though it proved unable to create an important economic basis by itself. Such caution appears in the conclusion of a study of New England: "The importance of being a center of government cannot be completely separated from the advantages of early settlement and favorable geography" (Daniels 1986, 21). The number of declining or small capitals proves this, especially since their decline—in relative terms—occurred during a period of national economic boom. The real GNP grew on average by more than 4% a year during the 1865–1914 period (Nouailhat 1982). The only capitals that improved their rankings were the ones that industrialized, indicating that capitals did not depart from the classic urban economic patterns of the 1850–1950 "industrial period."

Beyond Dunn's impressionistic statements, Indianapolis serves as an example of the relationships between governmental and other forces in the growth of a state capital (Bodenhamer and Barrows 1994). The main reasons behind Indianapolis's steady growth during the 1850–1950 period were the importance of transport, a diversified economy, and a mixture of public and private contributions to development. Only the last factor is linked to the presence of state government, which was "a source of considerable economic benefit from 1825 to the present."[19] This stems from the strong political interests that Indianapolis's citizens have always manifested and to the "linkage of politics and promotional boosterism."[20] Examples of public-private partnership appeared as soon as in 1828, when leading citizens formed the Indianapolis Steam Mill Company and, thanks to their close ties to the state, obtained cheap land and the permission to cut timber on state lands. The company was too big in a market that was too small and isolated, and it had to close in 1835. The state supported transportation endeavors from the beginning, and the coming of the first railroad in 1847—and then of numerous others, which later earned the city the title of "the crossroads of America," ensured growth through an enlarged trade area and the creation of banks. In 1853, the Indianapolis Board of Trade was created, which soon attracted a rolling mill. The processing of agricultural products eventually became the leading industry. In 1880, the capital employed approximately ten thousand industrial workers, and by 1900 it was ranked twentieth among manufacturing centers in the nation, partly because of the state's natural gas fields. The links between legislature and boosters took another turn in the early 1900s, when an intense campaign by local businessmen, especially the Merchant's Association of Indianapolis, led the legislature to set up the Indiana State Board of Accountants. Their aim was to improve the accounting practices of local authorities after a series of scandals in Indianapolis in 1903 and in Marion County in 1908 (Lambert 1959). Meanwhile, prosperity continued: no single industry was dominant, and crises were better managed. Around 1900, several automobile companies were founded, inducing the building of the famous Indianapolis Motor Speedway. Pharmaceuticals became important with the rise of Eli Lilly and Company. This shows that external factors were as much at work as internal ones: local entrepreneurs were important, but Indiana belonged to the manufacturing belt and was located in a dense part of the nation at the crossroads of the direct routes between New York and St. Louis and between Chicago and Atlanta. Indianapolis therefore followed the common trend of Midwestern cities, whose success relied not only on the positive response of their citizens to economic changes (Abbott 1978), but often also on "forces beyond the control of individual towns. . . . Interacting with the system did not insure success, but to avoid cultivating such interactions would insure failure and stagnation" (Mahoney 1990, 274).

Government helped, sustained, and controlled growth, but it did not create it single-handedly. Further evidence of this appears in a study published in 1943 by a Chicago geographer, Chauncy D. Harris. His aim was to propose a classification of American cities, but he was unable to put state capitals in the classes he had defined. His "political capitals" class included only Washington, DC. He had to create a specific category for the (then forty-eight) state capitals apart from his general classifications, "because no satisfactory criterion has been found to measure the relative importance of the political function" (Harris 1943, 86–99). Three classes were identified in his category:

1. The political function is clearly dominant (sixteen capitals).
2. The political function is "probably dominant" (twenty capitals).
3. Political activities are overshadowed by trade or industry (twelve capitals).

The conclusion is clear: thirty-six capitals (three-quarters) proved unable to broaden their economic bases much beyond state government during the first century (more or less) of their existence as capitals (see table 6.3). The reasons for this were multiple and interconnected. The first was the deplorable image of capitals, which did not help to attract as many prospective investors as city fathers hoped (hence the creation of Indiana's Board of Accountants). A second cause of the capitals' slow economic growth was the dichotomy that often existed between the political boundaries of states and the boundaries that were being defined by the industrial economy. Capitals ruled over areas that had not often been drawn according to economic logic. A third cause—partly the result of the first two—was that railroads often by-passed capitals, therefore hindering the processes of industrialization, the main engines of urban growth (or vice versa).

Table 6.3 Economic situation of state capitals, 1850–1900 and 1900–1950

Capital	Economy, 1850–1900*	Economy, 1900–1950
Montgomery	Trade and government	Trade and government
Juneau	Mining (1880s)	Mines and government
Phoenix	Service station of central Arizona; railroad hub	Service station of central Arizona, railroad hub
Little Rock	Center of the state	Center of the state
Sacramento	Agriculture; railroads (Central Pacific); trade	Agriculture; trade
Denver	Metropolis of the state	Metropolis of the state and beyond; Trade, tourism, banks, manufacture
Hartford	Insurance and manufacture (Colt, machines, air conditioning); richest city in America	Insurance and manufacture
Dover	Small center	1899 incorporation act attracted company headquarters
Tallahassee	Center of Florida's life, but not for long	Comparative decline (rise of sea tourism in other cities; lack of war industry during WWII)
Atlanta	Distribution center for northern products; banking and insurance center of the state	Uncontested center of the state
Honolulu	(Royal capital)	Major center
Boise City	Trade center near mines	Center; WWII air base
Springfield	Mostly government	Mostly government
Indianapolis	Railroad hub; diversified manufacture	Railroad hub and diversified manufacture (cars until 1937)
Des Moines	Insurance; overall center of the state	Insurance; overall center of the state
Topeka	Boom and bust agricultural center; government	Manufacture (agribusiness—but 3rd rank); air base; government
Frankfort	Government; tobacco and bourbon	Government; tobacco and bourbon
Baton Rouge	Secondary town	Secondary town; oil
Augusta	Trade (roads; inland port)	Decline
Annapolis	Naval Academy; government	Naval Academy
Boston	Metropolis	Metropolis
Lansing	Secondary center	Automobiles (Oldsmobile, then G.M.); Michigan State University
St. Paul	Railroads; trade and banks; linked with government	Twin Cities; overall metropolis of north-central United States (transition from agriculturally based economy)
Jackson	Secondary center; government	Became center
Jefferson City	Government (printing and shoe industry)	Government
Helena	Government; redistribution center for mines; banks	Government; small trade center
Lincoln	Government; universities; agricultural trade center	Universities and government; agricultural trade center
Carson City	Trading center near mines; government	Government; economy weakens (Reno and Las Vegas grow)
Concord	Government; small rural center	Government; small center
Trenton	Trade; manufacture	Manufacture
Santa Fe	Government; trade (decline of Santa Fe Trail)	Decline (Albuquerque expands: university; manufacture; military)
Albany	Transportation hub; manufacture	Relative decline
Raleigh	Government	Government
Bismarck	Railroad town in agricultural state	Railroad town in agricultural state (crisis)

Oklahoma City	Services to agriculture (Sooners, 1889)	Oil; service center; air base; government
Columbus	Balanced economy	Balanced economy in an industrial state
Salem	Government; university (1842); food processing; lumber	Government; university; food processing; lumber
Harrisburg	Railroad; industry; trade	Railroad; industry (metal); trade; convention; government
Providence	Manufacture (large growth in textile and metal); finance; university	Port and manufacture (but crisis since 1930s); finance; university
Columbia	Government; inland trade center; manufacture (1880s)	Government; army; textile; university (still small)
Pierre	Government (railroad); small agricultural center (since 1880)	Government (railroad); small agricultural center
Nashville	Financial and political center of the state; trade (railroad)	Distribution point; finance; DuPont; university
Austin	Government and trade (modest); university (1881)	Government; modest trade; university; but pales compared to other cities
Salt Lake City	Main center of Utah and Mormons	Main center of a poor state
Montpelier	Government; small trade center	Government; small trade insurance and tourism center
Richmond	Trade; manufacture (tobacco); government	Manufacture (tobacco); banks; government
Olympia	Government; small local center; port	Government; small local center
Charleston	Government; small manufacturing center	Government; small manufacturing center
Madison	Government and college city	Government and university
Cheyenne	Railroad; small center of an agricultural and mining region; army; government	Small center of an agricultural and mining region; tourism; army; government

Note: Government is indicated only if it has major importance in the capital's economy.

* Some capitals were founded after 1850, but almost all before 1900. Juneau became capital in 1900, and Oklahoma City in 1910.

Neanderthal Politics

Show your State Legislatures; show your Rings
And challenge Europe to produce such things
As high officials sitting half in sight
To share the plunder and to fix things right.

James Russell Lowell, "Centennial Ode"[21]

Although the apportionment system favored the countryside, state capitals were far from the incarnation of *rus in urbe*. We must first distinguish between territories and states. Before their composition, formation, and workings, territorial governments were far more exposed than state governments to influence, corruption, and unsavory schemes. When historian E. Robinson explained the reasons why corporations and influential individuals opposed statehood in North Dakota during the 1880s, he described all the drawbacks inherent to territorial power. The Democratic majority in Congress opposed statehood, because Dakotans were Republicans. Eastern investors refused to see their debtors control the loans. The railroads wanted to keep the rather lax territorial railroad laws and their power over local politicians. Finally, Bismarck wanted to remain the capital of a large territory (Robinson 1966, 202).[22] New Mexico provides a similar example. Here, big business succeeded, through corruption and ballot-box stuffing, in perpetuating territorial status for sixty years, although between 1849 and 1910 fifty acts providing for statehood were introduced in Congress. The reasons were manifold: slavery, struggles over Reconstruction, fear of Eastern domination, religious bigotry, and racial discrimination. The most important one, however, was certainly that big business feared that statehood would bring more taxes (Beck 1962, 226–27).

The powers of states evolved cyclically during that period. The period up to 1870 was one of apprenticeship in new states and of great local power. From 1870 to 1932, especially at the beginning of the twentieth century, state policies were progressive, establishing workers' compensation, unemployment insurance, and public assistance, as well as corporate regulation and programs for economic development that became the models for New Deal programs. In 1931, US Supreme Court justice Louis Brandeis famously characterized states as laboratories that can "try novel social and economic experiments without risk to the rest of the country" (Van Horn 1989, 20). Only during the 1932–1962 period were local governments almost completely overshadowed by the increase in federal powers and action in the context of wars. However, the image of state capitals was more stable: throughout the 1850–1950 period they were viewed as seats of factionalism, and, as Abraham Lincoln quite rightly forecasted, distrust of government was almost general: "As a result of the war, corporations have been enthroned, and an era of corruption in high places will follow, and the money power of the country will endeavour to prolong its reign by working upon the prejudices of the people until all wealth is aggregated in a few hands and the republic is destroyed."[23] Fifty-eight years later, a review echoed this sharp analysis: "This country is less democratic now than it was a hundred and forty years ago. . . . We are faced today with complete lack of control over nominations of candidates, which are made by irresponsible bosses, and with despotic courts that annul the will of the majority almost as they choose and take all the responsibility from law-making bodies" (Logan 1925–26, 394–95).

Larry Sabato aptly characterized pre-1970s state governments: "Neanderthal constitutions, nightmarish organizations, rural-based malapportioned legislatures,[24] inadequate and undiversified revenue systems, and shackled governors who often exercised little leadership" (Sabato 1983, 176). State expenditures remained small, compared to federal and local ones. According to estimates, between 1820 and 1930, federal expenditures accounted for 33% of the total, local expenditures for 56%, while state expenditures amounted to 11%. In 1970 the federal share was 62.5%; local, 18.1%; and state, 19.4% (Gilfoyle 1999).[25] In 1902, local governments raised almost 95% of their revenues themselves; since then, the percentage has continuously dropped. Conflict reigned before 1930 between state and federal governments, creating a "layer-cake" form of federalism, where national, state, and local governments were exclusive or autonomous (Wright 1988, 66–70). Although conflict was replaced by a cooperative phase during the 1930s and 1940s, all these factors explain why state legislatures were not very sought after. In 1939, legislatures met only biennially in all states except five, of which New Jersey was one (Rosenthal 1989).[26]

Local politics were not very democratic. Railroad and mining companies exerted an exorbitant power in many states from the 1850s until the 1900s or 1920s. Corruption had been almost standard since the Grant administration in Washington. The people seemed unable to act against corruption after the 1840s "Dorr conspiracy." Thomas Dorr supported popular sovereignty but was convicted of treason in 1843 in Rhode Island. In 1848, the United States Supreme Court decision that it was a political and not judicial matter "ended the possibility that there was any peaceful way for the people to assert their sovereignty against a government legally elected, however corrupt it might be" (McLoughlin 1978, 136). Although the Australian (or secret) ballot was first introduced in the United States in 1888 (in Louisville, Kentucky), South Carolinians had to wait the 1950s to use it (Kemp 1999, 49). Historians are therefore not very kind when explaining the workings of state government. In Pennsylvania, "inexperience, lack of party discipline, corruption, log-rolling, and constituent disinterest all influenced the legislature, which was characterized by untidy procedures and importuned by conflicting interests" (Harrison 1979, 334–55). Long legislative sessions were not seen as a sign of good and thorough politics, but as a means for legislators to garner more per diems. Tennessee's constitution of 1870 (still in use today, with a few amendments) set a limit on the number of days the legislators could collect per diem pay in order to prevent lengthy and expensive legislative sessions, which were typical in the postwar radical years (Bergeron, Ash and Keith 1999, 179). The same was true for municipal governments: personal enrichment, nepotism, and "exploitation of municipal wealth" were the usual activities of most city council members (Kemp 1999, 55).[27] This created the well-known boss-and-machine system that eventually induced the reform movement of the early twentieth century. The power of party machines lasted at least until the 1930s, when depression and the New Deal altered the pattern.[28]

The amount of corruption that plagued most state capitals during the 1850–1950 period is well known. However, some nuances have to be added to that bleak picture. Big corporations provided many local jobs, and their power had some limitations. First, railroad companies often compromised on taxation issues (Spalding 1964). Second, some railroads were in the hands of "public-spirited, useful citizens," such as the Boston and Maine Railroad, which controlled all the tracks in New Hampshire as well as the state's politics at the end of the nineteenth century (Heffernan and Stecker 1996, 156). Third, some states have maintained good reputations. Wisconsin has offered, since the early twentieth century, a progressive, efficient, and innovative image sharply contrasting with the usual state legislatures. The personality of Robert M. LaFollette (1855–1925), who dominated the state's political life from the mid-nineties to 1925, is certainly a reason for this, but he was supported by citizens. Historians Robert Nesbit and William Thompson tried to explain that case (Nesbit and Thompson 1989, 551–52). The main factor was the balance that was always maintained between the state's seventy ethnic groups, none of which was powerful enough to rule the state's affairs. Other reasons listed are the tolerance of Catholic theology for human error, Norwegian and German readiness to accept a "social service state," links between agriculture and industry, the medium size of most cities, and the fact that Milwaukee was never as dominant as Chicago or Detroit and had to vie for power with smaller Madison, the state's political and intellectual capital. A capital "town," as envisioned by the Puritans, would therefore not be an asset, since the capital must be able to contend with the primary city to ensure a healthy democracy. Despite Wisconsin's case, the overall image of politics and, by implication, of capitals, was a negative one. This situation widened the criticism of state legislatures, despite the progressive movement of the 1900s–1910s.

A second argument can be put forward to explain the relative decline of most capitals: the declining power of municipalities, which was also a factor hindering capitals' development. Capitals had a small initial advantage over most other cities. Indeed, while many towns had to wait long before being incorporated, capitals were soon incorporated, with the notable exceptions of Boston, which was not incorporated until 1822, and Indianapolis, which was incorporated in 1874 (Elazar 1987, 150). Even Washington had to wait until 1871 to be incorporated.[29] Their early incorporation shows that capitals were considered more important than other cities in the state, as emphasized in chapter 2. But incorporation remained mostly symbolic, for the relationship between states and localities was characterized by "open hostility" until the 1970s. Localities had been given near autonomy after 1776, but their powers were gradually overtaken by the states, especially after Iowa's 1868 Supreme Court decision known as Dillon's Rule. Later upheld by the US Supreme Court and still in use today, it stated that "municipal corporations owe their origin to, and derive their powers and rights wholly from, the legislature. . . . As it creates, so may it destroy. If it may destroy, it may abridge and control. . . . They are . . . the mere tenants at will of the legislature" (Sabato 1983, 175–76). There are regional differences: cities traditionally had more power in New England; in the South and the trans-Mississippi West, counties and states, founded before municipalities in the West held the power (Hollingsworth and Hollingsworth 1979, 49–50). Municipal elites and city government therefore often had little power when faced with "intrusion" by the state. The workings of legislatures in capital cities may have been almost similar, but municipal behavior differed. Partial proof appears in a study of all 278 American cities with a population between ten thousand and twenty-five thousand (among which were twelve capitals) during the late nineteenth and early twentieth centuries (Hollingsworth and Hollingsworth 1979).[30] The main finding is that the capitals' political structures widely differed: most were oligarchic around 1900; others were polyarchic or autocratic. Only four capitals had a high socioeconomic status and a high government activity, while governmental activity was low for Augusta and Baton Rouge. The only conclusion that can therefore be drawn is that state capitals behaved as did all other American municipalities.

State Boundaries and Economic Influence: Two Different Logics

Politics and economy do not have the same geographical limits, and political centrality and economic centrality are not the same: "State boundaries have been drawn by a capricious history, and only occasionally (and then by accident) does a state constitute the most logical economic unit for either making policy or delivering services" (Donahue 1997, 162). As a result, even if a state capital is the largest city in its state, it may be overshadowed by a metropolis in a nearby state; Indianapolis is indeed the uncontested metropolis of Indiana, but it is under the aegis of Chicago. Such a situation stems from the drawing of state boundaries, which are based more on ideology than on economics. Donahue was mistaken in stating that caprice was the main force behind their fixation. In addition, even if economy was an issue in defining boundaries, it was, for most states, the pre-industrial economy. This largely explains why rivers, which used to be the main arteries of the nation, were often sought as boundaries.

A comprehensive synthesis remains to be written to explain the processes at work.[31] The thirteen colonies were limited by their perceived economic returns and by the power of the future owners when they were proprietary and later crown provinces. They were limited as well by internal divisions and the potential of pioneer groups when they were founded under the aegis of religious belief.[32] This explains why northeastern states are generally smaller than southeastern ones. For new states, the first concern was the future political stability and balance of the Union, more than their future economic balance. The belief—based on Greek examples and the writings of Montesquieu—that republics could survive only in relatively small countries, where they could resist centrifugal tendencies, had to cede to the desire of coalesced interests to control large tracts of land. The Northwest Ordinance of 1787 therefore only provided for three to five new states, thus leaving the majority to the thirteen original ones. The situation was soon altered, however, as new territories were added to the Union, first by the Louisiana Purchase in 1804, then by the acquisition of western lands, either bought or violently colonized. According to the law, Congress had two obligations regarding the size of new territories: it had first to be amenable to government, and then to respect "the geographical affinities and dependence of its parts" (Meinig 1993, 440).

Each one of the first eight new territories—extending to the Mississippi—had a size roughly equivalent to that of the larger original states (40,000–50,000 square miles, roughly three times the area of the Netherlands). But their shapes varied, according to the different logics of their creation. The first three states carved out of the Northwest Territory were rectangles bounded by rivers and lakes. The lack of economic forethought is evident in Indiana, which was granted some lake frontage with no effective harbor. This led Illinois's delegate to negotiate a forty-one-mile extension to the south that added to the state the port of Chicago, Galena's lead mines, a future canal—and great prosperity. It is therefore not surprising that Meinig takes Indiana as "the abstract model of a constituent American republic" (Meinig 1993, 445). A quasi-rectangle (the Ohio River serves as its southern boundary) made out of rectangles (the townships), it had no immediate physical meaning. However, Congress did not forget the "geographical affinities" and reserved 5% of the net proceeds from the sale of federal lands for public transportation. As a result, the National Road crossed Indianapolis along Washington Street, making the new capital an important stop on the nation's main inland road, the first step of Indianapolis's uninterrupted growth.

On the other hand, boundaries in Kentucky and Tennessee were the "simple" continuation of old state lines. Economic considerations were not absent, but two groups with different conceptions fought. The first argued that rivers and their valleys were geographical units and should be located in the centers of states and not serve as borders.[33] The second wanted every state to be able to use the rivers, and thus have them as boundaries. Rivers were the fastest transportation routes at the time, especially after the coming of steamboats. Moreover, the Missouri River was the frontier with Indians. This is why Iowa fought to have both the Mississippi and the Missouri Rivers as boundaries. This is also why St. Paulites fought hard to have a north/south Minnesota with a frontage on Lake Superior rather than the east/west rural one that St. Peterites proposed (see chapter 5). Division could also come from previous political and economic logics. Mississippi and Alabama inherited the rivalry between the former Natchez and Mobile districts.

Later states were larger and could present more complex forms, such as Texas (which is larger than France) or California. But Texas came to the Union as an independent state, and all subsequent attempts to divide it never succeeded. Both states were located in very dry environments, and nobody predicted at the time that a century and a half later they would be the nation's most populated states. The argument that resources were scarce was used to explain why states of such size were admitted to the union: Minnesota was granted eighty-four thousand square miles on the grounds that it comprised many useless areas. The most common form was geometrical, since new states, even more than earlier ones, were the expression of republican principles and were delimited without particular attention to earlier—that is Native American or Spanish—forms of organization, with some attention given to local rivers as boundaries (such as the Snake and Columbia Rivers). Their size ranged from seventy-five thousand to a hundred thousand square miles, as much because of the scarcity of settlements as because of the enhanced mobility that railroads allowed.

It is therefore not strange that the national urban hierarchy is not based on state capitals, although some of them participate in metropolitan governance. During the westward expansion period, newly founded communities (from the western part of the Great Plains to the Mountain Region), whether they were capitals or not, "stood largely tributary to larger cities on the eastern edge of the Great Plains, such as Minneapolis, Omaha, and Kansas City. These centers were dominated in turn by Chicago, Saint Louis, and, ultimately, New York. . . . At best, the western city could hope to achieve a respectable but relatively low status in the general metropolitan hierarchy" (Stelter 1973, 189). This is well illustrated by the map of trade centers produced by Donald Meinig, in which trade areas are largely free from state boundaries (Meinig 1998, 298). For instance, South Dakota was divided into three trade areas: the north was controlled by Minneapolis-St. Paul; the center and west, by Sioux City, Iowa, and a small southern part by Omaha, Nebraska; Pierre, the capital, had no influence. Even a larger capital like Columbus had a trade area that did not come close to covering all Ohio, since it was situated between the trade areas of Cleveland and Cincinnati. Only twelve capitals had their own trade areas.[34]

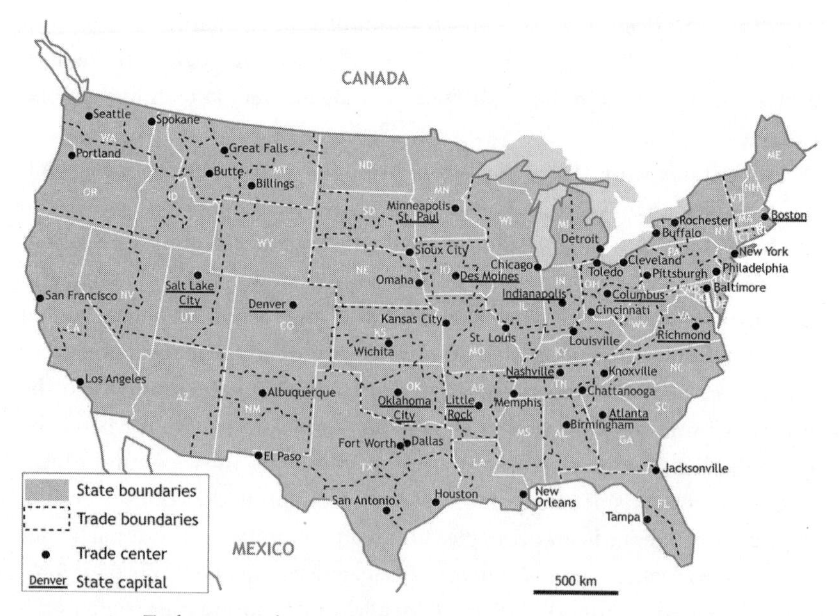

FIGURE 6.1 Trade areas and state boundaries in 1905 (Meinig 1998, 298). Credit: Christian Montès and M. L. Trémélo.

Successive Frontiers and Railroads

By and large, the continuing frontier is what keeps the U.S. and its cities vibrant and vital. Yet, not all cities benefit from the advance of the frontier. Some, economically linked entirely to the needs of earlier frontiers, are left in the backwash as the frontier passes on.
(ELAZAR 1987, XXI)

Successive frontiers were established, to which state capitals have been very unevenly adapted. In an era of shifting economic sands, some first capitals could not survive the economical stabilization following the opening of a merchant economy and of the connected industrialization process.

Colonial or territorial capitals were often important rural centers (with land offices and markets), as well as crossroads of an incipient road or canal system. During the second part of the nineteenth century, with the progressive development of a national economy, things changed. One major factor of change was the construction of the railroads. Railroads had appeared in the 1820s–30s, but only as small lines linking seaports and their close hinterlands, which remained mostly agricultural. The railroads were especially important compared to those in Western Europe, where railroads arose among numerous other means of transportation; in the United States, they grew "as a new development in the absence of any good prior transportation" (Vance 1995, xi).[35] The first phase of railroad development did not alter fundamentally the transportation geography but only modified the hierarchies, because the eastern seaboards used the railways to maintain their preeminence. Boston became the world's oldest "junction city," with seven railroad lines radiating from it. Likewise, Albany benefited from railroads from Massachusetts and from New York that joined the Erie Canal and Montreal. In the South, Charleston and Savannah competed to control the cotton trade and intercept the river traffic from cities like Milledgeville, Georgia's capital. Rail lines thus followed the initial river transportation system, a pattern that persisted. "The accordance between hydrography and railroad routing in the United States is very much a feature of the developmental nature of rail construction in North America" (Vance 1995, 81). Although rail connections were believed to be the best insurance of the Union, railroads did not connect Washington to all capitals: Lansing, Tallahassee, Baton Rouge, Little Rock, and St. Paul were still out of the system (Meinig 1993, 347).

During the second phase of railroad development, the crossing of the Appalachians from the 1850s on, hierarchies began to change with the rise of inland hubs, which modified the classic Atlantic pattern. Concurrently, a national network was rapidly constructed; the opening of the first transcontinental railroad in 1869 was one early symbol of the closing of the frontier.[36] Railroads could indeed mean the economic demise, salvation, or even creation of state capitals. Railroad companies found that creating towns along their lines often proved to be more profitable than transporting passengers and freight. Existing cities fought to attract railroads, which often functioned as economic boosters in addition to providing transportation. Part of their financing came on a pay-as-we-go basis (Lingeman 1980, 163). Public funds were used at municipal and state levels to pay for the railroads, although some small capitals, such as Tallahassee, were unable to raise the necessary payment.

Wisconsin is a good example of the overlapping of early and "modern" logic. To become Wisconsin's uncontested metropolis, Milwaukee relied on the power it had gained during the era of land and water transportation. Milwaukee was better located than the capital, Madison, as it was one of the few cities that offered good harbors for ships between the Door Peninsula and Chicago, and it was far enough from Chicago to be a competitor. Milwaukee also had clear land titles, as did Green Bay, but it was located far too north of the commercial flows. In a period of difficult internal transportation, controlling traffic was the key to metropolitan prominence. Milwaukee, which had built piers on the Lake Michigan shore as early as 1843, controlled the incoming and outgoing demographic and economic flows. As a result, banks came, waterpower was installed, pioneer railroads were built, and Milwaukee became a powerful interface between local commercial agriculture (mostly based on wheat) and a growing national economy. But population was ultimately more important than geography, and "the battle was really won by the people who were attracted to Milwaukee to start a great variety of enterprises and small businesses" (Nesbit and Thompson 1989, 188–89). Madison was indeed located on some major roads near the center of the state, but it served only as a point, not as a hub.

The third phase of American railroad geography was marked by "the projection of the various urban mercantile systems into the great agricultural areas of the North American interior" (Vance 1995, 104). The standardization of gauge and the concentration of the rail lines in the hands of huge companies controlled by eastern or midwestern capital (e.g., the New York Central, controlled by Cornelius Vanderbilt) helped to create a vast national market. In the Great Plains, railroad companies "defined the very network structure of the urban system that would serve the regions they plied" (Conzen 2001, 336). The new railroad hierarchy proved a mixed blessing for state capitals. The major hubs were in Chicago ("the Great Junction City") and St. Louis. The secondary hubs were Cumberland, Pittsburgh, Wheeling, and, farther south, Louisville (with the Louisville and Nashville Railroad), and Cincinnati (through the Cincinnati Southern Railroad, which built a direct route to Atlanta and New Orleans). Only two capitals appeared among the secondary hubs, Columbus and Indianapolis (only for the southern part of Indiana). Lansing lost all control over railroads, and Des Moines was only able to maintain a small north-south elongated hinterland between the influence of Omaha and Chicago (Conzen 1975, 373).

The construction of the transcontinental railroad induced great competition between towns (existing or on paper) to boost their economies. Of the four different routes considered, which ran along parallels, only two were really studied by a survey conducted in 1854. The first, along the forty-seventh and forty-ninth parallels, proved too costly to build. The thirty-fifth parallel was investigated by lieutenant A. W. Whipple, from Fort Smith, Arkansas to Los Angeles. The transcontinental railroad was seen as a line and not as a network of cities, because the West was at the time very sparsely populated. It was easy to choose the western terminus of the line, Sacramento, because the gold rush had made it the economic heart of the state. The rest of the Pacific coast was still too underdeveloped to contend seriously with California. The eastern terminus proved far more difficult to select, as St. Louis, the Iowa towns, and St. Paul contended for it. President Lincoln compromised (compare the processes of choosing capitals) by selecting Council Bluffs in 1862, across the Missouri river from Omaha, so that the company would have to build the bridge. As a direct side effect, the capital of the Nebraska Territory was selected with regard to the future route of the transcontinental railroad (Olson and Naugle 1997, 78–79). When the territory was created in 1854, the only settlement was Bellevue, which had fifty people living around the Indian Agency, the Presbyterian mission, and a trading post. That post belonged to Peter A. Sarpy, who also ran a ferry across the Missouri River to St. Mary, Iowa, where he conducted most of his trade. Sarpy created the Bellevue Town Company on February 9, 1854, intending the new town to be the capital. Bellevue's main opponent was successful Omaha City. It was founded in November 1853 by the Council Bluffs and Nebraska Ferry Company, which hoped to attract the transcontinental railroad by making Omaha the territorial capital. Omaha won the capital because of the Iowan leanings of the acting governor, who apportioned the new territory according to his own interests, with no regard for the population.

The fate of three capitals was to depend upon the route chosen by the transcontinental railroad: Golden, Denver, and Cheyenne. Studying them shows that a railroad town without a hinterland cannot forge a metropolis. Larimer, Denver's founder, wanted Denver to be on the transcontinental. The company refused in 1866, for two reasons: first, since the federal subsidy that enabled its construction was based on mileage completed, southern Wyoming was preferred; second, it could not find a suitable pass at Denver, but southern Wyoming offered more reasonable grades (Abbott, Leonard, and McComb, 1994, 84–86). This did not mean that Colorado was abandoned. The railroad wanted to fill its trains, and the traffic coming from the Colorado mines was welcomed through a branch line that would be built with Union Pacific's participation. The terminus of that line was sought by two of early Colorado's cities that had competed from the outset, each with a faction in the local Republican party: Golden, the first territorial capital, and Denver, which was soon to become the capital. The railroad question was vital because the 1867 decision of the Union Pacific to go through Cheyenne had led hundreds of people to move from Denver to Cheyenne (Leonard and Noel 1990, 34). In response, the Denver Board of Trade was created at the end of 1867, and the Denver Pacific Railroad Company in 1868. Golden set up a rival company, the Colorado Central Railroad, headed by the founder's associate, Loveland (whose store served as meeting hall for the legislature). The winner was the Denver Pacific; its 106-mile line between Denver and Cheyenne opened in 1870, and the Kansas Pacific reached Denver in the same year. The Colorado Central only built a 15-mile line to Denver one month later, but it had to wait until 1877 to reach Cheyenne and fell under the control of the Union Pacific in 1879. Behind that story are the territory's first governors. Colorado's first governor, William Gilpin (1861–62), had already envisioned Denver as a railroad hub. The state's second governor, John Evans (1862–65), had a background in railroads and secured money through his Chicago and Washington, DC connections to build the line to Cheyenne. After his discharge in 1865, he remained in Denver and was the mastermind behind the railroad network that was being built. He also created the Denver Seminary, which became Denver University (Colorado State Archives website). "Railroad[s] guaranteed Denver's survival, but the city's prosperity depended upon the regional economy" (Leonard and Noel 1990, 44). In fact, only 2,200 mining jobs remained in 1870. Denver became a major railroad hub for immigrants and tourists. Fifteen hundred miles of tracks had been built in Colorado and beyond from 1878 to 1883. In 1880, Denver ranked as the nation's fiftieth city; by 1890, it was twenty-sixth. Denver's economy was based on its role as the primary supply, banking, and industrial center for a large part of the Rocky Mountain West (Abbott, Leonard, and McComb 1994, 245–46).

While Denver thrived, Cheyenne never fulfilled its early hopes of becoming the major supply center for the central Rockies, despite having being built by the Union Pacific and serving as a point where train crews were changed and locomotives and other rolling stock were serviced. At the end of 1867, the very year of its creation, the city had about nine thousand residents. Despite violence and gambling (and vigilantes!), the city offered newspapers, seventy saloons (in the same ratio as for eastern cities), and variety theaters, with a Masonic society, a YMCA, two churches, and one school. After the departure of the construction crews, however, activity dwindled, and the town's population dropped to 1,450 in 1870. The major reason was that the Union Pacific's promises to make the city a great railroad center never really materialized. In 1868, Union Pacific's main shops were located in the new city of Laramie, inducing a big population flight. Cheyenne suffered a further blow when Denver reaped most of the commercial benefits of the branch line between both cities, which was completed in 1870 (Stelter 1967, 13). Then came a period of small booms and busts: in 1875 a gold rush to the Black Hills proved to be only a boomlet. Ten years later, Cheyenne became the market center of the northwestern range cattle industry. Cheyenne's population "jumped" from 3,456 in 1880 to 14,087 in 1900. "Throughout this early period the city's aggressive businessmen displayed a remarkable amount of energy and extended their commercial hegemony over a huge region" (Stelter 1967, 32–33; Stelter 1973). The region was indeed huge but underpopulated, as it was impossible to cultivate Wyoming's arid plains. Cheyenne therefore remained a small city, and its population even fell to 11,320 in 1910. Forty years later, it still was a mere 31,935. Without a real hinterland—Wyoming is the least populated state—significant growth proved impossible. The proximity of Denver also hindered Cheyenne's development.

The following transcontinental lines, seven of which had been built by 1910, mostly on the 1854 survey tracks, opened a second phase that "was most critically shaped by the burgeoning settlement geography of the West, and the desire to attach as many potential markets to a company's lines as possible" (Vance 1995, 204). The Northern Pacific Railway, completed in

1883, established in 1873 a new town at its bridge over the Missouri River, 446 miles west of St. Paul, named Bismarck. The last transcontinental line, the Western Pacific from St. Louis to Oakland, completed in 1911, passed by only three large cities that were already the capitals of their states: Denver, Salt Lake City, and Sacramento. This second phase thus did not give rise to a Denver-style hub, proving that the first line had been well chosen. Bismarck never rose to more than local prominence (as a supply center during the Black Hills mining boom); Helena is better known for the major obstacle encountered there by the Northern Pacific than for the great traffic it induced;[37] and Phoenix is located on a branch of the Santa Fe Railroad, which provided service to southern Arizona from Chicago and the Middle West via northern Arizona and the Pacific coast. The Southern Pacific transcontinental line had first bypassed Phoenix, but "enterprising landowners and merchants furnished the capital to build their own line—the Maricopa and Phoenix—to achieve the rail service necessary for survival and prosperity in the competition for Western urban supremacy" (Reps 1981, 89).[38] More generally, railroads proved essential for desert cities, allowing them to become focal points (Nash 1985). Railroads and their political power helped transform desert cities into regional capitals at the beginning of the twentieth century, with the help of federal funding that gave them stability during the boom and bust period. But they remained relatively small towns (in 1920, the largest one, El Paso, had seventy-seven thousand inhabitants, and Phoenix had twenty-nine thousand).

Atlanta is the only example of a capital that, almost entirely because of the railroad, was able to become a metropolis (Coleman 1991). Before its creation, around 1840, Georgia's most prosperous city was Savannah, the terminus of the Central & Georgia Railroad and the banking and industrial capital of the state. But, after the Georgia Railroad was built from Augusta (1834) to Athens (1841), and the Central from Savannah to Macon (1843), the Western & Atlantic Railroad (1843–47) founded a new town that was first called Terminus and then Atlanta. A decade later, Atlanta was becoming the hub of a railroad system that linked Georgia to eastern as well as western markets. Macon and Augusta were the other hubs, while the capital, Milledgeville, was situated on a thirty-seven-mile spur line to Eatonton. In 1860, although the largest city and main industrial center was still Savannah (22,292), Atlanta (9,554) was busy building a real manufacturing basis around its machine shops for railroad maintenance, including an iron plant. Atlanta's economy was more stable than that of Macon and Augusta because it was located outside the "king cotton country." Atlanta's excellent rail connections transformed it into a wholesale and retail distribution center for northern agricultural and industrial products. The city also became Georgia's banking and insurance center as well as the state's largest city in 1880, when it had 37,409 inhabitants (65,533 by 1890). Atlanta also gained a new university in addition to its religious universities; Oglethorpe College moved there in 1870 from the former capital, Milledgeville.[39] In 1940, Atlanta's population had grown to 302,288 inhabitants. Railroads were still the major reason for such growth, along with manufacturing (mostly textile and agricultural), banks, insurance, and conventions. Its connections also attracted branch offices of national firms and home offices of many southern firms.

Railroads did not always bring success to other capitals. One famous railroad company bears the name of two state capitals. Two years after the arrival of the first trains of the Kansas Pacific Railroad at Topeka, in 1866, the founder of Kansas's capital, Cyrus K. Holliday, started construction of what was to become the Atchison, Topeka & Santa Fe line (Reps 1981, 76). But Santa Fe and Topeka did not reap great benefits from the line. Santa Fe was the end of the well-known Santa Fe Trail (a carved stone in the city's square reminds visitors of that fact). But when railroads came in 1881, they favored an easier location, Albuquerque. Although it had been founded one century later than Santa Fe, in 1706, Albuquerque's role as military post gave it great importance, not only for defense but also for trade (it was at the center of irrigated agriculture), culture, and hospitals, as it had become a large health resort at the turn of the twentieth century. Its importance has increased because of federal military investment since World War II. Air bases and atomic production and research facilities were given to the city according to the usual spoil system, aided by the presence of the University of New Mexico (Beck 1962, 288–89). At the other end of the line, Topeka was certainly a railroad center in a rich agricultural country, but it was under the shadow of the real primary city, Kansas City, only fifty-six miles east.

Most capitals were bypassed by the railroads at first, and later became only secondary railroad centers, such as Lincoln. When the Missouri Pacific first went through Nebraska, it bypassed Lincoln and came there only in 1886 via a connection from Union and Weeping Water. Lincoln had at the end of the nineteenth century five trunk railways with eighteen diverging lines. The major hub remained Omaha, only forty-nine miles distant. Likewise, Olympia had been a hub of maritime commerce in Washington Territory since the 1850s, but the mainline railroads bypassed the capital in the 1870s, and residents had to build their own line to connect with the Northern Pacific mainline at Tenino, fifteen miles to the South. Because rail traffic remained modest, the city reverted to maritime efforts. In 1911–12, Olympia undertook a gigantic dredging project to create a deep-water harbor, but Seattle and Tacoma were already firmly established as the main ports and rail hubs.

Even a pioneer railroad did not guarantee success. In Austin, the Houston and Texas Railroad came in December 1871 and made the city the center of a vast trade area, because it had become the westernmost railroad terminus in Texas. Population almost doubled in five years, reaching 10,363. But although a second railroad, the International and Great Northern, arrived at Austin in 1876, new competing railroads diverted almost all trade to other towns, halting Austin's growth.[40] Indeed, if rail meant boom, it also brought competition for local banks and industries.

To conclude, the examples of Guthrie and Oklahoma City, the first and second (permanent) capitals of Oklahoma, show that in order to be useful to a city's economy, a railroad must be on an important route, and the city must become a dispatching point. The Atchison, Topeka & Santa Fe Railroad had created both cities, but Oklahoma City gained preeminence with the coming of a second railroad that bypassed Guthrie (Forbes 1938, 7). During the 1896–1902 period, the citizens of Oklahoma City paid for the coming of the Choctaw, Oklahoma, and Gulf Railroad, which brought inexpensive coal for industry, as well as connections to other railroads that made the city the regional rail hub. Packing companies from Chicago soon settled there (Welsh 1982), followed by numerous other manufacturing plants (Meredith and Shirk 1977, 297–300). This underlines the close links between industrialization and railroads: manufacturing plants need the best transportation available.

Industrialization and Capitals: Two Worlds Apart

Before the 1920s, when trucks began to compete with locomotives, railways reigned supreme in transporting industrial goods. The United States underwent an early process of industrialization, earning the second rank in the world in 1840, behind Great Britain, and the first rank at the beginning of the twentieth century. Historians call the 1860–1920 period "the industrial city era" (Weil 1992, 59).[41] Although the federal government's motto was "laissez-faire," state governments supported the economic evolution of their citizens. They did so by granting charters, sponsoring banks or transportation companies, investing in stocks (canals and railroads, for instance), and owning banks and enterprises (Pessen 1985, 103). But state capitals seldom benefited from such largesse. Only three eastern state capitals figured among the main industrial cities of 1860: Boston (ranked fourth), Providence (eleventh), and Richmond (thirteenth).[42] The concentration that occurred at the end of the nineteenth century created large companies that helped major cities by attracting capital and manpower. The only capitals able to attract factories were the ones that succeeded in enlarging their economic hinterland beyond agriculture. Denver concentrated on wool and ores; St. Paul-Minneapolis, on wheat. The presence of mines was an asset. Charleston, for instance, benefited from four successive industrial booms: salt (from 1797 to 1875); coal, oil, and gas (1875–1900, which earned it the nickname of COG City); chemicals (1920–37, with 7,407 workers in 1936); and the chemical industry's rediscovery of the use of salt brine in the production of chemicals in the 1960s (Goodall 1968b). Providence succeeded in becoming an industrial stronghold, first by acquiring important capital through maritime trade—especially the notorious Atlantic slave trade—then by "inventing" modern American manufacturing in 1793 (at nearby Pawtucket, nicknamed "the birthplace of the industrial revolution," although it was based on British innovations). It had added base metals, machinery, and jewelry by 1830, and strengthened its role as a rail hub with the early completion of several railroads. In 1900, forty-five thousand people were employed in manufacturing.[43] Industrial decline began after World War I and was more important during the 1930s. Generally, however, capitals were not as important as the well-known industrial centers: Detroit, Chicago, Pittsburgh, Cleveland, Kansas City, and Buffalo.

Among the numerous capitals bypassed by industry and modern trade, Annapolis is a good example. At the height of its prosperity, from 1763 to 1785, it was the social, economic, and political center of the Chesapeake Bay area, despite a population of only 1,280 in 1783 (Potter 1994, 76).[44] Even then, Annapolis traded mostly with the adjacent counties and never became the large tobacco center it was intended to be. Baltimore overtook Annapolis because the main agricultural staple had become grain rather than tobacco, and because Baltimore's harbor was larger and deeper. Even the genteel society that had nurtured Annapolis's luxury crafts and merchandising had left the capital for Baltimore (Shakel 1993, 62–69). Annapolis was therefore transformed into a small, local market town. The coming of the Naval Academy in 1845 helped Annapolis to fend off attempts to relocate the capitol to Baltimore, but the academy remained quite small during the nineteenth century and did not induce any boom. At the beginning of the twentieth century, the main feature of the town was the newly rebuilt (1899–1913) Naval Academy, which added large, granite beaux-arts buildings to the city, but the town was still far from the main commercial routes, and the corps of engineers refused to dredge the harbor. The navy itself viewed the city with contempt: "After the turn of the century someone at the Academy published an "Annapolis Alphabet," in which B for Busy pictured two cows walking up Main Street toward Church Circle . . . Annapolis town leaders considered confiscating all copies of the booklet" (Brugger 1988, 435–36). Rail service was discontinued between Annapolis and both Baltimore and Washington in the 1930s.

Harrisburg, Pennsylvania, provides an example of an industrialization process that remains as a quarter-century "interlude" in the capital's economic history, from the 1860s to the 1880s (Eggert 1993, xvii). A classic trade and governmental center of 7,834 inhabitants in 1850, it built its first plant in 1849, well after cities like Philadelphia (93,665 inhabitants in 1850) and Pittsburgh (21,115). Since the Civil War had made the city a major railroad center, iron and later steel became Harrisburg's leading industries, but without innovation. The population grew to 13,405 in 1860 and to 23,104 in 1870. Harrisburg was taking part in a movement that touched all of eastern Pennsylvania, where coal mining and steel production supplied the national market (Conzen 2001, 340). But Harrisburg never rose to the highest ranks. Philadelphia contributed 70% of the value added by manufacture in eastern Pennsylvania in 1910, more than Reading and Scranton. Industrial decline began for Harrisburg in the 1880s, and after 1945 the city was no longer described as an industrial center. The main cause was a lack of entrepreneurship and the effects of the concentration of the steel industry—with the huge US Steel company—which left almost no place for outsiders like Harrisburg's firms. The fates of Annapolis and Harrisburg were shared by the majority of state capitals, which never achieved major importance in the urban hierarchy that railroads and manufacture had established.

State Capitals and the Urban Hierarchy

This chapter has dealt with about a century of evolution, roughly from the end of the Civil War to the aftermath of World War II. That century began with state capitals that were full of promise. When it ended, some promises had been fulfilled—social ones, for instance, at least for the ruling minority—but most of them had not—mainly political and economic ones. This explains why capitals do not form a network. The fifty capitals had neither specific economic relationships among them, nor political ones (e.g., no association of state capitals was created). It is thus not surprising to see only a few state capitals on the maps of the American urban hierarchy drawn by Michael Conzen for 1840, 1880, and 1930, or among the trade centers of 1910 established by Donald Meinig (Conzen 2001, 335; Conzen 1977; Meinig 1998, 298). Capitals participated in the broader urban hierarchy, which was based on chains of wholesaling towns, instead of creating their own. In 1840, six capitals were listed by Michael Conzen: Boston was a major control point; Providence, an industrial center; Albany and Richmond, industrial centers and central places; Nashville and Columbus, central places. In 1880, eleven capitals were listed. Boston was still a major control point, along with Providence and Trenton (industrial places); Albany (industrial-entrepôt); Richmond, Nashville, Indianapolis, Columbus, and Atlanta (central places); St. Paul (central place/entrepôt); and Denver (the only western capital listed, a central place). In 1910, only twelve capitals had their own trade areas,[45] and only thirteen appeared in the banking hierarchy.[46] In 1930, the number of cities listed by Michael Conzen had climbed to seventeen, with the addition of Sacramento, Oklahoma City, Des Moines, and Lincoln (entrepôts), Salt Lake City (central place/entrepôt), and Lansing (industrial center). But this still only represented one-third of all capitals, close to the figure found in 1943 by Chauncy Harris. Moreover, with the great exception of Boston, capitals were not at the top of the hierarchy, most of them being subordinate to larger cities in their own states: Sacramento to San Francisco, Lincoln to Omaha, Columbus to Cleveland, and Lansing to Detroit. The fact that between 1880 and 1930 only six new capitals were added to the hierarchy is further proof of the fact that capitals were mostly ignored by the modernization processes that reached their fullest extent between 1880 and 1930, adding almost all the cities that today constitute the upper level of the national urban hierarchy.

The roots of the developmental delay of many state capitals are now known: they have been at once geo-economic (they have been bypassed by modernization, railroads, and manufacturing) and cultural (an often very negative image). Would their situation change with the coming of the postindustrial era? After the Great Depression, which halted development almost everywhere, and the war economy of the 1940s, most capitals would leave Purgatory, under the pressure of internal and external factors. Internal factors brought a reassessment of the role that states had to play and how they should play it, and external factors brought a new economic impetus that worked on the white-collar forces that were present in state capitals. The result was perhaps not always Heaven, but never Hell.

7

State Capitals since the 1950s

The Renaissance of Forgotten Cities

The following quotations provide three different assessments of the link between capitals and their economic and social development that translate the processes of transformation at work.

> Capital crimes: political centers as parasite economies. (Vedder, 1997, 20)
>
> It's a pain being a state capital. But it's an even bigger pain not being one." (Lemov 1993, 46)
>
> What used to be called "India-No-Place" is now a vibrant 24-h place to work, live, and play.[1]

In the 1950s, state capitals initiated an aggiornamento and have since been modeling their evolution on mainstream America's. As federal and state budgets expanded, capitals expanded more rapidly than other cities, from demographic, social, and economic points of view (Carroll and Meyer 1982). As a result, governmental centers grew in importance in the new division of labor that was emerging (Ross 1987). But this growth was linked to economic and demographic catch-up processes (as in the Sunbelt) and has not fundamentally modified the national urban hierarchy.

A First Demographic and Economic Assessment

When looking at the demographic evolution of state capitals, municipal and metropolitan areas must be distinguished. For municipalities, the growth pattern is not overwhelming. Whereas state capitals lagged behind comparable cities until 1930, they grew more rapidly after that date (Carroll and Meyer 1982, 569).[2] However, because the Great Depression of the 1930s and World War II introduced too many biases in the evolution of American cities to allow the identification of coherent trends between 1930 and 1950, the latter date will serve as a starting point for the study of capital city renewal. The list of America's one hundred most populated cities from 1950 to 1990 shows the rise, emergence, or revival of fifteen capitals. During the same period, nine capitals disappeared or declined, and some experienced a new decline between 1990 and 2010 (see table 7.1).

But such rankings suffer from an important bias: the difference between the areas of municipalities. While Annapolis and Juneau have ap-

Table 7.1 Rank of state capitals among the hundred largest US cities, 1950–2011, and MSAs, 2000–2010

City	1950	1960	1970	1980	1990	2000	2011	2000 (MSA)	2010 (MSA)
Montgomery	—	90	—	76	86	89	(103)	123	136
Little Rock	—	—	—	99	96	(112)	(117)	74	72
Phoenix	99	29	20	9	9	6	6	14	14
Sacramento	67	63	55	52	41	41	35	25	25
Denver	24	23	25	24	26	25	23	19	21
Boston	10	13	16	20	20	20	21	7	10
Hartford	54	78	86	—	—	—	—	42	45
Atlanta	33	24	27	29	36	40	40	11	9
Honolulu	—	43	44	36	44	47	55	56	53
Indianapolis	23	26	11	12	13	12	12	29	35
Des Moines	53	55	64	74	80	95	(104)	91	88
Baton Rouge	82	81	84	62	73	75	88	70	66
St. Paul	35	40	46	54	57	60	66	15	16
Jackson	—	85	91	71	78	(111)	(135)	96	95
Lincoln	—	96	92	81	81	77	71	147	154
Trenton	80	—	—	—	—	—	—	(in NY)*	(in NY)*
Albany	69	94	—	—	—	—	—	57	59
Raleigh	—	—	—	—	75	63	42	41	47
Columbus	28	28	21	19	16	15	15	33	32
Oklahoma City	45	37	37	31	29	29	30	49	43
Providence	43	56	71	100	—	(121)	(132)	40	38
Nashville	56	73	30	25	25	22	26	39	37
Austin	73	67	56	42	27	16	13	38	34
Salt Lake City	52	65	74	91	—	(113)	(123)	36	48
Richmond	46	52	57	64	76	97	(105)	51	44
Madison	—	97	78	84	82	84	82	98	89

Source: US Census Bureau, 1998, 2000, and 2010.

Notes: 1. Some rankings greater than 100 appear to show either the current ranking of capitals that once belonged to the hundred largest or the ranking of the MSAs of the capitals which belong the hundred largest cities.

2. Rank in 2010 of the state capitals that do not belong to the hundred largest cities but are included in the hundred largest MSAs: Columbia, SC, 70; Boise, ID, 85; Harrisburg, PA, 94.

* (in NY) means that Trenton is included in the New York City MSA.

proximately the same population (38,394 and 31,275, respectively), the first city covers 7.2 square miles, when the second, a city-borough, extends over 2,701.9. Moreover, their area is unstable, since capitals, like other American cities, frequently alter their boundaries by way of annexations. During the 1990–2010 period, some capitals experienced major land gains that sometimes largely explained the rise in their populations. For example, Tallahassee expanded from 63.3 square miles to 100.2, and Boise City, from 46.1 to 79.4.

Likewise, the suburbanization process largely explains the decline in rank of some cities and the growth of their conurbations through overspilling. The MSA rankings are quite different, often higher than the city rankings. Some smaller cities belonged in 2010 to the hundred biggest MSAs: Columbia, South Carolina (70th), Boise City, Idaho (85th), and Harrisburg, Pennsylvania (94th). For Hartford, Providence, Trenton, and Richmond, insertion in the megalopolis also biases the evolution. In 2011, when a majority of capitals (thirty-four) were not on the list of the hundred most populated cities, only twenty-four were missing from the hundred most populated MSAs. This means that more than a quarter (twenty-six) of the largest national metropolitan areas are state capitals (but only two are among the ten largest and six among the twenty-five largest), emphasizing the economic and demographic catch-up processes as well as their limitations. From 1990 to 2010, growth continued; no capital MSA/CSA lost population, and nineteen grew by more than 50% (for precise figures, see table A.3 in the appendix).

We must therefore look beyond the capitals' demographic achievements. Table 7.2 provides a summary of their economic situations. The growth of capital cities did not come from the three main factors in the national population shifts—region, industrial mix, or state population—according to a model cited in Carroll and Meyer (1982, 569). Among the factors at work, the Sunbelt environment certainly largely explains the expansion of southern and western capitals. In addition, the considerable modernization process of American politics that began in the 1950s was a prerequisite and strongly influenced the growth of capitals, although it was not the unique or even the main factor. This is why we examine it first. Two elements are emphasized: the rising importance of state bureaucracies and the rising importance of morality and professionalism in local politics, which brought more people to work in capital cities and modified the way capitals were perceived.

Table 7.2 Economic situation of state capitals in the early 2000s

Capital	Type of economy
Montgomery	Trade, cultural center of the state. Government (25%), army, some high-tech
Juneau	Government (57%), tourism, mining, fishing, logging
Phoenix	Metropolis, high-tech, tourism
Little Rock	Center of the state, hub, army, health, government (21%)
Sacramento	Metropolis, trade (hub), two air bases, air and space industry, state university
Denver	Metropolis (energy), federal jobs, high-tech, huge airport
Hartford	Major U.S. insurance center, general crisis (30% of pop. are poor)
Dover	Business headquarters (fiscal causes)
Tallahassee	Government (38.1%), two universities
Atlanta	Regional metropolis; Coca Cola, CNN, manufacturing, regional headquarters, and so on
Honolulu	Three-quarters of state's population, universities, tourism, navy, government (22%)
Boise City	Cultural, economic (high-tech), and political capital of Idaho
Springfield	Government first, health services, tourism, software
Indianapolis	Hub, trade, manufacturing (medical), university
Des Moines	Great Plains hub/insurance, overall center of the state
Topeka	Secondary hub, government, mental hospital
Frankfort	Government (53%), tobacco, bourbon, auto parts
Baton Rouge	Petrochemicals, port, universities, government
Augusta	Government (40%), local service center
Annapolis	Naval Academy, government, heritage tourism
Boston	Metropolis
Lansing	Government, manufacture (G.M.), Michigan State University, some finance insurance, and real estate (FIRE)
St. Paul	Twin Cities (MSA). Overall metropolis (from banks to high-tech and 3M)
Jackson	Main center of state (oil, gas), hub, manufacturing
Jefferson City	Government, some manufacturing
Helena	Government, tourism, local service center
Lincoln	Government, universities, agricultural center
Carson City	Government, tourism
Concord	Government, tourism, local center
Trenton	In New York CSA. Government, trade.
Santa Fe	Government, cultural and heritage tourism
Albany	Hub, trade, G.E., university, government (31% in county)
Raleigh	Center of the state (MSA). Triangle Research Park
Bismarck	Hub (rail), trade center, oil, government
Oklahoma City	Center of the state, hub, army, manufacturing, oil and gas
Columbus	Balanced economy
Salem	Near Portland MSA. Government, food processing, lumber, university, some high-tech
Harrisburg	Trade hub, government, some manufacturing (defense)
Providence	"City-state"; banking, universities, heritage, government
Columbia	Government, army, trade, University of South Carolina (25,140 students), high-tech
Pierre	Government, small agricultural center
Nashville	Government, finances, trade, manufactures (auto), university, music
Austin	Metropolis; high-tech, university, government (air base until 1993)
Salt Lake City	Metropolis of Utah, tourism, high-tech
Montpelier	Government (35%), insurance (FIRE 18%), and tourism
Richmond	In Megalopolis. Trade, finance, law, three universities, Philip Morris
Olympia	In Seattle MSA. Government (39%); small, active port; lumber
Charleston	Economic center of Kanawah Valley, chemicals, government
Madison	Government, University of Wisconsin
Cheyenne	Government, army, small center of an agricultural and tourist region

Sources: Local studies and websites.

Note: Government is indicated only if it has a major role in the capital's economy.

Political Purification: "One Person, One Vote" and the New Federalism

Almost unnoticed in Washington, there has been a revolution in state capitals from Albany to Santa Fe, from Olympia to Tallahassee, and from Richmond to St. Paul. No longer the province primarily of hangers-on and political hacks, most state governments today are remarkably sophisticated and professional, competent to address problems that only a decade ago seemed beyond their grasp.

D. P. DOYLE AND T. W. HARLTE, "A FUNNY THING HAPPENED ON THE WAY TO NEW FEDERALISM" (1985)[3]

Clear evidence of the new powers gained by the states is the annualization of legislative sessions (except for four states, one of which is Texas).[4] The rise of desegregation since the 1950s turned the focus on state politics. Modern, televised politics appeared, not as much based on local newspapers as before. A saner democracy grew when electoral fraud was abandoned. Concurrently, intergovernmental rules between the federal and state governments were redefined. Post–Civil War Reconstruction, the New Deal, both World Wars, and the Cold War had first considerably reinforced the role of the federal state. With the end of the fight against the USSR and communists, it began to lose importance, but the titles of recent books on intergovernmental relations prove that the process of state renewal is far from complete: for example, *The Rebirth of Federalism* (Walker 2000); *Can the States Be Trusted?* (Ferejohn and Weingast 1997); *Disunited States* (Donahue 1997); and *Governing Partners* (Hanson 1998). Recent events have also shown a surge of nationalism in the United States that favors the federal government. State and local spending increased between 1950 and 1990 as a share of the GNP from 9% to 14.8%, and federal government spending has expanded even more steadily, from 15.7% to 24.9% (Gray and Jacob 1996, 320). The trend has since continued.

The changes that occurred from 1961 to 1998 underline the contrasting characteristics of "conflicted federalism" (Walker 2000, 16). The main changes are that state governments are today more innovative and active than ever. They have regained the role of "laboratories of democracy" that they had from 1870 to 1932; they implement federal domestic aid programs more responsibly than they did from 1933 to 1960; and they are more caring toward their local governments than ever (Walker 2000, 12). These changes were the result of the emergence of a new political system. The traditional, decentralized party system that existed before the 1960s—founded on bossism, on party control over primaries, and on the white primacy in the South—was replaced after 1968 by far more powerful national parties, although their influence on federal officials has declined with the rise of pressure groups, consultants, and think tanks, along with the media's influence and the cost of access.

Two contrasting developments have occurred. On the one hand, federal importance has grown, through the rise of federal grants to states and social regulation, with the help of a Supreme Court that basically serves as a centralizing force. On the other hand, most conditions that helped create new regional institutions were abolished, and the comity between federal and state taxes was eroded. During this "quiet revolution," the states were revitalized but still reined in. As for the states themselves, about four-fifths modernized their constitutions. Citizens' rights were expanded, governors were given four-year terms (except in three states) and more power, the judicial system was integrated, and legislatures were streamlined and became more professional.[5] The state bureaucracy was overhauled in three-fifths of all states, and revenue sources were diversified. But the movement has not been total, owing primarily to the states' very diverse political atmospheres, which do not always work in the public interest. Second, although the states are more integrated in the operation of the federal system than ever before, "in the national political and pressure-group realm, centralization persists, and the states are still treated as not much more than one category of pressure group" (Walker 2000, 298–99). At the local level, urban municipalities gained new powers, since rurally controlled legislatures gave way to population-based ones after reapportionment and redistricting. For instance, Indiana's legislature failed to redistrict from 1921 to 1962, which largely favored rural interests (Bodenhamer and Barrows 1994, 168). The newly elected legislatures were keener to give cities home rule and to support urban political reformers. In the case of Indianapolis, this resulted in the creation of Unigov in 1970, which combined city and county functions and has since acted as an agent of economic development.

Tennessee epitomizes the changes brought about by pressure from citizens for better public services and facilities. The state of Tennessee's expenditures grew from $702 million in 1965 to $8.4 billion by 1990. The major question was legislative apportionment, which was still based on the 1900 census in the 1950s, although the constitution required a decennial reapportionment. As a result, rural Tennessee ruled an urbanized state, as had been the case in Britain's "rotten boroughs." Urban counties had up to one hundred times more voters than rural ones for the same number of representatives.[6] The system changed under legal pressure after a Tennessean challenged it in federal court in 1959, in what became a famous class-action suit, *Baker v. Carr*. In 1962 and in 1964, the US Supreme Court ruled that states must apportion their legislatures on the basis of "one person, one vote," under pressure from the federal courts, if necessary.[7] It was necessary in Tennessee, where the district court imposed its apportionment plan in 1968, after the General Assembly failed to meet its requirements (Bergeron, Ash, and Keith 1999, 330–31). Until the 1960s, the legislature remained weak because of high turnover and inadequate staff. Since then, the annual meeting of the legislature, the arrival of blacks and Republicans in the legislature and in the governor's seat induced changes in the legislative workings.[8]

Indeed, old devils are still alive, even though, contrary to Lincoln's forecast, the republic is not dead. Calls for more honesty in politics and for less influence by big business were common during the presidential campaign of 2000 (to no avail). A book review in *The Economist* on July 29, 2000, was even titled *Washington, Babylon?* The evolution of politics seems, according to an article in *The Economist* on August 26, 2000, to turn from national issues toward more and more local ones. The title of the article is "The Unbearable Localness of Politics." State and even metropolitan geographical levels seem less important than neighborhood or utility bill issues. Since political candidates spend a third of their time trying to raise money, the debate is about well-funded pressure groups (like the notorious National Rifle Association or the American Association of Retired People) and about campaign-finance reform. Getting elected is costing more and more money, and 2000 saw the first billion-dollar presidential election. The case of George Aiken, who used to hold down his expenses for reelection to Vermont's Senate to less than $20 (spending $17.06 in the 1968 campaign for instance), seems anecdotal and antediluvian (Morrissey 1981, 46). Morality still ranks high in the electors' minds, and rigid Puritanism is on the rise.

Local politics are still not always "pure." Commenting on West Virginia's political life in the 1980s, historians titled their chapter "An Era of Political Corruption" (Rice and Brown 1993, 290). Another interesting example is New Mexico, where nothing has really changed since World War II. The constitution is still that of 1910, which minimized government and fractionalized and constrained what was created. Reflecting conservatism and elitism, the constitution is very difficult to amend. The main features of New Mexico's political life are localism, a shortage of vision and leadership, and interests that support the status quo—"a little bit for everyone." The reason, according to F. C. Garcia, is that "such a high price is paid for the accommodation, if not the integration, of our political subcultures, that few resources are left for government to be responsive to the needs of the people of this state as a whole, the New Mexico commonwealth" (Etulain, 1994, 30). As a result, citizens hold the legislature in low esteem, and all the reapportionment plans proposed between 1964 and the 1990s were rejected by the federal courts. Short annual sessions that last for thirty days in even-numbered years and for sixty days in odd-numbered years restrict the power of legislators. They are not salaried but receive minimal per-diem expenses, which means that lobbyists—under very weak control—pay for much of the legislative "workings." Garcia concludes by stating that "ethnic-racial, religious, and regional differences have been so mutually reinforcing and so distinctive that they have posed a major threat to the stability of the system" (Etulain 1994, 54). New Mexico thus has the oldest capital in the nation (1609) as well as the most divisive political and social functioning. In this case, the lack of discussion about relocating the capital hides the division of the people. Santa Fe is still the capital partly (or perhaps mostly) because of conservatism and the stagnant political process. Discussing a new location would only open Pandora's box.

New Mexico's situation also shows that fragmentation results in part from the exploding number of groups and lobbyists who promote special interests in the legislature. The greatest increase has occurred among contract lobbyists, who represent multiple clients, reside in the capital, and maintain offices near the capitol (Bell 1986, 14–15). In Nebraska, nearly 400 lobbyists were located within five blocks of the statehouse in 1996—a year when more than 43,700 lobbyists were registered in the United States, significantly more than the 35,864 of 1988 (The Council of State Governments 2002, 484).[9]

In 2003, several local politicians were arrested in New Jersey, a state nicknamed "the Louisiana of the north." The offices held by the corrupt politicians were at the city or county level (mayors or county "bosses"), not at the state or capital level.[10] The contemporary increase in the power of governors is ample proof of the new importance of the states. Indeed, the state of New York has bequeathed five presidents and one vice president (Nelson Rockefeller) to the nation: Martin Van Buren, Millard Fillmore (he was a state comptroller), Grover Cleveland, Theodore Roosevelt, and Franklin Delano Roosevelt—but New York is an exception. "State-produced" politicians have attained national power only recently, especially in the Sunbelt states. The last few presidents—except for George Bush Sr. and Barack Obama—were former governors: James Earl Carter (1976–80) from Georgia, Ronald Reagan (1980–88) from California, William Jefferson Clinton

(1992–2000) from Arkansas, and George Walker Bush (2000–2008) from Texas.

However, the evolution of the capitals of California, Georgia, and Texas clearly emphasize the limitations of an exclusively political explanation for the renewal of state capitals. Other forces were at work to transform Atlanta in the last generation into a conurbation with a population of several million and to change Austin from a sleepy country and college town to the center of a metropolitan area with a population of almost two million.

State Government and the Economic Development of Capital Cities

There are no structural features of American democracy which force state governments to either favor or discriminate against capital cities or capital regions.

(BROMLEY 1990, 7)

Because of their location near the state capital and county seat, Nashville area businesses are fortunate to have access to state and federal government offices, as well as offices for Davidson County government.

(NASHVILLE AND DAVIDSON COUNTY CHAMBER OF COMMERCE WEBSITE, ACCESSED IN 2003)

Chapter 6 concluded that the influence of capital status on the economic growth of capital cities remained secondary, compared with outside factors such as railroads and industrialization. But the situation has changed since the 1950s, not only through the rise of state powers and administrations, but also through new economic patterns that have altered the relationships between "legislative hill" and the rest of the capital municipalities and MSAs, as the second epigraph above indicates. An initial approach to examining the influence that capital status might have on capital cities is to compare the current population of the fourteen cities created by the legislature (or the colonial governor) to become capitals (e.g., Washington, DC) to the population of the five cities created by land speculators to become capitals (table 7.3).

Only Annapolis remains really small, proving that the influence of legislatures never extends beyond their states, and only three of them more or less control their states. As for the five cities created by land speculators, although their population greatly varies, four of them hold the first rank in their states, and one (Madison) holds the second one. It therefore seems

Table 7.3 Population and rank in 2010 of cities created to become capitals

Capitals created by state legislatures	Population in 2010	Rank in 2010
Annapolis, MD	38,394	4
Austin, TX	1,716,000	4
Columbia, SC	805,106	2
Columbus, OH	2,071,052	3
Indianapolis, IN	2,080,782	1
Jackson, MS	539,057	1
Jefferson City, MO	149,807	6
Lansing, MI	534,684	3
Lincoln, NE	302,157	2
Raleigh, NC	1,749,525	2
Richmond, VA	1,258,251	2
Salt Lake City, UT	1,744,886	1
Santa Fe, NM (Spanish colonization)	184,416	2
Tallahassee, FL	367,413	12
Capitals created by speculators		
Boise City, ID	616,561	1
Cheyenne, WY	91,738	1
Denver, CO	3,090,874	1
Little Rock, AR	877,091	1
Madison, WI	630,569	2

Source: US Census Bureau, 2010.
Note: Sometimes a village or camp already existed. The rank is CSA's or MSA's, when applicable.

that speculators had more influence than legislators or that legislators had more than economics in mind when they selected the capital, as shown in chapters 4 and 5.

Beyond this first glimpse of the issue, the number of jobs provided by the states gives another answer. In state capitals, state government is often the major employer and brings to the city a stable economy at least, if not a large one. This was the quasi-unanimous answer in a small survey: "Historically, Madison has been insulated against sharp shocks to the local economy from boom and bust cycles that occur elsewhere."[11] This is not only true for small capitals, for an assessment of Sacramento's economy stated that "its commercial function is almost overshadowed by the administrative importance of being the capital of the most populous state" (McKnight 1992). Table 7.4 indicates the effects of capital status on the occupational structure of capital cities, especially in government. In 1978, capitals had 25% more jobs—in relative terms—in government than did other cities.

In absolute terms, the figures have risen, for state employment has grown substantially since the 1970s (table 7.5), even though the recent recession has meant public job cuts. States tend to employ more workers than municipalities: Delaware, for example, had about the same population as Indianapolis in 2000 (784,000 and 792,000 inhabitants, respectively), but offered 23,700 state jobs, while Indianapolis had about 6,000 municipal jobs.

Table 7.4 Nonagricultural employment in 1978

	National capital	State capitals	Other cities
Manufacturing	3.6	18.6	22.7
Construction	5.1	4.5	4.1
Services	23.9	19.3	18.3
Government	37.8	21.1	15.6

Source: Carroll and Meyer 1982, 575.

Table 7.5 Number of state government jobs controlled by state capitals, 1974–2000

State capital	Jobs controlled in 1974 (thousands)	Jobs controlled in 2000 (thousands)	State capital	Jobs controlled in 1974 (thousands)	Jobs controlled in 2000 (thousands)
Montgomery	—	79.6	Helena	—	17.9
Juneau	—	22.9	Lincoln	—	29.8
Phoenix	25	64.8	Carson City	—	22.4
Little Rock	—	48.8	Concord	—	18.8
Sacramento	195	355.3	Trenton	—	133.1
Denver	37	65.7	Santa Fe	—	47.8
Hartford	39	65.8	Albany	184	251.0
Dover	—	23.7	Raleigh	—	123.3
Tallahassee	—	185.2	Bismarck	—	15.8
Atlanta	57	120.0	Columbus	89	136.3
Honolulu	30	54.8	Oklahoma City	39	64.4
Boise City	—	22.7	Salem	—	53.3
Springfield	—	127.7	Harrisburg	—	149.7
Indianapolis	55	82.6	Providence	15	19.6
Des Moines	—	55.2	Columbia	—	78.7
Topeka	—	42.6	Pierre	—	13.4
Frankfort	—	74.1	Nashville	50	81.1
Baton Rouge	—	94.8	Austin	—	268.9
Augusta	—	20.6	Salt Lake City	19	49.3
Annapolis	—	91.5	Montpelier	—	13.6
Boston	65	95.6	Richmond	68	118.6
Lansing	—	141.9	Olympia	—	112.5
St. Paul	45	73.4	Charleston	—	32.0
Jackson	—	55.5	Madison	—	63.7
Jefferson City	—	91.4	Cheyenne	—	11.2

Source: Johnston 1982; for 2000, Council of State Governments 2002, 396.

The figures are impressive, but state capitals do not host all state employees. Diversity reigns, stemming from generally weak legislative provisions regarding the location of state employment. Washington State is an exception: all state offices not located in the capital were forced to relocate to Olympia in 1958 by order of the state's Supreme Court. Three categories can be created. The first includes hegemonic capitals, such as Phoenix, where 60,116 out of the 64,846 state jobs were located in 1999 (92.7%) or Honolulu, where 48,000 out of 54,832 state jobs were located in 2002 (87.5%—more than their already impressive share of the state's population). The second category includes capitals that have between one-quarter and one-half of the total number of state employees, such as Little Rock (50.6%), Charleston (34%), or Columbia (29%). In the third category are the capitals that control less than a quarter of state employees. This category naturally includes the smallest ones, such as Bismarck (2,963 out of 15,772, or 18.8%), Jefferson City (16,368 out of 91,425 in 2001, or 17.9%), and Pierre (2,143 out of 13,400 in 2003, or 16%). It also includes the capitals of densely populated states, where state jobs are often located mostly in the largest metropolitan areas; as a result, Tallahassee only hosts 24.6% of Florida's state employees, and Harrisburg hosts 14.7% of Pennsylvania's.[12] Naturally, the percentage of the workforce employed by the state (or, more generally, by government) is inversely proportional to the capital's population: Phoenix's workforce relies modestly on government, while government employees make up half of the workforce in Pierre, Frankfort, and Montpelier.

Such weight is sometimes resented, as it is in Cheyenne, where one-third of all jobs are linked to government—federal more than state in this case (the federal government still owns half the state). The main employer in 2011 was F. E. Warren air force base (3,820), followed by the state government (3,379), the Laramie County School District (2,157), federal government nonmilitary workers (1,804), and the Regional Medical Center (1,618). The company that founded the city, the Union Pacific Railroad (594), even lost the title of largest private employer to the Sierra Trading Post (595). The local Chamber of Commerce's statement is ambiguous: "While appreciating the stability of its government-based economy, Cheyenne's business community is actively working to diversify the economy and expand the economic base. The recent addition of several high-profile national businesses and industries is being hailed as a positive step toward that goal" (website of the Cheyenne Chamber of Commerce; text unchanged since the 2000s!). Among the new companies was EchoStar Communications, a satellite uplink center with 324 employees.

The presence of a large bureaucracy induced some analysts to compare state capitals to parasites living off their states' economies (Vedder 1997). Vedder based his assumptions on the unweighted mean per capita personal income in 1989 for counties with state capitals, which was 10% higher than in noncapital counties. For the richest states, the difference was less than 1%, while for the poorest ones, income was 17% higher in counties with capitals. He concluded that "redistributing income from the general population to public employees living in capital counties may have an impoverishing impact on the rest of the state." Table 7.6 shows the current situation at metropolitan and municipality levels.

Table 7.6 Gross product (GP) of states and capital metropolitan areas and municipalities, 2010

State	1. Gross product ($billion)	2. Capital metropolitan area	3. Gross product ($billion)	4. Percentage of state GP	5. Percentage of state population	6. Col. 4 as percentage of col. 5	7. Income (city)* (%)
California	1,701.9	Sacramento	92.9	5.46	5.77	94.6	82.4
Texas	1,149.9	Austin	86.0	7.48	6.83	109.5	123.2
New York	1,005.3	Albany	41.1	4.09	4.49	91.1	78.2
Florida	661.1	Tallahassee	13.4	2.03	1.95	104.1	88.1
Illinois	574.4	Springfield	9.7	1.69	1.64	103.0	101.0
Pennsylvania	479.1	Harrisburg	28.7	5.99	4.33	138.3	76.3
New Jersey	428.9	Trenton	26.7	6.23	4.17	149.4	54.1
Ohio	414.4	Columbus	93.4	22.54	15.92	141.6	97.4
North Carolina	378.1	Raleigh	57.3	15.15	11.86	127.7	123.7
Virginia	374.7	Richmond	64.3	17.16	15.73	109.1	84.8
Georgia	359.6	Atlanta	272.3	75.72	54.39	139.2	121.8
Massachusetts	341.2	Boston	313.7	91.94	69.53	132.2	90.0
Michigan	330.0	Lansing	19.6	5.94	4.69	126.7	80.9
Washington	305.0	Olympia	8.8	2.89	3.75	77.1	98.3
Maryland	262.0	Annapolis	n.a.	n.a.	n.a.	—	106.1
Minnesota	242.0	Minneapolis–St. Paul	199.6	82.48	61.84	133.4	87.1
Indiana	238.2	Indianapolis	105.2	44.16	27.09	163.0	106.9
Colorado	230.0	Denver	158.3	68.83	50.57	136.1	100.2
Tennessee	229.6	Nashville	80.9	35.24	25.05	140.7	113.5
Arizona	223.7	Phoenix	190.6	85.20	65.60	129.9	97.8
Wisconsin	219.2	Madison	35.6	16.24	10.00	162.4	110.5
Missouri	216.0	Jefferson City	6.0	2.78	2.50	111.2	106.7
Louisiana	204.8	Baton Rouge	39.4	19.24	17.70	108.7	109.5
Connecticut	197.5	Hartford	88.0	44.56	33.92	131.4	46.7
Oregon	177.8	Salem	12.5	7.03	10.20	68.9	91.4
Alabama	151.5	Montgomery	14.9	9.83	7.84	125.4	106.6
South Carolina	141.6	Columbia	32.0	22.60	17.24	131.1	100.3
Kentucky	140.5	Frankfort	n.a.	n.a.	1.63	—	113.4
Oklahoma	132.8	Oklahoma City	58.3	43.90	33.40	131.4	108.2
Iowa	126.2	Des Moines	39.5	31.30	18.70	167.4	98.9
Kansas	112.8	Topeka	9.4	8.33	8.20	101.6	95.4
Nevada	111.2	Carson City	2.8	2.52	2.05	122.9	95.2
Utah	106.2	Salt Lake City	66.5	62.62	40.67	154.0	114.1
Arkansas	91.2	Little Rock	33.0	36.18	24.00	150.7	137.3
Mississippi	84.9	Jackson	24.4	28.74	18.17	158.2	108.0
Nebraska	79.8	Lincoln	14.4	18.05	16.54	109.1	107.0
New Mexico	70.4	Santa Fe	6.9	9.80	7.00	140.0	147.5
Hawaii	58.1	Honolulu	51.3	88.30	70.07	126.0	112.4
Delaware	56.4	Dover	5.8	10.28	18.08	56.9	83.3
New Hampshire	55.7	Concord	n.a.	n.a.	11.12	—	92.2
West Virginia	53.4	Charleston	15.4	28.84	16.42	175.6	157.9
Idaho	51.2	Boise City	25.5	49.80	39.33	126.6	127.2
Maine	45.0	Augusta	n.a.	n.a.	9.20	—	98.0
Alaska	43.6	Juneau	n.a.	n.a.	4.40	—	117.9
Rhode Island	43.3	Providence	66.3	153.12	152.09	100.7	71.6
South Dakota	34.2	Pierre	n.a.	n.a.	2.45	—	116.5
Montana	32.0	Helena	n.a.	n.a.	7.56	—	116.7
Wyoming	32.0	Cheyenne	5.3	16.56	16.28	101.7	103.5
North Dakota	31.8	Bismarck	5.0	15.72	16.17	97.2	117.0
Vermont	22.9	Montpelier	n.a.	n.a.	n.a.	—	109.6

Source: US Bureau of Economic Analysis website (www.bea.gov); US Census Bureau, 2010.

Note: n.a. = figure not available.

* Column 7 corresponds to per capita income in capital municipalities as a percentage of their states' per capita income in 1999. For the precise figures, see appendix table A.5.

Six capitals have a share of their state's gross product that is lower than their share of the state's population.[13] This shows not only the comparative advantages of capitals over other cities in their states, but also that such advantages are not universal. Moreover, only twenty-six capitals have more than 10% of their state's gross product, underlining their comparatively low economic status in their states. Sixteen capitals have shares lower than 5%; fourteen have shares above 25%; and eight have more than 50%.

Vedder's conclusions were therefore too hasty: state capitals employ not only bureaucrats but also other highly paid managers in the private sector, especially in poor states where state capitals are often economic capitals. Vedder takes the example of the poorest state, Mississippi, where the capital county had an income level more than one-third higher than elsewhere in the state (in 2001, Jackson's MSA had a share of the state's gross product that was 139.2% of its share of the state's population). But he forgot that Jackson was a regional hub (railroads and aviation), home to twenty Fortune 500 companies, and home to thirty-five manufacturing firms. It also had facilities for eighteen international corporations and eleven hospitals, and had in 1999 a $508 million output from high-technology industries (e.g., telecommunications). The jobs these companies provide explain the difference in income between Jackson and Mississippi's other cities better than the 55,500 state jobs that Jackson controls, only some of which are located in the capital city. Table 7.6 further indicates that Jackson's per capita income is only 108% of Mississippi's. This argument is valid for most capitals.

Vedder also forgot that state governments have been accused of hindering the economic development of capital cities. States do not always pay taxes on property, for example, and their buildings cover large areas (e.g., 25% of the municipality for Harrisburg and 40% for Frankfort).[14] This is seen as a burden in Madison,[15] but the mayor of smaller Olympia dismissed it as "a minor matter." State employees tend to live in the suburbs, so we must distinguish between the central city and the suburbs. The capital city is only one part of the MSA or CSA, for the larger ones. Naturally, municipalities and the counties studied by Vedder have very different areas, so the bases of all these figures are not coherent. Both do not tell the same story, since twenty-one capital municipalities have lower per capita incomes than that of their states. Vedder should therefore have attacked suburbia. Moreover, capitals with a higher income than their state's are often not the richest cities in their states: for example, the figure for Lincoln, Nebraska, is 107%, but Omaha's is 110%; Oklahoma City's is 108.2%, but Tulsa's is 122%.

Would Vedder contend that the citizens of Omaha and Tulsa "have an impoverishing impact on the rest of the state"? In reality, capitals are not exempted from the woes of contemporary cities, and their downtowns often experience a marked "white flight" and significant commercial decline.

The worst situation was experienced by Hartford, Connecticut (Zielbauer 2002). Well-known as the national insurance capital, the municipality suffers from the second highest urban poverty rate in the nation, 30%, and the worst figure in table 7.6, with only 46.7% of Connecticut's per capita income, and only 84.7% of Mississippi's. Three major factors explain this situation: failing public and private leadership, disastrous racial divisions, and a home rule that favored affluent suburbs. Hartford lost more than a quarter of its factory jobs in the 1980s, and it lost twenty-five thousand insurance jobs in the 1990s.In addition, half of its land is untaxable park, church, and state property. The picture is not completely bleak, however; the city still offers about thirty thousand highly paid insurance jobs and "remains a city with a serious arts community that includes a ballet, a symphony and a world-class art museum. It is the capital of a wealthy state and the hub of a prosperous metropolitan area" (Zielbauer 2002). Smaller capitals have suffered similar problems. Baton Rouge and Hartford are resorting to similar measures to try to renew their failing downtowns. Inspired by the success stories of Baltimore and Boston, they intend to offer new waterfronts (Smolski 1990) and "festival marketplaces." Hartford seems to be succeeding, with the help of state financing, followed by the return of some out-of-town finance companies to a prettified center. Baton Rouge's early experience shows, however, that the expected benefits are sometimes hard to attain. The Centroplex, an administrative and cultural complex located in the vicinity of the capitol, opened in 1977 but proved to be a financial pit. The MSA is too small and too close to New Orleans, which undermines Baton Rouge's attempts to attract conventions, tourists, and casinos. Its waterfront is far smaller than Boston's or Baltimore's, and the returns remain highly uncertain. Mr. Katz, urban planner at the Brookings Institution, was certainly right when he argued that "if that's all you do, you end up with a Potemkin village that people can come in and visit on the weekends, but that's it" (Zielbauer 2002).

Vedder's arguments and Hartford's situation underline two important factors about state capitals: the difficulty in finding scientific criteria to assess the exact role of state government in the growth of state capitals, and the dichotomy often present between capital municipalities and capital MSAs, for capitals are, in that respect, evolving like other cities.[16] Five studies give more precise answers to these questions. They all argue that since the 1950s, capital status has become an asset in the economic competition between cities, not by concentrating the state's wealth in the greedy hands of state employees, as Vedder contended, but by attracting private jobs and capital. The expansion of state bureaucracies "exerts a major influence on the spatial distribution of resources. . . . Because state capitals have been regarded historically as backward areas, growth patterns favoring capital cities signal a potential restructuring of the intercity hierarchy" (Carroll and Meyer 1982, 565).[17] From 1949 to 1974, state capitals fared better than comparable cities in terms of income evolution, transportation and communication resources, and numbers of employees in the advanced and service sectors of the economy. For Atlanta, Columbus, Denver, and Phoenix, the main factors were the following:

- The development of strong complexes of corporate activities and producer services that increased their nodal role.
- The relatively heavy dependence of Atlanta, Denver, and Phoenix on their retail and consumer services sectors.
- The low importance of manufacturing, except in Columbus.
- The fairly important role of government, except in Atlanta. (Stanback and Noyelle 1982)[18]

Government influences the economy by means of an "overlap in government and business networks of influence relations similar to that of multinational corporations and national government networks" (Ross 1987, 258).[19] This is illustrated in the new geography of large law firms, for state capitals have become more prominent control centers (Lynch and Meyer 1992). However, expansion of state government was not in itself sufficient for a capital to become a large law center; it also had to be a nodal and financial center. Not surprisingly, the largest law centers are Boston, Atlanta, and Minneapolis, followed by Denver, Phoenix, Salt Lake City, and Hartford. Capitals gained from the contemporary building of branch networks, centered on Washington, DC, the position of which was reinforced by increased governmental regulations. Intraregional branching is focused on state capitals—from New York to Albany and Trenton; from Washington and Philadelphia to Harrisburg; from San Francisco to Sacramento; and from Detroit to Lansing. This movement was initiated in the 1950s and served as an advance signal of the transformation of the system of cities.

These findings were sustained by a later study of five capitals (Bromley 1990).[20] Although they had very different locations and evolutions, all experienced a dramatic change in their economic positions after since the 1960s, from quiet administrative towns to technopoles, that is, "metropolitan center[s] heavily oriented toward professional services, research and development, administration, education, health care, banking and finance" (3). Their assets, compared to those of other cities, were the following:

- More highly educated workforces
- Higher proportions of senior decision-makers
- Clean, "white-collar" image
- Greater economic stability and lower unemployment
- High level of legal activity
- Presence of liaison offices of major companies and institutions; thus, proximity to power
- Important tourist and educational flow; cultural and convention facilities
- Special transportation facilities
- Special business opportunities in lobbying, consultancy, news, computers, offices, environment, design and preservation, in a context of improving local business leadership

Their main drawback was that the private sector had to compete for employees with the public sector, which could offer better benefits. The balance is thus largely in favor of capitals. The main benefits concern image-building, business climate, state and city government policies, workforce, infrastructure, culture, education, and community life. Tallahassee, for instance, set up a lobbying task force in 1986 that united the Chamber of Commerce, the city, Leon County, Florida State University, Florida Agricultural and Mechanical University, and the Leon County School Board (Bettinger and King 1995). Sacramento established annual "capital to capitol" lobbying periods, both at Sacramento and Washington. Concerning infrastructures, the most important ones are the airports: Sacramento has developed a major new airport; Austin built its new one around the closed Bergstrom air force base; and Tallahassee improved its airport facilities with the help of state funds (Bromley 1990, 18).[21]

All five studies underline the transformations at work in state capitals. Indeed, with the more or less active support of state governments, most capitals underwent dramatic changes in the last generation. But we cannot generalize these studies' conclusions, because the first two only analyzed the pre-1980 period, the last two focused on very specific groups of capitals, and none studied the small capitals. An analysis of Oklahoma City revealed, for instance, that the Oklahoma City Chamber of Commerce, reorganized and rejuvenated after the 1980s oil crisis, was behind most economic development activities and that "the public development role [was] nearly non-existent" (Clarke 1995, 526).[22] We must therefore enlarge the analysis to include all state capitals to see whether they participated in the general process of economic modernization that the nation experienced since the 1970s.

The Age of Air, FIRE (Finance, Insurance, and Real Estate), and High Technology

> Given its existing functions as a wholesaling center and its development of service institutions in the 1920s, Nashville appears to have been well prepared for the postindustrial revolution that has characterized the American economy since 1930. The "new industries" of corporate services (banking, insurance, law), government bureaucracies, higher education, health care, communications, and tourism all had strong foundations in Nashville.
>
> (CHUDACOFF 1981, 12)

Does the preceding statement work for most capitals, further proving their current process of normalization? Factors of growth having changed since the industrial and railroad period, capitals seem better suited than other cities to take the postindustrial train, since most of them were not hindered by Rustbelt liabilities. Three factors are particularly representative: transportation, financial control, and high technology.

Transportation is still very important, but the means of transportation have changed. Whereas railroads were crucial until the 1950s, their importance is dwindling, except for heavy goods. Pierre and Salt Lake City offer examples of that trend. Although a railroad company made Pierre the capital of South Dakota, the city's mayor wrote in 2003 that "the two major issues that are on my desk at this time are the railroad bypass and air service."[23] To attract new businesses and people, Pierre is promoting an old means of transportation, the Lewis and Clark Trail.[24] Salt Lake City's mayor said in his "*State of the City*" address (January 14, 2003) that he wanted to reduce or even eliminate rail traffic on 900 South, near downtown Salt Lake City. The city has sued Union Pacific, but to no avail. Today's railroad tracks are bringing more problems than prosperity,[25] but a good airport is seen as the key to success. State capitals use airports more than railroads, but not much more, since, aside from Atlanta, Denver, and Phoenix, capitals' airports are not among the ten largest airports in the nation. The airport hierarchy nevertheless differs from the demographic one. For instance, Atlanta, the tenth largest national metropolitan area, has the largest single airport in the nation and is the fourth most important air hub after New York, Chicago, and Los Angeles. Salt Lake City has the twenty-fourth busiest airport, while its MSA is twenty-eighth, because of its role as a regional hub. The cause of the discrepancy in rankings is the same as the cause of the low influence of some capitals: the proximity of major metropolitan areas and airports (Harrisburg is close to Philadelphia; Sacramento, to San Francisco; Providence and Hartford, to Boston; and Boston, to New York). Some capitals are too close to have a "proper" airport, as is true for Olympia (near Seattle) and Santa Fe, which is only sixty-two miles from larger Albuquerque. Isolated capitals, on the other hand, have rather large airports compared to their size, such as Boise City, Des Moines, Tallahassee, and Honolulu. Other airports have less than 0.5 million passengers, such as Lincoln's (around 0.25 million), Bismarck's (0.2 million), and Springfield's (only 148,000 passengers; it is close to Chicago, St. Louis, and Indianapolis). Table 7.7 shows the numbers of passenger traveling through the nation's major airports that are located in capital cities. Capitals thus show mixed results concerning the "age of air."

When it comes to the control of the new economy, in 1972 (i.e., at the beginning of the movement), only sixteen state capitals were listed as major locations for company headquarters in metropolitan areas. With regard to hosting the top one thousand manufacturing firms and the top fifty firms in other sectors, Minneapolis was 7th, Boston 9th, Hartford 15th, Atlanta 17th, Richmond 21st, Denver 26th, Phoenix 28th, Columbus and Nashville tied for 33rd, Indianapolis and Springfield tied for 36th, Providence 38th, Oklahoma City 39th, Honolulu 44th, and Albany and Salt Lake City tied for 50th (Johnston 1982, 131–32). But the fortunes of the largest state capitals varied with regard to Fortune 500 industrial headquarters between 1950 and 1980: Atlanta was included in the list in 1970, Minneapolis added two more headquarters between 1950 and 1980, but Boston and Denver lost some of those they had, and Hartford had lost in 1980 the two it had in 1970 (table 7.8).

To understand the high-tech turn, we must acknowledge the prodigious role of World War II and then the Korean, Vietnam, and Cold Wars—the well-known military-industrial complex—in the growth of capitals, along with the more general Sunbelt renewal. Some capitals were bypassed for strategic reasons by this tremendous process of industrialization. Olympia, Washington gave way to Portland, Oregon, and Richmond was outdone by

Table 7.7 Traffic (million passengers) at major US airports in capital cities, 2002–11

	2002	2011	Population rank in 2000
Atlanta	76.9	92.4	11
Denver	35.6	52.8	19
Phoenix	35.5	40.6	14
Minneapolis–St. Paul	32.6	33.1	15
Boston	23.0	28.9	7
Honolulu	19.8	17.9	56
Salt Lake City	18.6	20.4	36
Raleigh-Durham	8.5	9.2	41
Nashville	8.0	9.6	39
Sacramento	8.0	8.7	25
Indianapolis	7.4	7.4	29
Austin	6.7	9.1	38
Columbus	6.7	6.4	33
Hartford	6.5	5.7	42
Providence	5.5 (2001)	4.0	40
Oklahoma City	3.2	3.6	49
Boise City	2.8	2.9	97
Little Rock	2.6	2.2	74
Richmond	2.5	3.2	51
Des Moines	1.8	1.9	91
Albany	1.5	2.5	57
Madison	1.5	1.5	98
Harrisburg	1.2	1.3	67
Jackson	1.2	1.3	96
Tallahassee	1.1	0.6	c.135
Baton Rouge	c. 0.8	0.8	70
Columbia	0.5	1.0	80

Source: Airport websites; FAA.

Table 7.8 Fortune 500 industrials for seven state capitals, 1950–80

SMSA	Classification	1950	1960	1970	1980
Boston	RN	4	5	4	3
Atlanta	RN	0	0	3	3
Minneapolis	RN	6	5	7	8
Denver	RN	4	3	1	2
Phoenix	RN	0	0	1	2
Salt Lake City	SN	0	0	0	0
Hartford	FN	0	1	2	0

Source: Lynch and Meyer 1992, 49.
Note: RN = regional node; SN = subregional node; FN = functional node.
SMSA : Standard Metropolitan Statistical Area (an earlier version of MSA, now discontinued).

Norfolk-Hampton Roads in Virginia because both Portland and Norfolk-Hampton Roads were far better placed to host huge naval works and navy yards. But many capitals profited from large army bases, partly due to the presence of the legislature: for example, Tinker Airbase in Oklahoma City (Crowder 1992) and the Curtiss-Wright aircraft plant in Columbus, Ohio (with up to twenty-five thousand workers). These bases drew manufacturing plants that fostered in turn the development of manufacturing in the region. The air and aerospace industries used high-technology components and therefore promoted high technology in construction as well as in research and development.

After a generation of changes, it is possible to draw some conclusions on the place of state capitals in the new economic order. Table 7.9 presents three indicators qualifying for the adjective *modern*: financial activities that indicate the level of control over the economy, professional and business services that emphasize the global environment of economic activity, and high-technology output and rankings that show the level of modernization in manufacture.

The results are once more mixed. Indeed, twenty-three state capitals appear in the hundred most high-tech MSAs in the country. Nine more appear in the second hundred, but twenty-seven belong to the hundred largest MSAs, and only seventeen (one-third) appear in MSAs with the highest share of high-tech in Gross Metropolitan Product. Moreover, only two state capitals belong among the fifteen largest North American techno-scientific metropolitan areas; Boston and Atlanta, respectively ranked fourth and twelfth (Alvergne and Latouche 2003).[26] Financial activities remain modest in most capitals (e.g., Madison pales compared to Milwaukee), except for the largest ones and notable exceptions such as Hartford and Des Moines, which host large insurance sectors. The same is true for professional and business services. Even recently praised "new economy capitals" are generally lagging: in North Carolina, Raleigh is second to Charlotte (the second national banking center by assets) and partially subordinate to Greensboro; in Texas, Austin is far behind Dallas, even in high technology. Table 7.9 nevertheless emphasizes the economic catch-up processes at work, since about half the state capitals participate in the upper levels of the new economy—more than during the previous period.

Two short case studies, at two different scales, will help us to understand table 7.9 better. The first deals with a metropolis at the state level, Austin, Texas, and the second with Salem, Oregon, a medium-sized capital gaining from the momentum of a metropolitan area.

Austin, Texas. Austin became an economic high-tech metropolis in the 1960s, after being an almost exclusively political town.[27] In 1860, although Texas had a population of 604,215, its first city, San Antonio had only about 8,000 people, followed by Galveston, with more than 7,000. Houston had about 5,000 inhabitants, and Austin, 3,500. Kenneth Wheeler described a

Table 7.9 State capitals' employment in modern activities

MSA	Total nonfarm wage and salary employment, July 2003 (thousands)	Financial activities (thousands)	Professional and business services (thousands)	High-tech output ($M, 1999)	Rank in nation	High-tech share of gross metropolitan product (%)	Rank in nation
Montgomery	162.0	10.6	17.2	346	206	—	—
Little Rock	313.3	19.8	39.8	1.704	94	10.1	98
Phoenix	1,576.6	130.6	259.2	18.610	11	18.4	35
Sacramento	744.5	53.9	88.8	13.277	21	22.8	21
Denver-Boulder	1,304.4	105.0	205.1	8.729	30	11.1	80
Hartford	598.7	72.4	60.6	8.660	31	14.6	45
Dover	54.9	2.5	4.1 (L 6.7)	—	—	—	—
Tallahassee	155.0	7.1	18.0	896	132	—	—
Atlanta	2,207.8	146.1	386.3	14.691	16	10.0	100
Honolulu	417.1	22.5	56.3 (L 58.9)	869	134	—	—
Boise City	226.7	12.2	32.7	4.489	46	36.0	7
Springfield	111.8	8.2	10.3	651	155	—	—
Indianapolis	853.1	62.3	100.3	4.020	56	—	—
Des Moines	283.4	44.6	30.0	985	123	—	—
Baton Rouge	301.0	16.3	35.9	1.271	109	—	—
Boston	1,940.6	171.2	315.0	44.401	1	20.6	29
Lansing	236.2	16.1	20.9	549	168	—	—
St. Paul–Minneapolis	1,703.1	136.2	239.8	14.859	15	13.4	55
Jackson	232.7	16.6	25.9	508	175	—	—
Lincoln	156.6	11.3	16.7	1.170	112	13.4	56
Trenton	225.2	17.7	30.6	2.742	70	16.5	41
Santa Fe	80.2	3.4	8.7 (L 10.5)	388	193	—	—
Albuquerque	366.5	19.2	59.0 (L 36.4)	5.209	42	22.6	22
Albany	456.6	26.1	52.3	4.244	51	12.2	66
Raleigh	681.4	33.1	99.6	13.433	18	33.6	10
Charlotte	820.5	68.5	116.5	4.159	54		
Greensboro	625.8	37.3	72.6	2.494	76		
Columbus	874.7	74.7	125.1	4.800	44	—	—
Oklahoma City	533.8	32.8	70.3	2.749	69	—	—
Salem	138.5	6.9	12.0	3.491	63	38.9	4
Harrisburg	370.8	26.1	35.7	1.687	95	—	—
Providence	524.0	34.5	53.5	2.444	78	—	—
Nashville	673.9	43.2	95.3	2.086	84	—	—
Memphis	577.9	31.3	74.0	1.631	97		
Austin	660.3	38.2	87.8	15.401	14	36.2	6
Dallas	1,914.8	167.6	274.4	27.427	6	19.1	33
Salt Lake City-Ogden	703.4	50.9	94.0	5.476	40	13.0	66
Barre-Montpelier	34.7	2.9	2.1 (gov. 8.8)	—	—	—	—
Burlington	107.6	5.6	10.5	2.445	77	38.6	5
Richmond-Petersburg	572.5	45.8	88.5	3.277	64	—	—
Madison	73.1	3.8	6.0	1.457	101	—	—
Milwaukee	828.7	60.5	102.9	4.450	47	—	—
Cheyenne	c. 43.0	?	?	—	—	—	—
Casper	34.1	2.0	3.2	—	—	—	—

Sources: US Bureau of Labor website; United Conference of Mayors and National Association of Counties website, 2001.
Note: L = leisure and hospitality. For some states, a comparison has been made with other cities.

political capital happy to stay so; its merchants were seeking no regional trade, contrary to the city founders' wish to create a great metropolis (Wheeler 1968, 165). The "official" description of 1940 was still focused on the political side of a quiet city:

> Although a commercial city of importance and the leading educational center in Texas, Austin's life revolves around the capitol, whose massive red dome dominates the physical scene. The course of the city's business runs close and confluently with the business of the State, and the speech of the man on the street is flavored strongly with reference to its affairs. Befittingly, Austin wears a mantle of dignity. It is a stately city, with broad tree-lined avenues and boulevards, and imposing public edifices set in attractive grounds; a city of institutions, its lines everywhere sobered and beautified by the design of its schools, churches, and State buildings. It is a tranquil city, with an air of serenity, decorum, and permanence, that dwarfs the temporary turbulences of its political life. (WWP Texas 1940, 167)

Austin had experienced its first industrial growth by 1940, linked to the building of dams and power plants on the Colorado River by the Lower Colorado River Authority, the aim of which was to harness the river and bring manufacturing to the capital. It worked: Austin had more than a hundred manufacturing plants in 1940, the largest of them linked to furniture and agriculture. The improvement of railroad and highway facilities increased its importance as a trade center. This growth was aided by the foundation first of St. Edward's University in 1878 and of the University of Texas in 1883, enabling Austin to become the state's educational center (with the great help of the oil discovered on the two million acres of public domain land set aside for the university's support).

Despite these assets, Austin's MSA had fewer than 200,000 inhabitants in 1960. In 1980 it had 585,000, and 1,716,289 in 2010, exceeding all forecasts. Manufacturing played a role, but in 1995 it offered only 68,400 jobs, 13.2% of nonfarm employment. Major firms were producing semiconductors (half for Motorola) and computers (IBM and Dell). Services represented 27% of employment (25% of which was in high technology). Administration was important: there were 55,000 federal jobs, and 17,000 were offered by the University of Texas and Texas A&M University. Ninety-three percent of its 1.8 million square meters of office were occupied. Airport traffic grew from 1.78 million passengers in 1980 to 5.34 million in 1995, and freight transportation from 53,400 tons in 1988 to 156,300 tons in 1995.

The university did more than offer jobs. In a 1957 study it recommended developing light industry in order to add a third sector to the local economy (besides government and academics). A plan was made in 1957, just when the Texas Instruments Company was founded in Dallas. In 1963, IBM came to Austin; Motorola came in 1974, and others followed (Dell is today the most famous), transforming Austin into a technopolis (nicknamed Silicon Hills) ranked fourteenth in the nation, with 36.2% of its gross metropolitan product (GMP) in high-tech industries (seventh rank) in 1999.

Austin only was a "far-away post" for the big high-tech firms until 1983, when Texas's economic crisis prompted the city to court all firms offering jobs linked to R&D, in order to transform itself into a "high-tech Mecca." Through "major government-business-university collaboration" (Bromley 1990, 15),[28] Austin succeeded in attracting in 1982 the federal project managed by the Microelectronics and Computer Technology Corporation, which employed 150 people in 1996. Twenty-two companies cooperate with the MCC to defy foreign competition in information and computer technology. Five years later, Sematech was launched in Austin to revive American leadership in semiconductors. With MCC's arrival, Austin's Chamber of Commerce ordered a new long-term economic study (the first since 1957) from SRI International, a California company whose optimistic report, *Creating an Opportunity Economy*, became the basis of Austin's economic politics. It stated that "the city possesses all necessary elements to construct an advanced information and knowledge economy and [is] able to offer a chance to every inhabitant in a unique and charming environment." Five sectors were put forward, the being first science and technology (R&D). Austin was not alone is this growth process, belonging to the "Austin-San Antonio growth corridor," which was eighty-two miles long (Davies 1986, 472). However, such mushroom growth did not proceed without a darker side. In 1995, one-quarter of students, mostly belonging to minorities, left high school without diplomas, and "in the early 1990s Austin was still seeking to balance the economic development it had long sought with the kind of life it had long treasured" (Texas State Historical Association website, http://www.tshaonline.org/handbook/online). The growth of the capital of Texas naturally owed much to its lenient regulations and mostly nonunion workforce, as was the case in Phoenix (Konig 1982, 1984), but its development was largely helped and controlled by government, through the efforts of all its components: federal, state, municipal, and academic.

Salem, Oregon. When situated near a metropolitan area, a capital can easily take part in its growth process. This is the case for Salem, Oregon. The arguments stated by its Chamber of Commerce's website are clear: "Because of its proximity to Portland, its outstanding outdoor recreation, and the addition of high-technology industries, Salem is in the midst of sustained, steady growth." The chamber adds to Salem's advantages a family-friendly ambience and an easy commute to the Portland metropolitan area. A recent study by Cleveland State University ranked Salem number one in manufacturing growth in an "exurban area," for the reasons stated above, plus excellent transportation, availability of land, quality of the workforce, an excellent education system (with Willamette University), and economic incentives. The role of government is not mentioned. Another Oregon city, Eugene, experienced similar growth during the recent period, without government, but with the University of Oregon (since 1876). Salem's largest employer, besides state government, is the food product industry, followed by Salem Hospital and the Spirit Mountain Casino. Willamette University is rather small. Agribusiness has nothing to do with legislatures (except perhaps through pork-barrel policies), and high-tech growth came from Portland, making Oregon's capital the fourth most technological city in the nation (in relative terms), with 38.9% of its workforce employed in that sector in 1999. One of the reasons for growth in the areas of high technology and FIRE is the presence of a university. The specific relationships between capitals and universities partly explain the capitals' mixed results concerning high technology.

Higher Education and Capitals: A Late Marriage

For any would-be metropolis, the presence of a good university is a real asset. But, although "emerging business-government-university linkages in capital regions"[29] have been witnessed in some capitals, state capitals are not the best-placed cities in the American educational system. The compromise system of capital selection often distinguished capital city and university town (e.g., Harvard, founded in 1636 outside Boston). This system did not correspond to the Jeffersonian ideas that based democracy on education as well as on private property, with the aim of creating a Republic of Letters. In fact, the Northwest Ordinance (1787) stated that "religion, morality, and knowledge being necessary to good government and the happiness of mankind, schools and the means of education shall forever be encouraged." But Congress did not provide any substantive support for education, and most new territories did little to fulfill that pious wish. New states had to wait for the 1862 Morrill Act as well as for the 1890 Land Grant Act, which resulted in the establishment of eighteen universities (e.g., in Frankfort, Kentucky).

There are exceptions: Yale was created in New Haven in 1701, and Brown was transferred to Providence in 1770. Vincennes University (Indiana) was established in 1806, because without it "our excellent government . . . is liable to be assailed by the various arts of cunning and intrigue, of designing, ambitious, and desperate Individuals" (Cayton 1996, 224). Tuscaloosa, Alabama's "permanent" capital from 1826 to 1845, was selected on December 29, 1827, by the general assembly as the location of the state university. In the case of South Carolina, the location in 1801 of South Carolina College (now the University of South Carolina) at Columbia originated in the political will to unify the state, which was sharply divided between antagonistic regions along "ethnic," economic, and political lines. The removal of the capital from Charleston to Columbia in 1796 had the same purpose (WWP South Carolina 1941, 92).

Faced with the lack of higher education facilities, capitals often created a university or received a branch of the state university in the twentieth century. Oklahoma City had to wait until 1910 for a branch of the University of Oklahoma and until 1961 for a branch of Oklahoma State University, and it is still ranked only fourth in the state for higher education. Students can exert an important influence, especially in smaller capitals, where they add population and lower the average age. Enrollment figures were 48,906 (fall 2012) for Michigan State University in Lansing (42.8% of the city's population—and 62% if Lansing Community College, with an enrollment of 22,000, is included); 30,721 (fall 2011) at the University of South Carolina in Columbia (23.8%); 42,595 (fall 2010) for Wisconsin State University in Madison (18.3%); and 24,593 (fall 2011) for Nebraska State University in Lincoln (9.5%).

But, as a result of belated university foundations among already existing institutions, state capitals seldom host the "best" universities, except for Boston, which has two of the most prestigious American universities: Harvard and the Massachusetts Institute of Technology. Table 7.10 shows the results of two rankings of universities (best universities and best research universities).[30]

Only seven capitals are listed in the first ranking (Boston having three

Table 7.10 State capital universities that are among the fifty best American universities, 2010

Best universities	Best research universities
1. Harvard	2. MIT
5. MIT	5. Harvard
10. Duke (Raleigh MSA)	7. Duke
15. Brown (Providence)	12. University of Wisconsin–Madison
17. Vanderbilt (Nashville)	14. University of Minnesota–Twin Cities
20. Emory (Atlanta)	15. University of Texas–Austin
31. Boston College	24. Ohio State University–Columbus
36. Georgia Tech (Atlanta)	25. Vanderbilt (Nashville)
42. University of Wisconsin, Madison	39. Georgia Tech (Atlanta)
45. University of Texas, Austin	46. Arizona State University
	47. Brown (Providence)
	50. Boston University

Source: http://colleges.usnews.rankingsandreviews.com/best-colleges/rankings/national-universities; and http://mup.asu.edu/research.html (accessed July 11, 2012).

universities and Atlanta two), and ten in the second (Boston having three universities). Moreover, except for Boston and Raleigh, their rankings are not very high. Larger capitals have nevertheless experienced significant growth in their enrollments since the 1960s. To take only the best-ranked ones (fall 2012), Arizona State University in Tempe and Phoenix had an enrollment of about 60,000 students; Ohio State University in Columbus, 56,000; the University of Texas in Austin (founded 1883), 52,000 (it is the best Texan university); the University of Minnesota in the Twin Cities, 51,800; Indiana University (1916)/Purdue University in Indianapolis, 30,500; the University of North Carolina in Chapel Hill (founded 1789), 29,400; Emory University (1836), 13,400; and Georgia Tech (1885) in Atlanta, 20,500. Prestigious—and private—ones have smaller enrollments, such as Duke University (founded 1926, with about 14,700 students) in Durham, Vanderbilt University in Nashville (founded 1873, 12,860), and Brown University in Providence (founded 1764, 8,454).

It is clear that by entering the "normal" path of economic evolution, capitals also had to experience the darker sides of these processes. Although capitals profited from the so-called new economy, they are suffering from the current sharp crisis. State capitals are not immune to economic crises, a further indication of their "economic normalcy." Indeed, even if government, sometimes considered a backward area during boom-time, is positively reassessed during hard times because it plays the role of a social safety net, the dire straits of most state budgets also meant public job cuts. If capitals have more "normal" economies, what about other aspects of success—image and tourism? The study of the current image of state capitals sheds new light on their record.

A Revivified Image

The image of the state to the outside world is especially conditioned by the impressions gained by distinguished visitors to the capital city.

(GOVERNOR BLANCHARD OF MICHIGAN, QUOTED IN BROMLEY 1990, 15)

One way to assess the image of capitals in contemporary America is to look at major television programs.[31] Apart from their political news (see chapter 2), state capitals do not stand out as television magnets. Popular programs, such as *E.R.*, *Dallas*, *Friends*, or *Desperate Housewives*, are set in economic metropolises. A competitor to *Dallas* was indeed set in Denver, but its name was *Dynasty*—although the company portrayed was called Denver-Carrington. The only popular television series actually set and shot in a state capital is *Hawaii Five-o*, but it is sponsored by the Hawaii Tourist Department and is intended to draw more tourists to the entire archipelago and not specifically to Honolulu. *Profiler* was set in Atlanta only because of the presence of the regional FBI headquarters. Likewise, *Boston Public*, set in a high school, *The Practice* (which had the same producer), and *Fringe* use Boston as a metropolis and not as a state capital. *Providence* was really about the city—with beautiful aerial views in the credits—but the heroine, a doctor, lives in a better-off suburb, East Providence, and the shooting was mostly done elsewhere. Finally, the town where the characters in *The Simpsons* live, Springfield, does not apply, because the state indicated on the cartoon's car-plates, NT, does not exist (except in Australia, for Northern Territory). Springfield is said to have been chosen because it is one of the most common town names in America.[32]

Another—perhaps more scientific—way to look at the image state capitals have in the American mind is to study them as tourist magnets (table 7.11). Although the situation is improving, much remains to be done to make them a part of the growing national and international tourist industry.[33]

The United States is ranked first in the world for tourism (8.7% of the GDP in 2011). Its tourism, mostly domestic (81% of total expenses), comprises business trips (29% of spending) and pleasure trips (mostly short trips). Heritage and cultural tourism remain modest with regard to the country's major attractions—natural beauties, metropolises, and gambling cities. Indeed, inner cities, with their well-known problems, are not the first places chosen by Americans searching for a place to travel. The federal state instituted National Monuments long before creating urban historic

Table 7.11 Tourism GMP of metropolitan capitals in 2002

Pop. rank	Capital (MSA)	Tourism GMP ($million)	Rank	Tourism share of GMP (%)	Rank	Employment in tourism
11	Atlanta	9,770	5	6.7	7	156,361
7	Boston	6,302	10	3.0	51	100,977
14	Phoenix-Mesa	5,768	13	6.0	10	102,303
15	Minneapolis-St. Paul	5,113	14	4.8	18	87,363
19	Denver	3,729	20	4.4	23	57,238
56	Honolulu	3,670	21	13.9	2	57,238
28	Indianapolis	2,888	28	5.0	16	51,020
35	Salt Lake City	2,629	32	5.5	13	45,656
38	Nashville	1,992	40	4.3	24	37,750
40	Raleigh	1,492	46	3.4	44	26,198
24	Sacramento	1,503	45	2.3	66	20,358
37	Austin	1,476	47	3.1	48	24,546
32	Columbus	1,341	52	2.3	67	22,838
74	Little Rock	884	62	4.5	20	18,506
41	Hartford	834	63	1.4	86	10,190
50	Richmond	812	64	1.9	74	12,628
57	Albany	746	66	2.0	71	10,509
67	Harrisburg	699	68	2.7	58	10,963
80	Columbia	527	74	2.9	55	9,894
39	Providence	491	76	1.4	85	7,569
91	Des Moines	465	78	2.6	60	8,404
97	Boise City	440	79	2.9	56	7,808
98	Madison	394	84	2.2	69	7,855
93	Lansing	300	90	1.8	76	5,681
70	Baton Rouge	275	94	1.5	82	5,226

Source: DRI-WEFA 2002: *The Role of Travel and Tourism in America's Top 100 Metropolitan Areas.* Consulted online at http://www.usmayors.com.

districts (1906 for the former and 1966 for the latter). Nevertheless, the country has renewed interest in the value of its heritage since the 1960s. State capitals are firmly anchored in the American identity. Their capitols symbolize the states, and their historical museums display their history to the pupils and adults who come to investigate the past of their home states. Logically, they should attract tourists, as do European capitals, which offer a rich and well-preserved heritage. In fact, they are often left aside by tour operators and foreigners. Only thirteen capitals out of the twenty-five in table 7.11 are among the fifty largest tourist cities in the nation and have a higher tourist GMP ranking than their demographic ranking. Although they have become destinations for the pleasure trips of Americans, their development of tourist attractions linked to their political status remains low, and the appeal of capitals is mostly based on other assets. This is partly because they have long had an appalling image. Tourism specifically aimed at capitals is therefore based upon student visits to the capitol and upon business trips by legislators, journalists, and lobbyists. The population of Montpelier, the capital of Vermont, for instance, grows from eight thousand to fifteen thousand during legislative sessions, bringing important economic returns.

The most important French tourist guide, the *Guide Bleu*, grants stars to destinations: one star means a place is worth the detour; three stars mean it is worth the journey (Collective authors 1988, 1989). Twenty US capitals (40%) receive no star, and fifteen of them (30%) lack a specific article. Apart from metropolitan capitals, only Richmond, former capital of the Confederate South and rich in colonial houses, obtains two stars. The same lack of stars occurs on the web.[34] Most capitals are not even listed, and others are mentioned only for attractions that have nothing to do with their history or their monuments (rodeo for Cheyenne; auto racing for Indianapolis). Since eleven capitals host fewer than fifty thousand people, and only seventeen are the metropolises of their states, marvelous skyscrapers are seldom seen.

The overall image revealed by the *Guide Bleu* is one of peace and natural beauty. Likewise, *The Rough Guide to New England* appreciates Vermont's capital for its "low tourist profile . . . that makes Montpelier a refreshing counterpoint to the cultivated rural quaintness that pervades the rest of the state" (Hull and Keeling 1999, 310). But quiescence does not suffice to render these cities interesting. According to the *Guide Bleu*, "Jackson does not deserve a full visit; but one must absolutely see Vicksburg, situated less than 60 miles away" (Collective authors 1988, 457). For Hartford, Connecticut, the *Rough Guide* proves faithful to its name: "The only exception to the slow picturesqueness is Hartford, the state capital, an unattractive and largely dull city" (Hull and Keeling 1999, 273).

Political changes since the 1960s have improved the image of capitals, although much more remains to be done. The image of state capitals oscillates between praise for their quality of life and their human scale, and reproach for lack of animation. Indeed, although they are outstanding as historic cities, their tourist offerings are often limited. Heritage protection appeared rather late in the United States, after the 1960s, and local people saved only individual buildings, mostly located in the South, where there was appreciation of its romantic past.[35] The bicentennial of US independence in 1976 helped Americans to realize that their past was not restricted to the War of Independence, the Civil War, and the nineteenth-century conquest of the West. Capitals have not been at the forefront of this movement, either because they feared that their economic development would be frozen by heritage protection or because they were happy with their status as administrative cities. Heritage protection thus often came in extremis, following massive urban renewal in larger capitals, such as Denver, as well as in smaller ones, such as Helena, where the past has mostly been recreated through the rehabilitation of their main streets (a federal program in place since 1980), which is linked to the nostalgia for nineteenth-century Main Streets, the ones of lost communities. Capitals also designated historic districts,[36] but these abound in the United States. There are 13,500 of them, plus the 85,000 places listed at local and state levels, and the majority are not even tourist destinations (Hamer 1998).[37] Capitols are the only original tourist attractions of state capitals, although they are seldom the highest buildings in town, as they once were. Local pride in capitols reflects pride in the workings of local democracy, and capitols nurture a memorial heritage of the many famous men (and seldom women) who worked and declaimed in them. But the valorization of their presence proves delicate. First, they rarely had the notoriety of future President Lincoln, who spent twenty-six years in Springfield, Illinois.[38] Second, their traces are often immaterial: a long disappeared eloquence, a hotel room, a house, a desk (often sold and transformed). One must therefore reconstitute and evoke—processes that do not easily attract crowds.[39] To make matters worse, most capitals undervalue the polity. Size does not really matter: one would have thought that the smaller capitals would boast of their capital status, though the larger ones would have a more common metropolitan discourse, but this is not really the case. Indeed, larger capitals seldom mention their capital status in their tourist leaflets or on their websites. Atlanta's brochure, *50 Things to Do*, does not even mention the capitol among the historic buildings and districts it highlights. Some small cities are also almost mute on the subject, such as Carson City, Nevada, which rather dwells on its "long" history (it was founded in 1858) in a promotional leaflet. "The Old West reappears around every corner. Ghosts of explorers, mountain men, miners, settlers. . . . Challenging the Frontier and inviting you to do the same. On horseback, in a hot air balloon, at a casino, on the golf course, or the ski slopes."[40] With a population of 55,274, Carson City offers 1,700 rooms to tourists. A pedestrian trail, marked by a blue line, allows the visitor to see about thirty historic houses built from 1859 to 1914, twenty-four of which are "talking" (they tell their stories via podcasts), as well as official buildings, among which is the capitol. Carson City benefits from a very rich history (Zauner 1984).

This partial indifference is nowhere more flagrant than in Oklahoma City, which shows that most capitals first wanted to use their status to become economic metropolises (Robertson 1996). The tall derricks (123 feet) of the fifteen oil wells located since 1942 less than three hundred feet from its south entrance overshadow the capitol. They are no longer in use, but still stand as reminders and advertisements of the state's petroleum industry.

Hagiographic commentaries on capitals appear only in the capitol brochures. Rhode Island's capitol was praised on the first page of the 1998 visitor's leaflet: "It is built of white Georgia marble and is one of the four self-supporting marble-covered domes in the world, the largest being the dome of St. Peter's in Vatican City, followed by the dome of the Minnesota State Capitol in St. Paul and the Taj Mahal in Agra, India." Not bad company, is it?

Aside from capitols, for which hyperbole is common, capitals do not really dwell on their capital status for tourist purposes. Are capitals not tourist cities? Guides seem to concur for Augusta, Maine, about which the *Rough Guide* writes that "you can take a self-guided tour, though unless you have a particular interest in politics or architecture, this should kill no more than twenty minutes of your time" (Hull and Keeling 1999, 449). Events celebrating their heritage remain rare. The only events regularly staged at capitols are Fourth of July ceremonies and Christmas candlelight tours of capitols. One might add the moving exhibitions during the centenary year of a capitol. But boards, cheerleaders, and candles are not going to make capitals suffer from the syndrome of "museification-ossification" that some European historic cities are experiencing.

If capitals attract tourists, the reasons lie elsewhere. Honolulu draws tourists for its beaches, for the Pearl Harbor Navy Base, and for Hawaiian tropical landscapes. Only a few old structures were left after massive renewal programs, among which is 'Iolani Palace (1882), the only Royal palace in the United States that served as a capitol (1900–1969).[41] Likewise, Salt Lake City attracts people for its Mormon heritage and for the nearby ski resorts (it hosted the 2002 Winter Olympics). Nashville is a music Mecca. Montgomery's tourism draws upon its history as the birthplace of the Civil War and civil rights. Not all capitals profit from similar assets. Nebraska's major attractions are Omaha's zoo and the state parks, but not the capital, Lincoln, where the capitol attracts fewer visitors than the city's zoo or botanical gardens.

Likewise, one of the only examples of "museified" capitals, Santa Fe, owes little to its capital status and much to its ancient Native American culture and to its Spanish foundation (c. 1609).[42] The archetype of the cultural capital, Santa Fe has even created a new architectural style, "adobe and pueblo," for new structures built between listed buildings. The municipality has demolished many turn-of-the-century buildings.[43] Even in Santa Fe, however, the entire heritage has not been valorized: the country's oldest church and oldest house have not been restored. On the other hand, in Annapolis, restoration has been so thorough that the Historical Annapolis, Inc. association, founded in 1954, has been rechristened by its opponents Hysterical Annapolis (Brugger 1988, 614). Annapolis has fully profited from its poor economic record throughout the nineteenth and the first half of the twentieth century. The lack of economic growth allowed it to avoid demolishing beautiful old buildings to make room for new office buildings and condominiums. The city preserved fifty-five buildings from its eighteenth-century heritage, more than any other city in the nation except Williamsburg. Gentrification from nearby Washington and Baltimore and yachting also contributed to the renaissance of Maryland's capital (Potter 1994).[44]

Whereas European national and regional capitals generally rank among the cultural "giants," American state capitals are often left behind in artistic matters. For instance, Columbus, Ohio, had to wait until the 1950s to emerge as a small cultural center. Before that, sports were foremost for the city's inhabitants (Bureau of Business Research 1966, 36). Only a handful of state capitals appear in the rankings of American cities according to their cultural achievements. The best American symphony orchestras, for instance, are those of Cleveland, New York, Los Angeles, and Boston. Cincinnati and Minneapolis supersede Columbus or Indianapolis. With regard to museums, Atlanta's Museum of Art is certainly a beautiful contemporary building (by Richard Meier), but its collections—although well chosen—cannot match those of smaller Cleveland. Augusta's Maine State Museum (which boasts an "always free admission") is nothing more than an interesting display of local memorabilia ("12,000 Years in Maine," "Lumbering," "Made in Maine," "Quarrying," etc.). Albany's Museum (lodged in governor Rockefeller's grand plaza) only provides reproductions of generalities about New York State and City. To know Arizona's early history, one is better off visiting Sharlot Hall Museum in Prescott, a former capital, which houses Fort Misery (1863), the Governor's Mansion (1864), and Fremont House (1875), than the memorabilia displayed in Phoenix's former capitol (although it has recently been carefully restored to its former glory). The Phoenix (municipal) Pride Commission's 1999 survey listed as first and second "points of pride" the America West Arena and the Arizona Biltmore Resort (built 1929), although the city offers twenty-five historic districts.

Sport largely outweighs culture and the capitol was not mentioned. Tourism ranked second in Phoenix's economy (about 16 million yearly visitors). Places that attracted tourists were not the former capitol, but golf courses. With more than 190 golf courses, the city is one of the five major world destinations for golfers. The city also offers the country's largest botanical park dedicated to desert plants.

The modest role of the polity in tourism also concerns small capitals. Juneau, Alaska (31,275 inhabitants), welcomed in 2008 around 1.1 million visitors, of which 1 million came on cruise ships, making it one of the major destinations in the world. The main tourism claims of a city that describes itself as "one of the most beautiful cities in the world," are linked to landscape, culture, and shopping, but not to history (although the city has a historic district) or to politics. The website (www.juneaualaska.co) was very clear in 2003: "Often referred to as cosmopolitan for its size, Juneau carves its place in the world on a small strip of land between mountains and sea. A creator's paradise, the town is home to artists, actors, poets and writers who draw their inspiration from the untamed land and sometimes untamed people of Alaska."

I take Sacramento as the concluding instance of the under-valorization of the polity in a capital. More than two million people live in California's fourth metropolitan area. The capitol almost disappears behind new skyscrapers, although the city has tried to retain its majesty by passing a 1959 law restricting the height of nearby construction (Hansen 1967, 253). The capitol's valorization nevertheless remains largely internal. A brochure available inside the building states, "Fully restored to assure earthquake safety and to preserve California's rich history, the Capitol is a unique example of modern technology used to recreate the past, where today's Legislature carries on the democratic process in turn-of-the-century surroundings." The offices of major politicians (of 1902–6 and 1933) have been recreated on the first floor. The brochure ends with this phrase: "The Capitol is truly the most important building in the State of California. Even more important than the beauty of the building or the craftsmanship that went into its restoration is the work that goes on inside, which affects every one of us." The capitol is first perceived as a workplace, a functional place: tourism is secondary. It is thus not surprising that the city's major tourist attraction has nothing to do with its capital status. The Old Sacramento neighborhood, located between the Sacramento River and an elevated section of Interstate 5, is seen as "the premier resource for interpreting our brief western metropolitan history with a semblance of tactile authenticity" (Foster 1976, 21).[45] Although it offers three museums, the motto of its website, "Shop, dine, play," links it with the numerous shopping-cum-entertainment places that America has specialized in, especially on waterfronts. Sacramento remains secondary compared to San Francisco, Los Angeles, and the state's natural parks (e.g., Yosemite). While Sacramento ranked as the nation's twenty-fourth MSA, its GMP share of tourism was only at the sixty-sixth rank (table 7.11). Sacramento is indeed a capital, but it remains above all a business center (57.8% of the trips are business-linked), despite its nickname of "Disneyland North."[46]

State Capitals in the National Urban Hierarchy

In 1982, Carroll and Meyer concluded their study of the evolution of state capitals with a strong statement: "Thus, we are describing the beginning of a process that might result in significant restructuring of the urban hierarchy when it attains equilibrium" (Carroll and Meyer 1982, 577). But when Ray Bromley categorized state capitals in 1990, he still distinguished four types: "principal city capitals," "emerging technopoles," "classic government towns," and "small town capitals" (Bromley 1990, 6).[47] The upward movement of state capitals seems therefore far from complete, since the principal cities are "principal" for quite a long time (with the exception of Phoenix). To understand the current situation, we need to see the influence that capitals have in the nation and to go beyond the hierarchy established by traditional geographical overviews (McKnight 1992).[48]

Until 1950, capitals, except for Boston, remained at best secondary or local centers in a hierarchy still largely based on the eastern metropolises. Today, things have changed, with four more capitals near the top of the hierarchy (Atlanta, Phoenix, Minneapolis-St. Paul, and Denver) and nine others that strongly relay the economic influence of the former into their own hinterland (Columbus, Salt Lake City, Sacramento, Indianapolis, Oklahoma City, Raleigh, Austin, Nashville, and Hartford), all with MSAs of more than one million inhabitants (see figs. 7.1 and 7.2). But this situation did not induce any new urban hierarchy influenced by capitals that would replace the old one. The current urban hierarchy is based on new economic processes, be they called postindustrial or hyperindustrial, whose managers found in numerous state capitals the features they needed: white collars, proximity to decision making, and heritage, although they had to look elsewhere for the benefits of universities, since capitals were seldom the best centers of higher education.

Figures 7.1 and 7.2, like Carroll and Meyer's categorization, put the emphasis mostly on size and overall influence, and not enough on the processes of evolution. Figure 7.3 is an attempt to take these processes into account and to show the long-term evolution of capitals. The four categories

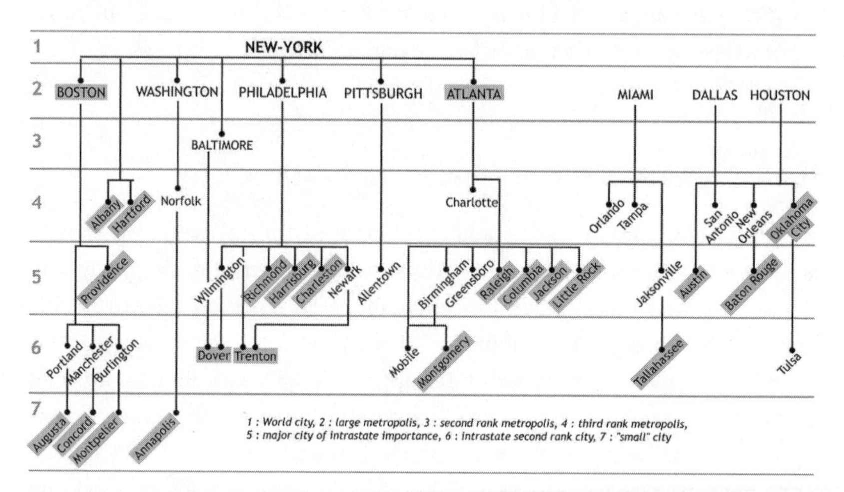

FIGURE 7.1 State capitals in the national urban hierarchy (top rank: New York). Credit: Christian Montès and M. L. Trémélo.

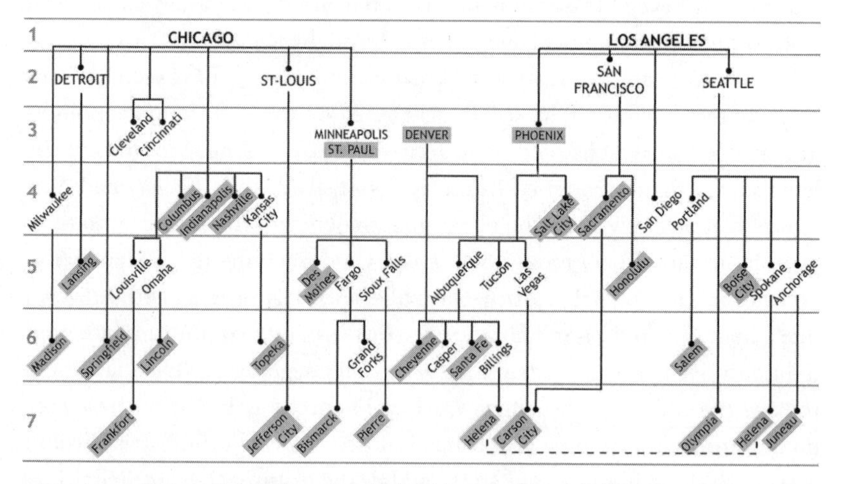

FIGURE 7.2 State capitals in the national urban hierarchy (top rank: Chicago and Los Angeles). Credit: Christian Montès and M. L. Trémélo.

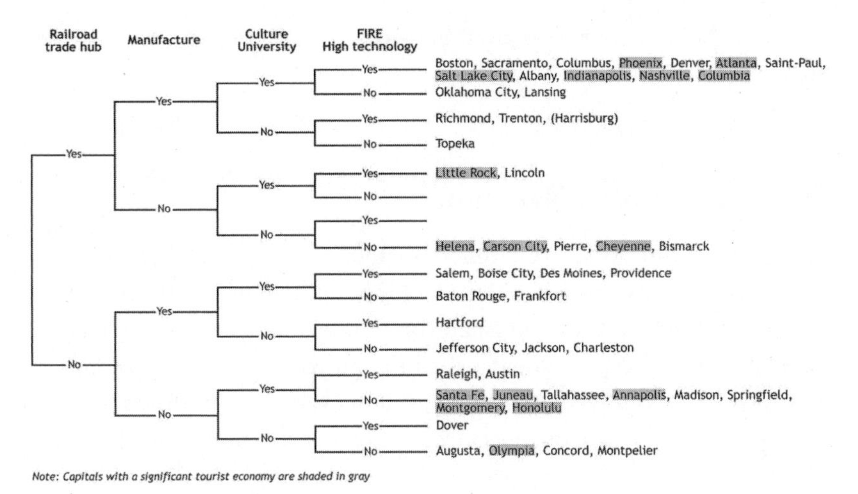

FIGURE 7.3 Processes of evolution of state capitals. Credit: Christian Montès and M. L. Trémélo.

represent the successive engines of growth in the American urban system: the first aim was to be a mercantile gateway, the second was to become a manufacturing stronghold, the third was to benefit from cultural and educational facilities, and the last was to develop high-technology industries and FIRE. The label *no* in the figure means that the process was absent or never reached an important level.

Some capitals never implemented their first promises, such as Harrisburg (see chapter 6), while others waited until the end of the twentieth century to begin their growth, such as Austin or Raleigh.[49] The second finding in figure 7.3 is that twenty-four capitals are participating in the FIRE/high-tech development (i.e. in the modern economy). If we include the eight capitals that are tourist magnets without fully participating in the former movement, the total is thirty-two. The majority of capitals take part in the post- or hyperindustrial hierarchy, whereas the industrial hierarchy bypassed most of them. But Carroll and Meyer's predictions have only been partly translated into reality: while capitals are at the top of the political hierarchy, they are still not at the top of the economic one. For instance, Austin's growth has been important, but the growth of Dallas and Houston has been far greater, and San Antonio is still a larger metropolitan area. As a result, capitals seem as hard to categorize as when Chauncy Harris made his classification sixty years ago (Harris 1943). The fifty capitals are scattered in almost all sixteen categories of figure 7.3, and Columbia, South Carolina, seems to be a surprising intruder in the "best" category, from the economic point of view at least. It is therefore necessary to try a complementary approach.

Main Factors of Explanation

Three groups of capitals were differentiated, each experiencing a specific set of evolutionary patterns. The typology is based on eight factors that have been weighed to explain the evolution of state capitals: the date of foundation of the capital, the density of population, the type of economy, centrality, the recent evolution of the economy, the size of the capital before the economic turn, the presence of federal institutions, and chance.

1. The date of foundation is not a major influence. A capital created after other cities may become the leading city: Des Moines proves this.
2. Density of population is a better indicator. It is easier for a capital to lead the urban system in a sparsely populated state like Iowa that in a densely settled one like Ohio (cf. also Phoenix, Salt Lake City, or Boise City). Boston is a special case: it was the colonial capital and thus had a long record of primacy, as well as an excellent port.
3. The type of economy is significant. If manufacturing dominates at the regional level, state capitals are at a disadvantage, because they tend to attract white collars more than blue ones. If agriculture dominates, centrally located capitals can easily become supply and trade centers. With a diversified economy, the case is more complex. Diversity is a characteristic of most capitals and enables them to profit from a stable and growing economic basis through public employment (around one-half of the total workforce for the smallest capitals and one-quarter for medium-sized ones). Being diversified in a mono-industrial state enables the capital to react better against economic downturns (e.g., Columbus or Cheyenne).
4. Centrality is favorable only if state centrality also means regional centrality. Austin is more central than Houston or Dallas in Texas, but it is far from the cattle and cotton centers, and out of the oil fields, whereas Atlanta is less centrally located than Macon, which vied for the capital well into the twentieth century, but far better located when it comes to controlling trade with the whole southeast. Political centrality may be an asset, but it is not necessarily an economic one.
5. The recent evolution of the economy is important because it is based on the traditional assets of capitals: white collars and urban amenities. To

reap the full benefits, however, requires access to academics, but state universities are often awarded to cities that had lost the capital fight. Capitals that are also academic centers thus fare better (e.g., Austin or Raleigh).

6. The size of the capital before the economic turn is also significant. Below a certain population threshold, it seems impossible to turn a "village-capital" into a "new economy" boomtown. In 1950, Austin already had 132,459 inhabitants, and Phoenix had 106,818.[50] The figures were not impressive, but their populations were far greater than those of Carson City (3,082), Pierre (5,715), Montpelier (8,599), and Frankfort (11,916). By 2010, Montpelier had 7,855 inhabitants; Pierre, 13,646; Frankfort, 25,527; and Carson City, 55,274, while the Austin MSA had 1,716,289, and the Phoenix MSA had 4,192,887.
7. The presence of military and civilian federal institutions makes a great difference. Despite the rise of state powers, federal influence remains high through federal land ownership (still important in the western states), the federal banking system (Boston and Richmond), other federal services (Denver), and through army and navy bases (Annapolis, Columbia, Little Rock, and Montgomery), which attract high-technology firms (Columbus, Phoenix).
8. We must also consider chance, a far less quantifiable factor. Why did Hartford and Des Moines become the nation's insurance capitals? Many other state capitals host insurance companies, which seek proximity to legislators.[51] The insurance business soon left New York—with the important exception of reinsurance—because, as a repetitive and normative business, it does not require constant contact with exclusive sources of information. But to become an insurance capital, a city must offer more than friendly local corporate laws. Such laws might help, but entrepreneurship has been most important. Likewise, the previous factors do not fully explain why numerous capitals remain small cities, a status not deliberately chosen. The local political climate could be a strong factor, since capitals were often considered as sites of corruption (as Frankfort was until the 1960s) far more than as sites of economic development.

The typology that follows is still a blueprint, because the criteria cannot be perfectly scientific, and some capitals followed ambiguous processes of evolution.

First Type: Modern State Capitals. The number of modern state capitals (30, or 60%) is close to the number of postindustrial capitals (fig. 7.3). Five subcategories exist, according to type of growth pattern and type of relation with the regional environment. In the first subcategory are capitals that were major centers from the beginning and profited from the new economic deal: Boston, St. Paul-Minneapolis, and Denver at a very high level in the metropolitan hierarchy; and Salt Lake City, Oklahoma City, and Indianapolis at a lower level (with between one and two million inhabitants).

Denver is the nodal center in the Mountain States region. It controls air, truck, and railway transportation, along with wholesale trade, and it is the only regional banking and insurance center. The discovery of coal and then oil and gas increased its importance. The influence of high-tech firms is relatively small, and the economic power of tourism is growing. Denver is therefore a classic metropolis, its capital status playing only a secondary role in its economic evolution, although many federal agencies are based or have offices in the MSA, making it the second-ranked federal center in the nation.

Salt Lake City's situation has greatly changed since 1981, when Utah had the fifth lowest per capita income in the nation and was led by the mining industry (Alexander and Allen 1984, 308–10). Utah's capital—still the largest city between Denver and the Pacific coast—became a high-tech city (ranked fortieth) thanks partly to the relocation of California firms (especially computer software and biomedical firms). It is also an air hub, a banking center, a higher education center, a tourist magnet (4.5 million people a year visit Temple Square, more than Waikiki Beach, Hawaii). Banking services employed more people than manufacturing in 2001 (15,742 vs. 14,267 out of a total of 211,742 jobs (see the city's website). Since 1987, the need for partnership between the various local actors, private and public, to foster economic growth has been recognized. The Economic Development Corporation of Utah is a result of that recognition, with sixteen private and sixteen public members, including the capital municipality. The corporation "is working quite well, even though some of the members, like Salt Lake City, are somewhat concerned how many direct benefits they are getting" (Weatherby and Witt 1994, 108–9). The municipality of Salt Lake was faced with the classic dichotomy between its population (20% of the county's) and its economic importance (it employed half the county's workforce in 1989), which resulted in recurring fiscal problems.

Indianapolis hosts high-technology firms and financial institutions larger than the size of its MSA would imply.[52] The city underwent a decline in its manufacturing sector after the 1950s, particularly in the durable goods (automobiles), but high-technology facilities replaced older plants (Bodenhamer and Barrows 1994, 65–71). In the 1970s, risk-taking leaders used public as well as private and charitable resources to foster downtown renewal (Walcott 1995). New buildings rose, the campuses were developed, and new sports facilities made Indianapolis known as "Sportsville, USA." The economy is still diversified and based on Indianapolis's position as the "crossroads of America," with four major interstate highways (65, 69, 70, and 74).[53] Waves of mergers in the late twentieth-century meant the growth of outside control. After a 1985 law allowed banks to cross county lines, Indianapolis's big three fell under out-of-state ownership. In 1983, only one Fortune 500 Company—Eli Lilly, inventor of Prozac—was headquartered in Indianapolis, the 2012 figure climbed to four (WellPoint, Eli Lilly, Bright Point, and NiSource). From the social point of view, which is excluded from such rankings, problems still faced the city: African Americans and women lag behind, pollution is still high, and the city has a poor mass-transit system.

In the second subcategory are capitals undergoing a contemporary metropolization process, which is strong (and well known) for Atlanta and Phoenix, strong but at a lower level for Sacramento, Columbus, Raleigh, and Austin, which are not at the first rank in their state, and for Nashville, which shares primacy with Memphis. Nashville is a subregional node. Despite its smaller size, it is still an important center of government (which generates employment and income) and of nonprofit activities (education and medicine). Nashville is also a well-known cultural center, as the capital of country and western music (Grand Old Opry). In 2003, Nashville's economy was described by its Chamber of Commerce as diverse and therefore balanced. Manufacturing represented 14% of the workforce (automobiles), more than government (13%). FIRE's share of employment was 6%, higher than that of Memphis. In 2006, health care provided ninety-four thousand jobs.

Columbia and Boise City are interesting cases. Both belong to the third category but are on their way to reaching the second (see figures 7.1 and 7.2). Columbia was ranked in 2010 as South Carolina's second CSA, after Greenville. With only 129,272 inhabitants in the municipality, Columbia is certainly supposed to belong to the "small capital" category. However, it was ranked in 2010 as the seventieth American MSA, with 805,106 inhabitants, a 77.3% rise since 1990. High-tech output represented 11.9% of its metropolitan gross product in 1999. Located in the center of the state, the capital had become an important cotton market and railroad town by the middle of nineteenth century. After the completion of a canal, it added textile manufacture at the end of the century. In 1917, the settlement of Camp

Jackson, a training facility three to four times larger than the city (they merged in 1968), added an important source of work (Moore 1993, 438). Another asset is the University of South Carolina.[54] State and federal government jobs are therefore still important (one-quarter of the workforce), although Columbia is also an important health and insurance center.

At first sight, Boise City would be considered a classic central place in a sparsely populated state. Idaho had only 1,567,582 inhabitants in 2010, and Boise City, with 205,671 inhabitants (616,561 in the MSA), was only the eighty-fifth largest conurbation in the nation. However, its recent growth has been tremendous, and owes little to its capital status. Almost half of Idaho's population growth occurs in the capital. Some of the reasons are classic. Boise City, as the central place in a rural state, hosts the headquarters of several timber and food companies, such as Boise Cascade, a large paper and wood products company founded in 1957. But the main reason is the transformation of the city into a high-tech stronghold. It held in 1999 the forty-sixth rank for value added and the seventh for the share of GMP in high-tech industry (36%), far better than its rank for population. Several Hewlett-Packard divisions have their headquarters in Boise, as well as Micron Technology (Weatherby and Witt 1994, 39–41). Boise also benefits from Boise State University, Idaho's largest. In 2011, the major private employers were Walmart (7,136), Micron Technology (5,000), Hewlett Packard (3,000), and J. R. Simplot (3,400).[55] Boise City's economic pattern relies heavily on the last three companies.

In a third subcategory are the capitals integrated into a metropolitan area: for Olympia (linked to Seattle) and Salem (linked to Portland). Providence and Richmond are in-between cases. Each has one million inhabitants but is located inside the megalopolis—close to Boston for the first, and to Washington for the second. Both had a strong manufacturing basis that is now in crisis and participated in the megalopolitan processes (with the presence of banks, high-technology firms, and universities). Their downtowns suffered important crises followed by recent impressive renewals. Two billion dollars were invested in Richmond to redevelop the riverfront, including the largest convention center between Charlotte and Washington, DC, and the creation of Virginia Biotechnology Research Park. These were modern additions to the city's numerous historic districts, evidence of its long and prestigious history. The city also gains from the federal presence, with the US Fourth Circuit Court of Appeals (the second largest US court), and a branch of the Federal Reserve Bank that has attracted other financial institutions.[56]

In the fourth subcategory are "niche" capitals: Madison and Lincoln (universities), Honolulu (tourism and government employ 20% of the workforce), Hartford (insurance capital of the United States), and Dover. Dover became Delaware's permanent capital in 1781 and was chartered in 1829. Its growth was inconspicuous, shadowed as it was by the great industrial city of Wilmington, which hosted half the population of the state in 1920 (110,168 out of 223,003), when Dover had only 4,042 inhabitants. The city nevertheless greatly profited from the state's main asset (besides the E. I. Du Pont Company). Its 1899 general incorporation law was "a gold mine. Lenient requirements, low fees, proximity to financial centers, stable political institutions, and, in time, a bench and bar experienced and competent in corporate litigation gave Delaware an advantage over other states that sometimes sought to capture this business" (Munroe 1993, 186). In 1919, 4,776 companies were incorporated in Delaware, and that number rose to almost 80,000 in 1975. The Financial Development Act of 1981 reinforced the corporate importance of Delaware by attracting the credit-card operations of out-of-state banks and by eliminating usury laws that restricted interest charges. "The results, in terms of employment, were fantastic" (258). Financial services became the second largest industry in the state in 1990, although by this time it mostly benefited Wilmington. The last way the state found to attract businesses was to pass its very lax bankruptcy laws.

In the fifth and last subcategory are small capital cities that initially experienced small growth linked to a stable economic basis (government, with services to the surrounding community) followed by strong contemporary growth due to the heritage movement, the leisure economy, and the economic boom of the 1990s. Capitals in this subcategory are Santa Fe, Juneau, Carson City, Annapolis,[57] Helena (Victorian city), and Concord.

Second Type: Seemingly Unchanged Capitals. This second group includes the fourteen capitals (28%) that have been unaffected by the growth processes or continue to follow mostly older economic logic. Three subcategories exist. In the first one are the central places of mostly rural areas (but of intermediary level only): Bismarck, Cheyenne, Montgomery, Little Rock, Jackson, and Des Moines. The first three share centrality, but the last three reign supreme in their states.

Although Cheyenne is the first-ranked MSA of Wyoming, its small size (91,738 inhabitants), its out-of-center location, and its proximity to Denver prevent it from controlling the whole state. Casper, with 75,450 inhabitants, plays a similar role for central Wyoming. Moreover, half the state belongs to the federal government (Yellowstone National Park is only the visible part of the federal iceberg). Cheyenne remains primarily a government-based city with important tourist attractions.

Bismarck is a classic central place of a sparsely settled state; North Dakota has only 672,591 inhabitants. The city is a distribution point for the rural resources of the area (wheat, livestock, and dairy), a small financial center, and the largest medical center between Minneapolis and the West Coast (one of the least populated areas in the nation!), according to its Chamber of Commerce.[58] Bismarck's MSA, with 108,779 inhabitants, is now ranked second in the state, after Fargo's (208,777), which benefits from its location on the Minneapolis-Winnipeg interstate highway and gains from Minnesota's economy and population. Fargo, a thriving community, hosts North Dakota State University (Grand Forks, the third-ranked MSA, hosts the University of North Dakota, established 1883 as a consolation prize in the capital contest). Bismarck's institutions of higher education are much smaller: Bismarck State College (established in 1939) and the University of Mary, the only private Catholic university in the state (established in 1955). The railroad that created and largely influenced the life of the city, the Burlington Northern, employs only 325 people.

Montgomery (427,691 inhabitants) is larger than Bismarck and even rose to second rank in Alabama in 2001, outranking Mobile. A distribution point, a central place in Alabama, and a tourist center, it nevertheless ranked only 206th in the nation for high-tech firms in 1999, although it has profited from the Air Force Data Systems Design Center at Maxwell/Gunter Air Force Base since 1971. Besides state government (25% of the workforce), the military has a huge impact, with twenty-two thousand direct and related jobs and nine thousand retirees. Montgomery nevertheless has to share centrality with Huntsville and Birmingham (the state's largest industrial city) in the north, and with Mobile in the south.

Little Rock is an in-between case that is influenced by more modern processes than the previous cases. It is also the capital of a poor southern state and relies heavily on government jobs: 32,200 for the state, 9,200 for the federal government, and 4,500 for the air force base. But Little Rock is also a major medical center (more than 50,000 jobs), a telecommunication center (Verizon and AT&T), and a tourist center. Manufacture remains secondary (avionics).

In the second subcategory are industrial cities that are declining today: Charleston, Trenton, and Harrisburg. Charleston, West Virginia, suffered from major losses of population between 1990 and 2010, losing 19.6%, but its MSA gained 21.7% (only six capitals had lower rates of demographic growth in their MSAs). With 51,400 inhabitants (85,796 in 1960), it is a small capital in a medium-sized MSA of 304,282 inhabitants. Although its Chamber of Commerce website boasts that it profited by the "influx in West Virginia of companies specialized in health care and manufacturing," the figures tell another story. Located in a (declining) coal and oil area, Charleston does have chemical plants (Dow Chemicals and E. I. Du Pont), but manufacturing employs only 5.1% of the MSA's workforce, while government employs 17.9%. As for health care, the city hosts only two companies of significant size.

The situation is different for Trenton and Harrisburg, which are both interesting cases of ambiguous contemporary evolution. Trenton is a classic example of industrial decline (with the fall of the pottery industry) and inner-city problems, although it does have a new high-tech center (Cumber 1989; Stansfield 1998). With 16.5% of its GMP in high-tech, Trenton's MSA ranked forty-first in the nation and seventieth for the high-tech output, although its MSA had only 366,513 inhabitants in 2010 (ranked 140th). This illustrates the metropolitan processes—Trenton is an edge city of New York—and the dichotomy between capital city and capital MSA, the latter drawing the main benefits from the new economic order. Trenton as a municipality belongs therefore to the group of declining former industrial cities; as an MSA, it is integrated in metropolitan processes and belongs to the first type, modern capitals.

Likewise, Harrisburg still hesitates between the old and the new economic and urban patterns. The city began its industrial decline as early as the 1880s. As a municipality, Pennsylvania's capital lost population, plunging to 49,528 in 2010, near half of its 1950 population of 89,544. The proportion of African Americans has risen from 7% in 1920 to 52.4% in 2010, a clear indicator of inner-city crisis. The poverty rate was 30.2%, and in 2012 the city was on the verge of bankruptcy. State employees (21,885) represent today 44.2% of the resident population. But the population of Harrisburg's MSA rose to 683,043 in 2010, owing to its position as a relay point between the megalopolis and the Midwest. The major private employer in the MSA is the Hershey Foods Corporation, based in its model town of Hershey (6,500 jobs, plus 8,848 at the Penn State Hershey medical center, and 7,500 in its amusement parks). Dauphin County has a high-tech base (e.g., HP Enterprise Services). The city of Harrisburg has been undergoing a "dramatic economic resurgence" since 1981, when it was the second most distressed city in the nation (according to its website). The number of vacant structures has been reduced by 85% (825 remain); the number of businesses has increased by 260%; crime rates dropped by 53% between 1982 and 1999; and $2.65 billion were invested. But the capital still lacks a big university, and the municipality's per capita income was only 66.6% of the state's in 2006–2010. One-third of the city's offices belong to the state capitol complex, showing the primary importance of state government.

Small cities without any hope of important growth constitute the third subcategory. In Pierre, still a small town, the state government employed 2,380 persons in 2010 (about one-quarter of the county's workforce), well ahead of the town's second-ranked employer, St. Mary's Healthcare Center, with 450 persons (Pierre Economic Development Corporation website). Montpelier, the smallest state capital (and the only one without a McDonald's), lost population between 1990 and 2010, dropping to fewer than eight thousand inhabitants. The town is located outside the megalopolis, and commercial competition comes from nearby Barre. History is preserved in some fine nineteenth-century buildings, but there is no coherent and comprehensive historic district. Vermont, like New Hampshire, offers scores of better-preserved and charming New England towns. This leaves government and insurance (National Life Group is headquartered here, but this is not sufficient to put Montpelier in the "niche capital" category). Other examples could be Frankfort (see chapter 8), Jefferson City (state government is the major employer), and Augusta, a self-styled "center of post-secondary education" in a μSA of one hundred thousand inhabitants. Its population is still lower than it was in 1990.

Third Type: Second-Rank Capitals. In the third group are six capitals (12%) that still have a secondary rank in their states but belong to medium-sized MSAs (225,000–950,000 inhabitants): Albany, Baton Rouge, Lansing, Springfield, Tallahassee, and Topeka. All are the result of compromise or big city rejection. Baton Rouge, Springfield, and Albany already existed when they were selected as capitals, while the other half were created specifically to become capitals. All immediately experienced competition from already established large cities, and none was able to compete with them. In all of them, government remains a leading force. In the first subcategory are the growing southern cities of Baton Rouge (ranked second in Louisiana) and Tallahassee (ranked twelfth in Florida). In a second subcategory are the northeastern and central capitals, where the municipalities are declining and the MSAs slowly growing: Albany (fourth behind New York, Buffalo, and Rochester); Lansing (third behind Detroit and Grand Rapids, despite its university); Springfield (fifth behind Chicago, Peoria, Rockford, and Champaign-Urbana), and Topeka (third behind Kansas City and Wichita).

Until the 1930s, Topeka was the center of a typical "medium-sized Midwestern area dependent primarily on its agricultural base," according to its Chamber of Commerce. Built where a ferry crossed the Kansas River on the Oregon Trail, it began to thrive with the arrival of railroads (the famous Atchison, Topeka & Santa Fe) but went through the boom-and-bust history of Midwestern towns. World War II was a turning point; the installation of Forbes Air Force Base attracted a Goodyear plant in 1944. Others firms followed suit, such as Hallmark Cards. After the closure of the air base in 1974, ten thousand people left Topeka, inducing voters to accept tax increases in favor of economic development, for which they also changed the form of their city government. A new airport and a convention center were built, and some firms came. The state's universities are located in Lawrence (twenty miles east) and Manhattan (fifty miles west), although Topeka founded the only municipally owned university in the nation, Washburn (smaller than both the others). As a result, the MSA remains small (230,824 inhabitants in 2010, owing to the extension of its limits to five counties instead of one). Kansas City is far too close, fifty-six miles away, for which Topeka can only be a relay point. This explains why government employs 20.2% of Topeka's workforce (MSA); manufacturing, only 7.6%; and FIRE, only 6.5%. The state of Kansas is the largest employer, followed by transportation (trucking first, and then the Burlington Northern Santa Fe Railroad).

To conclude this chapter, we must shift attention from the usual criteria, because, contrary to the booster model, size is no longer the main factor in deciding whether a city has succeeded. We have to see how state capitals participate in the "urban quality" network of the United States. Quality of life in a successful economic environment is today a major factor for most Americans. If concentration allows economies of scale, it also brings traffic congestion, pollution, and stress. The state capitals that fully participate in contemporary growth processes (about two-thirds) are at the top of the demographic hierarchy, but some are at the bottom and happy to stay there. Eight capitals benefit from the blooming leisure economy, without losing their souls, by treasuring their modest and pleasurable size and atmosphere (Juneau, Carson City, Olympia, Helena, Cheyenne, Santa Fe, Annapolis, and Concord). College capitals also dwell upon their pleasant, youthful atmosphere (Madison and Lincoln). Whether located inside a metropolitan area or splendidly isolated in grand mountain surroundings, capitals boast of their family atmosphere, far from the evils of large cities (Pierre, Salem, Olympia, and Boise City). The discourse on capitals' websites is a clear demonstration of this trend. High among the numerous rankings they all display with pride are the ones dealing with quality of life. Topeka, for instance, was proud to announce its nomination in May 2003 as a "five-star community" in the annual Quality of Life Quotient survey by *Expansion Management Magazine*. Quality of life is interestingly defined there as "being able to afford to take part in the 'American Dream.'" In the same issue, Nashville was ranked second for "Best Metros for Standard of Living" and given four stars. Nashville was also the seventh most generous metropolitan area and the fourth among Ten Great Places to Spend Christmas.[59] It is therefore not surprising that ten state capitals were listed among the twenty named as "America's Best Cities" in 2012 on the Bloomberg Businessweek website, with Boston, Denver, Austin, and St. Paul in the top ten. Larger capitals are therefore not excluded from the movement. This situation is fully in line with the findings of chapter 2, which emphasized the assets of most capitals: careful planning, attractive monumentality, presence of public places, and important symbolic attractions. The setting and the atmosphere of capitals are seen as strong positive factors for attracting new permanent inhabitants. Many Americans long for the quality of life they think was found in nineteenth-century small towns. This is, of course, a mythical view of the past, but it is heavily exploited on many capitals' websites, along with signs of the most up-to-date modernity. Smaller capitals thus see themselves as the perfect blend of modernity and tradition. They claim to have succeeded in what the urban thinkers of the nineteenth century dreamed of when they presented Lowell as the ideal city, with modern manufacturing in a still small and moral town. According to their websites (2003), Bismarck, North Dakota, "combines the progressive technology of a large city with the sincere friendliness of a small town, making it a true home"; Jackson, Mississippi, is "a big city with a small town feel"; Charleston, West Virginia, "has all the amenities of a big city with small town ambience"; Salem, Oregon, prides itself on a "family-friendly ambience"; and Lincoln, Nebraska, provides "the ambience of a friendly small town."

As usual, there are caveats. Before praising their "small town atmosphere," webmasters should read Tracy Kidder's *Home Town* (1999), which revealed the reverse of the coin.[60] Likewise, rankings are mostly aimed at "mainstream" Americans and not at left-aside ones. Table 7.6 shows that only nineteen capitals have per capita incomes above that of the nation.

One should therefore think twice about moving to Hartford, Trenton, Harrisburg, or Providence. Indeed, Hartford is still suffering from the acute racial segregation fuelled by the "Bishops," a group of insurance leaders who, in the 1960s, devised a plan to displace all blacks and Puerto Ricans in order to make Hartford a white city. With the second lowest home ownership rate in the nation (22.4%), a racial "Balkanization," and a low percentage of high school graduates (67.5%) due to one of the worst school systems in the nation, the quality of life of the majority of Hartford's (62.5% minority) residents has a long way to go before reaching the level of the capital's wealthy and almost entirely white suburbs, such as Avon and Farmington (Zielbauer 2002). Even in capitals with a per capita income above the state's, the situation is far from perfect. The actual situation in Baton Rouge (109.5% of Louisiana's per capita income) is not very appealing. In a short inquiry, made in 1995, the city appeared smallish, calm, conservative, and boring (Montarry 1995, 149–84).[61] People liked the "amicable persons" and the setting of the River City Capital but reproached its unplanned growth, its lack of activities and animation (80%),[62] its crime rate (sixth out of the 272 major cities in 1994), and its puritanism (alcohol sales were forbidden on Sundays). Baton Rouge's image was nevertheless positive for two-thirds of the people interviewed, owing to the presence of the university more than of the capitol. When asked to compare Baton Rouge and New Orleans, 50% answered that they were like "night and day," but 30% preferred the calm and tidiness of Baton Rouge, which they veiwed as a city to live in, while its rival was a city to visit. But we should be skeptical of this view, because the people interviewed mostly lived in the suburbs and seldom went to the city center. The image of the city as a whole is therefore quite different from the image of the capitol grounds or the governor's mansion. People are certainly keener to visit most capitals than to settle in them.

8

Validating Models through a Chronological and Concrete Analysis

Three Case Studies

After analyzing the processes of choosing current state capitals, and of their evolution (i.e., after having looked for explanatory models), it seems useful to go back to specific spaces in order to test the models and render them more concrete, while allowing for the reintroduction of time in its most classic form: chronology. Since the last chapters have underlined the multiple temporalities at work, this return to chronology must be understood not as a return to some linear time continuum, but as a writing mode, which helps to revive the individual and the subject.

Using case studies means making choices: among the fifty capitals, why single out the three discussed below? They were chosen because they shed light on the three categories of state capitals that were defined in chapter 7 after a precise analysis of their evolution since the 1850s. They encompass the processes of choice as well as the subsequent evolution of the capitals. The first, Columbus, Ohio, belongs to the category of capitals undergoing a contemporary metropolization process. It is also a new town built in the wilderness to be a capital that grew to become one of the leading cities in its state. The second case, Des Moines, Iowa, is the central place of a mostly rural area that became its state's leading city while remaining medium-sized. The last case, Frankfort, which has never succeeded in rivaling the metropolises of Kentucky, is an example of the small capital cities that have no hope (or, today, desire) for important growth.

Columbus, Ohio

Ohio is one of the nation's major economic strongholds. Its population in 2010 was 11,536,504 (seventh rank), but its economy is divided among several leading crops and several leading industries, and is based in several leading cities. The state's political fragmentation and factionalism echo these divisions, with county leaders acting as feudal kings (Knepper 1989, x). In such a fragmented, competitive—with competition balanced between several competing cities—state, Columbus, a city created to become the state capital in 1812, did not become Ohio's leading city precisely because the state does not represent a unity. Columbus did not lack advantages, as the city's Board of Trade stressed in 1904. The board appeared anxious to convey the idea that Columbus was close to the Garden of Eden: "To describe the happiest people on earth, where climate, fertility of soil, and congeniality of a contented citizenship conspire to create a veritable paradise in the center of the great State of Ohio; to describe its many busy factories and marts; . . . to set forth in just terms its exceptional educational facilities; its unparalleled transportation advantages; would require pages vastly more ample and extended than at our command in the restricted space here allotted us" (Columbus Board of Trade 1904, 1–2).

Reality proved quite different, however, as Columbus was mostly seen, around the mid-twentieth-century, as a conservative and sluggish city, the "largest small town in America." It was judged to be so representative of America as a whole that it was used as a test city for new products. Table 8.1 provides a first look at Columbus's evolution in terms of population.

After a period of slow growth until the Civil War, Ohio's capital ended the century with a steady, if not booming, rise in its population, but it remained Ohio's fourth city. Indiana, Illinois, and Michigan each experienced the rise of a leading metropolis that prevented the formation of a full, local urban system, but Ohio saw the growth of several competing or complementary cities. Columbus is one of them but not the most important, because, although located at the geographical center of the state, it was slightly out-of-center with regard to the main economic flows, illustrating the dichotomy between economic and political boundaries. Columbus's story is therefore the story of a city that has, until today, seen others deprive it of economic prominence.

When Columbus officially became Ohio's capital in 1816, thirty years of settlement in the Northwest Territory had already resulted in booming

Table 8.1 Population of Ohio's major cities, 1810–1900

	1810	1820	1830	1840	1850	1860	1870	1880	1890	1900
Chillicothe	1,369	2,426	2,847	3,977	7,100	7,626	8,920	10,938	11,288	12,976
Columbus	—	1,450	2,435	6,048	17,882	18,554	31,274	51,647	88,150	125,560
Cincinnati	2,540	9,642	2,4,831	46,338	115,438	161,044	216,239	255,739	296,908	325,902
Cleveland	200	606	1,076	6,071	17,034	43,417	92,829	160,146	261,353	381,768
Dayton	383	1,139	2,954	6,067	10,976	20,081	30,473	38,678	61,220	85,333
Toledo	—	—	—	1,222	3,829	13,768	31,584	50,137	81,434	131,822
Zanesville	1,154	2,052	3,094	4,766	7,929	9,229	10,011	18,113	21,009	23,538
Akron	—	—	—	—	—	3,477	10,006	16,512	27,601	42,728

Sources: Moffat 1992; Andriot 1980.

cities with which it would have to compete. The most important was Cincinnati, in the southwest. Founded in 1788, it became the capital of the Northwest Territory two years later, because it was centrally located within the territory's 260,000 square miles, nearer than other cities to the outposts at Detroit, Vincennes, Kaskaskia, and Cahokia.[1] The influx of pioneers remained steady, and Ohio became a state only three years later, on February 19, 1803, with Chillicothe, a Republican stronghold, as its capital. As Ohio's population multiplied, from 231,000 in 1810 to 581,000 in 1820, the general assembly decided to create a new capital city closer to the geographical center of the state and close to a navigable river. A new town, named Columbus, was platted on the east bank of the upper Scioto River, on the site of a former Wyandot village.

Lost amid dense woods, in the glacial drift plain of the central low plains and close to the Appalachian plateau, Columbus had slow economic beginnings. The surroundings were still wild, and the village was crudely built, often muddy, and prone to epidemics as well as floods, which were severe in 1832, 1834, and 1847 (Cole 2001, 25–26).[2] But Columbus slowly began to add facilities and jobs: 1813 saw its first sawmill, school, tavern, and bridge over the Scioto. The first newspaper, *The Western Intelligencer and Columbus Gazette*, moved from Worthington to Columbus in 1814. A market house and two Protestant churches were also added, as were a jeweler's shop and four lawyers in the following year. In February 1816, the population was 700, and the town was incorporated under the name "The Borough of Columbus." The Franklin Bank of Columbus was also incorporated. In 1821, when the population was 1,450, the federal court arrived. In 1824, Columbus replaced Franklinton, across the river, as the seat of Franklin County.[3] Columbus experienced serious economic troubles in the 1820s, as three out of the four members of the original land syndicate ended up in bankruptcy. In 1830, Columbus had only 2,435 inhabitants out of Ohio's 937,903. It was officially chartered as a city on March 3, 1834, with a population of 3,500. Only Cincinnati was significantly larger at that time, but Columbus was still a fragile pioneer town.

The 1830–1860 period was crucial to Ohio, when the bases of its economy were being established. Ohio became during that period a transportation hub of national importance and a manufacturing stronghold. In 1860, it ranked fourth in the nation for the value of its industrial products, a rank it would retain until the end of the twentieth century. People kept flocking to the state, doubling its population from 937,903 in 1830 to 1,980,329 in 1850. These factors fostered the growth of several cities, their rate of growth stemming from their respective location and from the degree of local entrepreneurship. Controlling the main commercial flows, which were tied to the pioneer movements, was crucial. As the major routes went through the north of the state, Cleveland became Ohio's second city in the 1850s. It controlled trade through all transportation modes: Lake Erie, the Ohio-Erie Canal, roads, and railroads, which made it a prime crossroads on the way to Detroit and Chicago. As for river trade, Cincinnati still ruled over Ohio, superseding Louisville. It soon became one of the prime cities of the United States, ranked eighth in 1830, sixth in 1840 and 1850, and ninth in 1890. In 1860, with 161,044 people, it was four times more populated than Cleveland and nine times more than Columbus (18,554). The latter's centrality in Ohio was not the best location with respect to the western expansion of the nation, which limited its market. Although not at the forefront, Columbus participated in this expansion. It became one of the crossroads opening on the unpopulated West. Under the leadership of an active merchant group, major improvements in transportation followed, making Columbus a transportation and market center (beef). A wharf was built on the Scioto River in 1830; the Ohio and Erie Canal arrived in 1831; the National Road (today's US 40) came through in 1833, making Columbus a hub for stage transportation, and the Columbus-Cincinnati railroad arrived in 1850, followed by the Columbus and Cleveland the next year. But contemporary journalists bemoaned the city's "lack of enterprise and public spirit,"[4] and later analysts lamented the fact that local capital was "more interested in land speculation—even then—and in transportation, with investments in both stagecoach lines and railroads" (Hunker 2000, 51). As a result, Columbus remained at the beginning of the Civil War essentially a political and commercial center, though some manufacturing activities had developed. The city was still quite small (although ranked forty-ninth in the nation). Columbus's citizens wished from the start to become the exclusive center of political power in the state. They first wanted to have the US courts removed from Chillicothe, the former capital, to Columbus. To that effect, they raised the transfer-money by subscription and, despite great protestation, succeeded in 1821.[5] In 1824, Columbus also seized the county seat from older and nearby Franklinton. As early as the 1830s, "she was accused of putting on metropolitan airs" (Studer 1873, 37).[6] These "airs" were resented so much that Columbus was accused of every conceivable evil, such as being very unhealthy, owing to poorly drained land and hogs in the streets, and insufficiently central. The senate therefore voted in 1842 to establish a committee to look for "the permanent establishment of the seat of government," but the proposition was narrowly defeated in the house on March 6, 1843. This settled the matter at last, but the capital's economy did not fare as well as was hoped. Economic uncertainty and the western pull took their toll during the 1850s, as Columbus's population grew by only eight hundred, far from the thirty thousand predicted by the city's leaders.

After the Civil War, new economic factors once again benefited other cities more than Columbus. First, Cincinnati remained the major transportation hub of the area, although it was surpassed by St. Louis and Chicago, which were better located to control trade with the newly settled West. Cleveland became the leading city, benefiting from big industry and oil—the huge Standard Oil of Ohio Company had its headquarters there. Other competitors also rose. The Dayton Electric Company (Delco) enriched Dayton. Toledo already dominated the northwest part of the state. It was an export point for agricultural products by river, railroads, and roads, and became an oil-refining center. Thanks to a supply of cheap fuel and the proximity of Detroit, it earned the nickname of "Glass City," by manufacturing glass for automobiles.

Columbus benefited to some extent from the economic and demographic booms of southeastern Ohio. Its position as the capital earned it a federal arsenal in 1864, the Deaf and Dumb Asylum, the Blind School, and the State Lunatic Asylum. Manufacturing benefited from the timber and mineral resources of the southeastern part of Ohio. With the creation of the Columbus Iron Company in 1870 and the arrival of Carnegie Steel in 1894, the city became less dependent on agriculture, but other industrial cities could have been better qualified for Columbus' nickname, the "Birmingham of America." Certainly more important for the future was the new Ohio Agricultural and Mechanical College (now Ohio State University), established in 1870 by the Ohio General Assembly under the provisions of the 1862 Land-Grant Act.[7] Columbus also became a rail hub of national importance: seven trunk railroads with eighteen divisions connected it with the whole country. By 1900, these boomlets had made it the twenty-eighth largest city in the country, with 125,560 inhabitants, but it was still referred to as an "overgrown county seat town" (Knepper 1989, 361) and "a conservative community with little risk capital available to support new industrial endeavors" (Hunker 2000, 53).

The twentieth century was more profitable for Columbus than the nineteenth century. Ranked third among Ohio's cities between 1910 and 1960, it rose to second (1970–1980), and has been first since 1990, and it is the only city that never experienced a decline in population (see table 8.2). We

Table 8.2 Population of Ohio's major cities, 1910–2010

	1910	1920	1930	1940	1950	1960	1970	1980	1990	2000	2010
Akron	69,067	208,435	255,040	244,791	274,605	290,351	275,425	237,177	223,019	217,074	199,110
Columbus	181,511	237,031	290,564	306,087	375,901	471,316	539,677	564,871	632,910	711,470	787,033
Cincinnati	363,591	401,247	451,160	455,610	503,998	502,550	452,524	385,457	364,040	331,285	296,943
Cleveland	560,663	796,841	900,429	878,336	914,808	876,050	750,903	573,822	505,616	478,403	396,815
Dayton	116,577	152,559	200,982	210,718	243,872	262,332	243,601	203,371	182,044	166,179	141,527
Toledo	168,497	243,164	290,718	282,349	303,616	318,003	383,818	354,635	332,943	313,619	287,208

Sources: Andriot 1980; US Census Bureau, 1980, 1990, 2000, and 2010.

should note, however, that Columbus is only ranked third among Ohio's CSAs, even if it is closing the gap with Cincinnati, and that its steady population rise owed as much to an ambitious annexation policy, begun in 1954, as to economic growth. Columbus was in 2010 the fourth largest capital city in the nation, behind Phoenix (1,445,656), Indianapolis (820,445), and Austin (790,390), but it was only the eighth largest capital MSA.

Ohio's economy largely benefited from the coming of the "industrial age," as well as from the adoption of home rule in 1913, which "encouraged creativity, inventiveness, and efficiency of hard-working teams, willing to bring prosperity to their jurisdiction within the state" (Thomas 2000, 190). As a result, Columbus's competitors remained strong. Diversification was at work in Cincinnati, with the creation of the Procter & Gamble Company, for instance. A newcomer, Akron, prospered initially from the Cleveland to Portsmouth river route, which was very important until 1913, and its abundance of iron, coal, wool, and grain. Its greatest benefit came because of Mr. Goodrich, who created the Goodyear Company in 1898, inducing a rubber boom (Akron's population grew from 69,067 inhabitants in 1910 to 208,435 in 1920) and the growth of truck freighting. Cleveland consolidated its industrial primacy until the 1960s. Dayton added to its Delco Company the Wright-Patterson air force and research complex.

In such a manufacturing context, politics and trade remained the main staples of Columbus's economy. With the official "closing" of the frontier and the populating of nearby states, its location became excellent,[8] which enabled it to attract the distribution centers of major retail store chains; in August 1965, Port Columbus became an official United States Port of Entry. But manufacturing should not be discounted. Local labor was qualified, reliable (nonunion), and relatively low-wage. The federal government stimulated the economy by establishing in 1918 a huge military depot (Hunker 2000, 54). Likewise, and as elsewhere in the nation, World War II acted as a catalyst to industrial expansion. Ohio's other prime cities had industrial contracts; the main benefit for Columbus was the location there of the Curtiss-Wright aircraft plant in 1941, which attracted migrants from the Appalachians and the South and opened Columbus to a new generation of industries. When one adds ideal transportation, excellent supplies of raw material, market accessibility, and available suitable industrial buildings and sites (due to the an active annexation policy), the attraction of Columbus is easy to understand (OSU Bureau of Business Research 1966, 29–38). Following its famed buggy industry, the city became a big supplier of auto parts, and heavy industry was important. Big companies settled there: General Motors (1946), the Columbus Auto Parts Company, Westinghouse (1952), Western Electric (1959), and North American Aviation. "By 1950, Columbus was well established as a center for national industry and its pattern of industrial development, so long dominated by local firms, was changed" (Bureau of Business Research 1966, 15). Columbus developed a durable manufacturing economy with at least seven different industries in 1969, electrical machinery and transportation equipment being foremost. Employment in manufacturing, stable between 1914 (c. 35,000 workers) and 1940 (32,784), rose in 1960 to 67,209 in Franklin County, the core of the 540-square-mile MSA. As total employment grew from 138,662 to 256,684, manufacturing represented 26.2% of the total. That relatively low figure (for Ohio) contributed to the dubbing of Columbus as "the 20% economy," but its diversity prevented crises. Cincinnati presents a similar balanced economic profile.

That balance explains why neither city was as affected as other Ohio cities were by the post-1973 industrial crisis. Indeed, although manufacturing is quite important in all of them, contrasting economic subregions exist that induce specific development paths.[9] The new pattern of economic evolution favored Columbus. Cleveland is still suffering from the de-industrialization process. It was overtaken by its competitors—Chicago and Detroit to the west, Pittsburgh and the megalopolis to the east. It never reached the status of a more than regional center. Dayton and Toledo also suffer from de-industrialization. While manufacturing represents 31% of the state's gross product (30% for Dayton and Toledo) it only amounts to 17% for Columbus, which lost "only" 11% of its manufacturing workforce between 1965 and 1996, while gaining 411% in its service sector.[10] In absolute figures, Cleveland gained 228,000 service jobs, Columbus 186,000 and Cincinnati 160,000. Ohio's location quotients (LQ) indicate that only Cincinnati lacks specialized industries.[11] Columbus has two of them: ceramics and glass ($LQ = 3.1$) and banks ($LQ = 2.6$). The corridor from Evansville (Indiana) to Louisville-Lexington (Kentucky) and to Cincinnati-Columbus is called today a "new activity area." Not considered essential during the Fordist period, its economic structures are adapted to the post-Fordist phase, and it has prospered significantly since the 1970s. But the post-Fordist transition and de-industrialization cannot fully explain the economic changes that the area experienced, according to Isabelle Thomas (2000, 87–88). First, growth occurred not only in new firms. Endogenous forces, arising from Columbus's economic history, proved resilient. Efficient systems of production, local natural resources, and a qualified workforce, allowed "old" firms to contribute more than half of all new jobs. Second, contemporary economic dynamics—the growth of state powers and offices, the university, and finance—are to be linked to Columbus's status as capital. In 1976, state government was already the largest employer in the MSA, with 20,400 employees. The federal government employed 12,000 more people, and the Columbus campus of Ohio State University, one of the largest state university systems in the nation, employed 16,000 people (Stanback and Noyelle 1982, 105). In 2011, the figures were 26,778 for OSU and 26,728 for the state of Ohio.

In the private sector, Bank One created the Visa card in the 1960s and developed its banking services. Cardinal Health, formerly a small wholesale grocery, became an international health-service provider. This helped the city to become a nodal center of corporate activities, hosting divisional head offices of a large number of Fortune 500 firms. The Chemical Abstracts Service (1,200 employees in 1982) and several insurance companies are also located in Columbus. A 1982 study explained the processes at work: "Columbus took an early lead in the service transformation because of its function as state capital and its early development of a strong research orientation. At a time when both the federal-state government connection and R&D functions were increasing in importance for the large corporation, Columbus appeared to have been well positioned to gain an important share of the expansion of the central administrative auxiliary establishments of large local firms and to propel the development of a number of associated corporate services" (Stanback and Noyelle 1982, 104).

But these assets are shared and sometimes undermined. They are shared by Cincinnati, which also became an attractive center for financial and health activities, and undermined, because Columbus lacks a strong research and development basis. Although Battelle Memorial Institute, the largest independent research organization in the nation, is headquartered in Columbus,[12] the city's research facilities were considered weak by a 1966 study that also stressed its "reluctance to contend for government contracts for military products, EDT&R, NASA research, etc." (Bureau of Business Research 1966, 66) Local banks were also analyzed as conservative, with little venture capital. In fact, as early as the late 1950s, the city had turned down proposals by Ford and International Harvester to locate new branch plants there. Columbus did not want to become a blue-collar town. One of the main reasons, according to Henry Hunker, is that the city leaders have been largely conservative and have resisted new business development. The city was controlled by a few influential families until the

1990s. The Wolfes were the foremost, owning shoe factories, banks, land, and newspapers. The Lazaruses and Galbreaths were also important. Today, the Wolfes are still influential in media, and a new player has come, Leslie H. Wexner.[13] The mayors did not play a great role, except for Mayor Sensenbrenner (a young scion of the family is still on the city council). "Unlike many other U.S. cities, political leadership in Columbus does not necessarily provide dynamic civic leadership, due in part to politicians' historically close ties to the Wolfes" (Hunker 2000, 82). The main growth factors tend to be exogenous. Since the merger of Bank One with First Chicago NBD Corporation in October 1998, the headquarters have left Columbus for Chicago. Outside firms control manufacturing, and Columbus has lost some important plants that left brownfields, such as the huge, abandoned McDonnell-Douglas units near the airport, the Conrail Storage Yards, and Armco Chemical.

Among contemporary metropolization processes, Columbus tends therefore to become a relay—not a relay of Cleveland or Cincinnati, the three cities having always followed parallel paths, but of the real regional metropolis, Chicago. As was the case one century ago, concentration favors outside centers. Table 8.3 compares Columbus with Cincinnati and Cleveland.

Being a strong relay has enabled Columbus to experience steady growth in the past decades. Columbus's CSA population climbed to 2,071,052 in 2010. It remains Ohio's third CSA, after Cleveland (2,881,120), but is almost catching up with Cincinnati (2,172,191). Columbus is ranked fifteenth among US cities and twenty-fourth among MSAs. The 2011 municipal credit rating would please Detroit: AAA for Standard and Poor and Aaa for Moody. In 2011, the unemployment rate was 7.6%; Nationwide Insurance provided 19,572 jobs in its headquarters and in its children's hospital; and JPMorgan Chase employed 18,000 people. In the MSA, FIRE employment was higher than employment in manufacturing—68,100 (7.6% of the workforce) vs. 63,100 (7.1%)—and 158,000 government workers accounted for 17.7% of the total workforce.

As a result, when most central municipalities, especially in the northeast, lost population, the population of Columbus grew quickly; since 1980, it has gone from 564,871 inhabitants to 787,033 in 2010, a 39.3% rise. Downtown Columbus remains quite attractive for public and private jobs and cultural activities. Renewal is fostered by the high architectural quality of the city center (fig. 8.1), and the presence of Ohio State University.[14] The city is home to the Columbus Symphony Orchestra, Opera/Columbus

Table 8.3 Columbus, Cincinnati, and Cleveland: Economic comparisons

MSA	Percentage of state's GMP, 2001	Percentage of state's population	GMP/pop.	Airport passengers, 2002 (millions)	Employment, 2003 (thousands)	Financial jobs, 2003 (thousands)	Professional and business jobs, 2003 (thousands)	High-tech rank, 1999
Columbus	16.10	13.56	119	6.7	874.7	74.7	125.1	44th
Cincinnati	15.93	16.12	99	20.8	878.0	59.3	127.0	43rd
Cleveland	21.32	25.94	82	10.8	1,108.7	80.1	134.8	38th

Sources: U.S. Bureau of Labor; airport websites; http://www.usmayors.com.

(celebrating its thirty-second season in 2012–13), and six theaters downtown. *Places Rated Almanac* in 2003 ranked Columbus as one of the best metro areas for the arts, ahead of Miami, Phoenix, and San Diego.

Could today's Columbus be called a "vibrant metropolis," as its website claims? Certainly. Being a capital has helped Columbus by diversifying its economic basis: politics and academics provide thousands of jobs. But capital status was not enough to bring Columbus to the first rank. This was certainly due to the success of manufacturing in Ohio. The state did not look primarily for an administrative place, but for ores and markets. Being located in the middle of a densely populated region with a strong urban system inhibited the emergence of a single metropolis in Ohio: the excellent location of the state gave rise to several cities that were as well situated as Columbus and had competing hinterlands (fig. 8.2). The post-Fordist economy, more based on mind and amenities, is currently benefiting Columbus, with its architectural quality and the presence of a good university (ranked twenty-fourth in the nation among research universities).

FIGURE 8.1 The skyline of Columbus, Ohio. This view from new North Bank Park shows a rather impressive skyline. However, as someone from City Hall noted, since "no one can name Columbus when you show its skyline" (June 2012, personal communication), the municipality hopes that a Fortune 500 company will come and build a "unique" building. Photo by Christian Montès, June 2012.

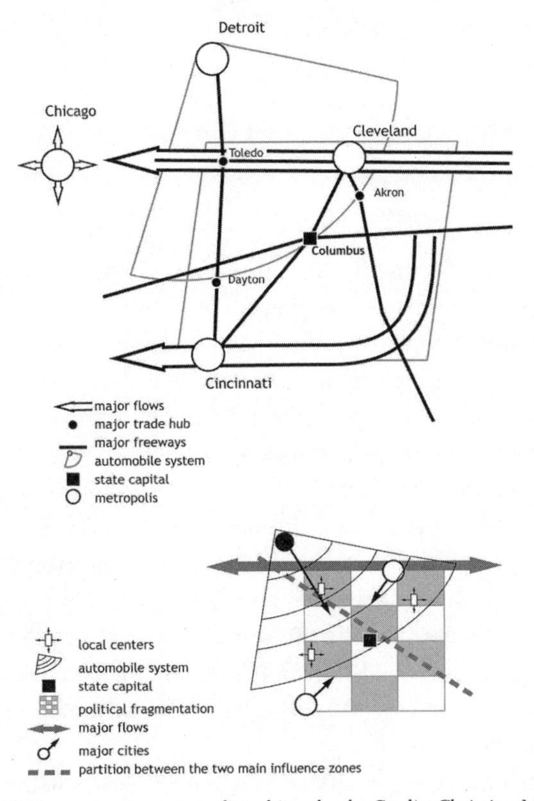

FIGURE 8.2 Ohio's competing metropolitan hinterlands. Credit: Christian Montès and M. L. Trémélo.

Columbus's future seems linked to FIRE, education and health care, and the capital could become one of Jean Gottmann's "transactional cities" (1983). However, in his 2000 study, geographer Henry Hunker still described Columbus as a government center with other activities induced by its capital status.[15] The proximity of competitors inside as well as outside the state may well limit the ultimate growth of Columbus, which is now driven largely by external forces and capital (Chicago).

Des Moines, Iowa

Though not a mushroom growth, the result of speculation and manifest destiny, Des Moines has rapidly, yet substantially, grown into prominence as one of the leading cities of the West.

(PETERSEN 1970, 223)

Gaining the State Capital was not the major factor in the growth of Des Moines.

(PETERSEN 1970, 217)

The city's selection in 1857 as Iowa's Capital had guaranteed longtime economic development.

(SCHWIEDER 1996, 175)

Des Moines could exemplify three processes. First, the case of capitals that have become their state's leading cities. Second, because it is only the twenty-third most populated capital MSA and the twentieth most populated capital city, it also exemplifies the "medium" capital. Third, because its MSA gained 44.9% between 1990 and 2010, it exemplifies the new momentum that state capitals experience today.

Tables 8.4 and 8.5 show that even though Des Moines was not among the first settlements in Iowa and had a slow start, it took the leadership, first timidly in 1880 (with two close contenders), and then strongly since 1890. The city did lose some inhabitants during the 1970s, but it has regained some since. Des Moines, despite this primacy in its state, has never figured among the biggest cities or MSAs in the nation. In 2010, the municipality held the 104th rank; and the MSA, the 88th rank, with 569,633 inhabitants, up from 392,928 in 1990.

W. Christaller and A. Lösch's central place theories help us to understand Des Moines's situation. Des Moines is the central place of an almost flat agricultural area, as are, on a lesser scale, other Great Plains state capitals such as Bismarck, North Dakota, or Pierre, South Dakota. This situation enabled Des Moines's political and geographical centrality to become an economic centrality, a fact enhanced by its designation as capital in 1857, in line with the settlement pattern of the state.

Iowa's first capital, Burlington, was located on its eastern fringe, where people were first settling in the new territory, which was founded in 1838.

Table 8.4 Population of Iowa's major cities, 1840–1920

	1840	1850	1860	1870	1880	1890	1900	1910	1920
Burlington	1,172	4,082	6,706	14,930	19,450	22,565	23,201	24,324	24,054
Cedar Rapids	—	—	1,830	5,940	10,104	18,020	25,656	32,811	45,566
Davenport	640	1,848	11,267	20,038	21,831	26,872	35,254	43,028	56,727
Des Moines	—	502	3965	12,035	22,408	50,093	62,139	86,368	126,468
Dubuque	1,059	3,108	13,000	18,434	22,254	30,311	36,297	38,494	39,141
Keokuk	493	2,478	8136	12,766	12,177	14,101	14,641	14,008	14,423
Iowa City	491	2,262	5,214	5,914	7,123	7,016	7,987	10,091	11,267
Sioux City	—	—	—	3,401	7,366	37,806	33,111	47,828	71,227

Source: Andriot 1980.

Table 8.5 Population of Iowa's major cities, 1930–2010

	1930	1940	1950	1960	1970	1980	1990	2000	2010
Burlington	26,755	25,832	30,613	32,430	32,266	29,529	27,206	26,839	25,663
Cedar Rapids	56,097	62,120	72,296	92,035	110,642	110,243	108,772	120,758	126,326
Davenport	60,751	66,039	74,549	88,981	98,469	103,264	15,333	98,359	99,685
Des Moines	142,559	159,819	177,965	208,982	200,587	191,003	193,189	198,682	203,433
Dubuque	41,679	43,892	49,671	56,606	62,309	62,374	57,538	57,686	57,637
Keokuk	15,106	15,076	16,144	16,316	14,631	13,536	12,451	11,427	10,780
Iowa City	15,340	17,182	27,212	33,443	46,850	50,508	59,735	63,027	67,862
Sioux City	79,183	82,364	83,891	89,159	85,925	82,003	80,505	85,013	82,684

Sources: Andriot 1980; US Census Bureau, 1980, 1990, 2000 (with 2001 correction for Iowa City: http://www.silo.lib.ia.us/datacenter), and 2010.

After a few years, however, the frontier line had moved ninety miles inland from the west bank of the Mississippi River, and a new town, christened Iowa City, was built in 1841 to host the capitol in a more central location. By 1857, Iowa City had already become too off-center, and Fort Des Moines was designated as the capital of the state and incorporated under the shortened name of Des Moines. Des Moines belongs to the second stage of westward movement in the state. It was created in May 1843 as a fort in the prairie wilderness, at the confluence of the Racoon and Des Moines rivers. In a second phase, a trading post was settled one mile west, on the site of three Indian villages. As pioneers continued to arrive, a town was surveyed in 1846 (Iowa's population had reached 96,088 by then) on a healthy plateau overlooking the rivers. This town of 127 people became in that year the seat of Polk County and was incorporated in 1853. When it became Iowa's capital, after only fourteen years of existence, the post in the wilderness had been transformed into a city. Not quite a boom town, Des Moines had only 3,965 inhabitants in 1860, up from the 502 in 1850, but far from Dubuque's 13,000 or Davenport's 11,267,[16] and still behind Keokuk, Burlington, Muscatine, and Iowa City.

Twenty years later, Des Moines's 22,408 inhabitants made it Iowa's leading city. The gilded dome of the new capitol, 275 feet tall, has been visible since 1884 for many miles. Built by the French architect A. Piquenard, who also designed the Illinois capitol, it was modeled on the dome of the Hôtel des Invalides in Paris. But Des Moines was not yet Paris! Until about 1875, its streets were still unpaved (in 1888 came the first electric streetcars). The growth of Iowa's cities was in fact far slower than the rise of its overall population: 192,214 people in 1850; 674,913 in 1860; and 1,624,615 in 1880. Iowa was a mostly agricultural state and could only induce the growth of medium-sized cities, while the cities that controlled Western expansion through transportation and manufacturing were located outside Iowa (St. Louis and Chicago). Furthermore, when the urbanization process began after the 1880s, settlement slowed, and population grew from 2,231,853 in 1900 to 3,046,355 in 2010. Although that does not amount to much in a low-density state, Des Moines's MSA now represents 18.7% of Iowa's population.

Des Moines benefited from the classic concentration of population and industry, such as John Deere and meat packing, resulting from the recurring agricultural crises (1930s, 1950s, 1980s) that devitalized small towns. The first territorial capital, Burlington, the first point of entry for immigrants, became the "Porkopolis of Iowa"—an important railroad center for shipments of lumber and cattle. But it is located at the edge of the state and remains a small city. Iowa City suffered a big blow when it lost the capital, although, to compensate the city for this future loss, the state university was established there in 1847 and opened in 1855. Its enrollment remained paltry throughout the nineteenth century, from 124 students in 1856–57 to 1,542 in 1900. Iowa City is still primarily a college town, despite some manufacturing (calendars), and it is a major medical and university center for the whole state. It has fine buildings, such as its 1840–42 Greek Revival capitol, "perhaps the most interesting architectural work in Iowa" (WWP Iowa 1959, 160). But the university and hospital are the contemporary growth factors that explain Iowa City's revival since the 1950s.

Sioux City is the leading city of the northwestern part of Iowa. It is far enough from Des Moines to grow, but it is located in a sparsely populated part of the state, which explains its standstill since the 1940s. Cedar Rapids, on the contrary, is centrally located in the east central and most populated portion of Iowa, and was able to expand (not in finance but in high technology). Such was also the case of Davenport that with Bettendorf, Moline and Rock Island form Quad Cities, on the Iowa-Illinois boundary. All of them of them are today strong relays of Des Moines's influence, as shown by table 8.6 and figure 8.3.

The central place model does not suffice to explain the evolution of Des Moines. Dealing with the question in the 1970 special issue of *The Palimpsest,* William Petersen rightly stressed that no unique factor could be singled out. The rich agricultural hinterland was important, but it was also available to every city in the state. Likewise, in 1878 there were six railroads that fostered the continual growth of agriculture, but other cities in the state also experienced a railroad boom: Cedar Rapids, Fort Dodge, and Waterloo. One of the supposed advantages that had been put forward to designate Des Moines as the capital—river transportation—remained unfulfilled (Hubler 1968). The Des Moines River ran from the north (Minnesota) to the south of the state and joined the Mississippi, and although it became the greatest commercial river in the state, its traffic, which peaked in 1859 at about one hundred steamboats, never attained the expectations of the town leaders, who wanted to make it the "gateway to the West." This failure was partly due to the cost of the works involved (twenty-eight dams and locks had been planned, but only six were completed) but mainly to the arrival of the railroad. Settlement had been too late for river navigation, as the railroad era was already coming to the West. Des Moines was reached in 1866. As a result, the Act of 1846, which declared the Des Moines River a public highway, was repealed in 1870.

Table 8.6 Des Moines, Cedar Rapids, and the Quad Cities: Economic comparisons

MSA	Percentage of state's GMP, 2001	Percentage of state's population	GMP/ pop.	Airport passengers, 2002 (millions)	Employment, 2003 (thousands)	Financial jobs, 2003 (thousands)	Professional and business jobs, 2003 (thousands)	High-tech rank, 1999
Des Moines	21.39	15.58	137	0.8	283.4	44.6	30.0	123
Cedar Rapids	8.61	6.60	130	0.8?	115.7	8.5	13.2	169
Quad Cities	13.78	12.24	113	?	170.1	8.9	20.8	—

Sources: US Bureau of Labor; airport websites; http://www.usmayors.com.

Note: Because the Quad Cities are located on both sides of the Illinois-Iowa boundary, the figures about GMP and population are to be taken only as broad indicators.

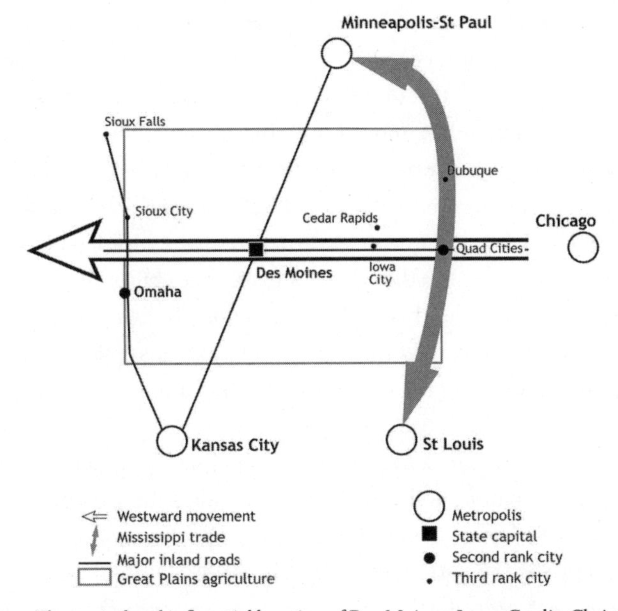

FIGURE 8.3 The central and influential location of Des Moines, Iowa. Credit: Christian Montès and M. L. Trémélo.

To explain Des Moines' success, other assets have therefore to be added, among them the proximity of productive coal mines. From 1870 to 1930, Des Moines also benefited from manufacturing growth in relation to Chicago and the East, although Iowa's industries have been located in medium-sized towns more than in major urban centers (Schwieder 1996, 351). Manufacturing was based on agriculture and could use the eight-foot falls of its six-hundred-foot-wide rivers for power.[17] Like every city in an agricultural state, Des Moines was also a trade center. It added academics with the opening in 1881 of Drake University, a private college relocated from Oskaloosa by the Disciples of Christ of Iowa.

But this still tells only part of the story. Two more factors were singled out, both linked to the winning of the state capital. According to Dorothy Schwieder, being a capital certainly "guaranteed longtime economic development" (1996, 175) to Des Moines. Although to quantify the exact weight of that guarantee proves difficult, we know that capitals attract newspapers and printers for state papers; this can explain Des Moines's powerful printing and publishing industry. Publishing first expanded by focusing on farming, then on country lifestyle magazines, dominated by the powerful Meredith family.[18] The links between insurance and state capitals are also well known (see chapter 7). But here too, this business went far further than in every other capital, except Hartford. The insurance boom at the end of the nineteenth century earned Des Moines the nickname of the "Hartford of the West." The first insurance company, Hawkeye (nickname of Iowa), was organized in 1866. Equitable of Iowa (1867) and the Bankers Life Company (1879) soon followed suit. In 1938, forty-seven insurance companies maintained home offices in Des Moines, with a payroll of more than 16,000 (WWP Iowa, 1959, 234). In 2002, Des Moines boasted on its website to have become the third largest insurance center in the world, with nearly sixty life, health, and casualty companies. In 2011, it had the "highest concentration of metro employment in financial services in the United States," with a growth of 20.5% between 2000 and 2011, when Hartford had lost 13.1%. The city's major employers were Wells Fargo (12,900 employees), Mercy Medical Center (6,900), Principal Financial Group (6,547), and Iowa Health (5,005), followed by Nationwide Allied/Insurance (4,396). What William Petersen deemed the most important factor was "the aggressive quality of the people who were lured to Des Moines in its formative decades that pointed the way to an ever-expanding industrial, commercial, and agricultural economy, liberally interspersed with a religious, educational, social, and cultural heritage" (Petersen 1970, 221). Indeed, the *New York Daily Graphic* closed an 1878 article on Des Moines with the following praise: "Energy and enterprise are visible on every street. It is no marvel that what has already been done is a matter of pride to every citizen or that their future aims are set high" (Petersen 1970, 247).

But, although the progressiveness of Des Moines's entrepreneurs has been real, the traditional "booster model" does not work. William Ferraro, in his study of the brothers of General William Tecumseh Sherman, could not find any clear relationship between the city's political leaders and entrepreneurs. Ferraro preferred Darwinism: "Why competitiveness flourished so vibrantly in Des Moines during its formative years defies precise explanation, but clearly the forceful personalities and shrewdness of men such as Frederick M. Hubbell [president of the Equitable Life and a speculator] and Jefferson S. Polk played a role larger than any structural features that differentiated Des Moines from Midwestern cities with more evidence of cooperation among their entrepreneurial layer" (Ferraro 1998, 273). One can add Hoyt Sherman (1827–1904), one of the sixteen founders of the Equitable Life Insurance Company in 1867, and its president from 1874 to 1888. But, by pointing principally to individuals, Ferraro tends to forget the broader principles and processes that can be found in nineteenth-century Midwestern and western boomtowns. His thesis does not explain why the shrewdest and most important personalities appeared here and not elsewhere. The availability of capital, like the location of the city where its owners settled, cannot be underestimated. Des Moines was in the center of the state, far enough from Chicago, St. Paul, Omaha, and St. Louis not to be overwhelmed by these cities. Iowa's rich agriculture contributed the capital necessary to attract insurance companies that had in turn to insure agriculture (machines and harvests for instance). Being the state capital brought at least the atmosphere of permanence that new cities needed in that mobile era of westward movement. Capital status certainly lured Drake University,[19] the printing and publishing industry, and a few industries. The Sherman brothers, for instance, came there because the new city seemed a good place to build fortunes. They perhaps stayed because, as the capital, Des Moines seemed more stable and attractive than other cities. Even if they did not ultimately succeed, they participated in the growth of the city.[20]

Had Des Moines not been selected as capital, it would probably still be Iowa's largest city, but it would not have the same skyline and perhaps not the same economic contours that it now has. The state is the city's second largest employer, providing seventy-five hundred jobs. In 2010, the Des Moines municipality had only 203,433 people in its 80.9 square miles (66 in 1938). It nevertheless puts on metropolitan airs (see fig. 8.4), although its official website contends that "metropolitan Des Moines provides the advantages of a big city with the atmosphere of a small town." A "small town" it may be, but it has buildings in the central business district of forty-four stories; a 2.8-mile, climate-controlled (but aging) skywalk connecting twenty-eight blocks; and an international airport that handles 1.8 million passengers and is ranked among the top fifty airports for air cargo tonnage. But it is also a small town with an average commuting time of only fifteen minutes, where the official goal of a "vibrant downtown" is still not entirely fulfilled, even after investments of almost $3 billion during the last decade.[21] (Seventy thousand people work in downtown Des Moines, but only sixty-five hundred people live there.)

Frankfort, Kentucky

Frankfort exemplifies the case of the ever-small capital. Although Frankfort was initially (1800) the second village of Kentucky, its rank in the state slowly receded as other cities grew. It had to wait for the twentieth century

FIGURE 8.4 The metropolitan skyline of Des Moines, Iowa. Taken from the steps of the statehouse, this photograph shows that a medium-sized, central place in a rural area might also wear some "metropolitan clothes" because of its importance in the financial sector. Photo by Christian Montès, August 2011.

to have more than ten thousand inhabitants, and did not grow again until the 1960s, after half a century of stagnation. Even so, it was only Kentucky's seventh city in 2010, with 25,527 people, far fewer than the metropolises of the state, Louisville and Lexington-Fayette (table 8.7).[22] Frankfort became the capital, not out of grand considerations, but because the commission had been instructed by the legislature to choose the town that pledged the largest contribution toward the construction of a statehouse.[23] Frankfort is not geographically central, but it is close to the demographic and economic heart of Kentucky, the Ohio River. But Frankfort was only slightly touched by the successive frontiers, making it a perfect example of dissociation between political centrality and economic centrality. This does not mean that Frankfort was happy with her *aurea mediocritas*. The capital's history is one of successive attempts to board the train of economic growth, a train it almost always missed because it is an "in-between capital," located between the state's most important cities, Lexington (twenty-two miles to the east) and Louisville (fifty miles to the west). In 2010, Frankfort had 0.06% of Kentucky's population.

Table 8.7 Louisville and Lexington, Kentucky's metropolises

MSA	Percentage of state's GMP, 2001	Percentage of state's population	GMP/ pop.	Airport passengers, 2002 (millions)	Employment, 2003 (thousands)	Financial jobs, 2003 (thousands)	Professional and business jobs (thousands)	High-tech rank, 1999
Louisville	31.85	25.38	125	3.5	569.4	38.5	62.6	85
Lexington	14.31	11.86	121	1.0	272.3	11.2	27.6	91

Sources: U.S. Bureau of Labor; airport websites; http://www.usmayors.com.

Seen in the eighteenth century as a "Western Eden," Kentucky was the main channel of pioneer flows into the Mississippi Valley and the first important English settlement beyond the mountains. After the creation in December 1776 of the County of Kentucky in Virginia, its population quickly increased to twelve thousand in 1783 and to a hundred thousand in the early 1790s (table 8.8). With westward flowing rivers and a flowering economy, Richmond, Virginia's capital, seemed another world. Statehood was soon asked for and obtained in 1792. The same year, Frankfort (established as Frankfort, Virginia, in 1786) was selected as state capital over Danville, the first important political city, after a six-month interlude at Lexington. Since then, the several attempts to remove the capital—Louisville having been the most prominent in its efforts—have failed.[24]

Kentucky had a good agricultural basis: hemp (until 1865), tobacco, and horse breeding are well known, as is the famous bourbon.[25] The state was well located in relation to New Orleans. The city that would control the overland routes, the most important being from Ohio to Mississippi, would therefore become the leading city of that part of the nation. Because of geography, instead of a fight between Frankfort and other cities, the fight raged between Lexington (organized in 1781) and Louisville (founded in 1780), and between both of those cities and Cincinnati (founded in 1788).

Lexington was the first to lead, because in the 1780s it became the land transportation hub of the state. The European war and the credit bubble had created some local industries. It was even briefly the largest "city" in the West, with 1,795 inhabitants in 1800. But its position as the main commercial center was challenged in the late 1810s, when the river began to concentrate all the trade. Louisville, the "River City," led the movement and supplied the southern plantations. It gained more influence with the opening in 1829 of the first canal around the falls of the Ohio River. Kentucky's population meanwhile grew quickly, reaching 570,000 in 1820 and 1,132,011 in 1870. Lexington could not compete and remained a small city until the 1960s. Without a riverfront, its turnpike and railroads were unable to break its isolation. Its local university tells the same story; after brief growth in the 1820s, which earned Lexington the nickname of the "Athens of the West," Transylvania University declined with the city. Louisville itself was passed by in the 1850s because the new railroads favored the northwest, reorienting trade toward Cincinnati and Chicago. But Louisville retaliated with the Louisville & Nashville Railroad, paid for with municipal bonds and subsidized by the city, which gained a monopoly in the Ohio and upper Mississippi River Valley market (Kentucky, Tennessee, and Alabama).

Table 8.8 Population of Kentucky's major cities, 1800–1910

	1800	1810	1830	1840	1860	1870	1880	1890	1900	1910
Bowling Green	41	154	821	1400	?	4574	5144	7803	8223	9173
Covington	—	—	743	2,026	16,471	24,505	29,720	37,371	42,938	53,270
Danville	270	432	849	1,223	4942	2542	3074	3,766	4,285	5420
Frankfort	628	1,099	1,682	1,917	3,702	5,396	6,958	7,892	9,487	10,465
Lexington	1,794	4,326	6,404	6,997	9,521	14,801	16,656	21,567	26,369	35,099
Louisville	359	1,357	10,336	21,210	68,033	100,753	123,758	161,129	204,731	223,928
Newport	106	413	717	1,400	10,046	15,087	20,433	24,918	28,301	30,309
Owensboro	—	—	—	—	2,308	3,437	6,231	9,837	13,189	16,011
Paducah	—	—	105	1,000	4,590	6,866	8,036	12,797	19,446	22,760

Sources: Moffat 1992; Andriot 1980.

The construction of a direct rail connection from Cincinnati to eastern Tennessee, bypassing Louisville, could not break the economic and political power of the Louisville & Nashville. In 1900, Louisville had 10% of the state's population.

When young General James Wilkinson founded Frankfort in 1786, his reason was that he saw its potential as "a collecting point for goods intended for export by way of the Ohio and Mississippi rivers" (Kramer 1986, 21). But although Wilkinson had obtained from Spain a commercial monopoly in the Spanish trade, and rich agricultural lands lay around it, Frankfort proved unable to fulfill Wilkinson's dreams. According to Carl Kramer, the reasons are not clear: among them are the crude surroundings of the town and the cheap land available at outlying settlements and stations. But the main reason was certainly that Frankfort was located far too close to the state's largest cities and lacked their assets. This prevented it from succeeding in its successive attempts to broaden its economic basis. Another reason was that it suffered from a bad image. The city was viewed as the state's capital of political corruption. According to Steven Channing, during the nineteenth century, Kentucky experienced political and cultural provincialism, parochialism, and general corruption. In 1900, the Democratic candidate who had run for governor in 1899 was assassinated. The period that followed was no better: "apathy, raids on the public trusts, and reactionary moral crusades," Democratic gerrymandering, and the "intellectual bankruptcy of the Republican" reigned until 1945 (Channing 1977, 165, 177). The common saying was, "An honest politician was one who when bought, stayed bought" (Kramer 1986, 252). Unfortunately, Kentucky was far from being alone in such muddy waters.

To succeed economically, Frankfort had first to become accessible. Its civic leaders tried therefore to promote turnpikes, steamboats, and railroads. Even after the completion of several turnpikes by 1840, however, river trade remained foremost and was firmly in the hands of Louisville. When the railroad reached the capital in 1852, Frankfort was only a stop between the state's largest cities. Moreover, the Lexington, Frankfort, and Louisville Railroad never achieved the aim of its promoters to provide a direct rail connection between the bluegrass and the cotton kingdoms. It remained a simple link in the national railroad network (Kramer 1986, 133). The white-collar dreams of Frankfort proved equally disappointing. As the state capital, it attracted attorneys, but in 1850, the bar counted only twenty-six members. Frankfort's attempts at finance were short-lived. Its banking monopoly lasted only six years, from the chartering in 1806 of the

Bank of Kentucky until 1812. Half a century later, the Frankfort Fire and Marine Insurance Company and the Frankfort Building and Loan Association, approved by the General Assembly in 1869 and 1872, remained paper companies, their organizers having failed to sell the necessary stock.

The economy of Frankfort still rested on the success of agriculture, and it was not brilliant. It nevertheless allowed the formation of several manufacturing companies, such as the Frankfort Canning Company, or the Frankfort Modes Glass Work, established in 1907, which employed 350 people (half the city's industrial workforce). The other big plant was owned by the Hoge-Montgomery Company, founded in 1889, which manufactured shoes and offered three hundred jobs, aside from its use of prison labor.[26] To promote the city, economic leaders organized themselves, first under the aegis of the Commercial Club (until 1912), then as the Frankfort Business Men's Club (1908), which became the Frankfort Chamber of Commerce in 1915. As was the case one century earlier, the priority was better transportation. "The chamber lobbied successfully to have three major interstate highways—the National Midland Trail, from Washington, D.C., to San Francisco; the Jackson Highway, between New Orleans and Niagara Falls; and the Boone Way, which crossed five states—routed through the capital" (Kramer 1986, 276). It consequently promoted the capital as a tourist and convention center that "capitalized" on its political status.[27] But the city suffered a major blow from Prohibition, which crushed its distilling industry and hindered further growth.

Since 1945, Kentucky's agriculture has experienced several ups and downs: the Tennessee Valley Authority, after an initial impetus, did no more succeed in transforming the local economy than did coal. The foundation of Kentucky Fried Chicken created jobs, but not enough to reverse the downhill tendency. On the brighter side, the state benefited from the southern migration of manufacturing, essentially to its major cities. Louisville, for instance, saw the arrival of General Electric and Ford plants. Kentucky's population grew from 3,685,295 in 1990 to 4,339,367 in 2010. Nowhere were these processes clearer than in Lexington. After the arrival of an IBM plant in 1956, the city shed its farming and college-town clothes to put on metropolitan ones.

Frankfort also began to see a rise in its population, but it probably was too late. With 11,916 inhabitants in 1950, it was a small town, lacking the urban atmosphere and resources sought by the new economy that was coming into being. It had no significant workforce and no local capital, and all these elements augured badly for the city, although it had one of the

Table 8.9 Population of Kentucky's major cities, 1920–2010

	1920	1930	1950	1960	1970	1980	1990	2000	2010
Bowling Green	9,638	12,348	18,347	28,338	36,253	40,450	40,641	49,296	58,067
Covington	57,121	65,252	64,452	60,376	52,535	49,585	43,264	43,370	40,640
Danville	5,099	6,729	8,686	9,010	11,542	12,942	12,420	15,477	16,218
Frankfort	9,805	11,626	11,916	18,365	21,356	25,973	25,968	27,741	25,527
Henderson					22,976	24,834	25,945	27,373	28,757
Hopkinsville					21,395	27,318	29,809	30,089	31,577
Lexington	41,534	45,736	55,534	62,810	108,137	204,165*	225,366	260,512	295,803
Louisville	234,891	307,745	369,129	390,639	361,472	298,694	269,063	256,231	741,096**
Newport	29,317	29,744	31,044	30,070	25,998	21,587	18,871	17,048	15,273
Owensboro	17,424	22,765	33,651	42,471	50,329	54,450	53,519	54,064	57,265
Paducah	24,735	33,541	32,828	34,479	31,627	29,315	27,256	26,307	25,024

Sources: Andriot 1980; US Census Bureau, 1980, 1990, 2000, and 2010.

* After 1970, it is Lexington-Fayette (174,323 inhabitants).

** After 2003, the City of Louisville merged with Jefferson County.

1890 land grant universities. If Frankfort more than doubled its population, to a modest total of 27,741, it was largely due to annexations. The growth of state government mostly explains the city's workforce expansion. In fact, government employed only 900 people in Franklin County in 1940. That number exploded to 2,300 in 1950 (853 in Frankfort), and to 4,700 in 1970, 3,400 of whom resided in Frankfort. The city's leaders nevertheless complained of the city's excessive dependence upon state government and sought once more to diversify the economic basis. To attract new manufacturing plants, the city created a park and recreation system and fostered urban renewal around a new complex called Capital Plaza. It included new federal buildings and was centered on a twenty-eight-story state office tower. The leaders also tried to revive the tourist and convention goals of the 1920s. Since that date, the Capital Hotel had closed (1962), and the city had to wait until 1979 for the designation of the central business district as a historic district, and until 1984 to see the opening of the 189-room Capital Plaza Hotel (Kramer 1986, 386–88).

Today, although Frankfort has participated, according to its scale, in the southward movement of automobile plants, automotive parts manufacturing accounted for only 7.6% of Franklin County's thirty thousand jobs in 2011.[28] Government is still the county's major employer (38.5%), followed by services (27.8%). Thus, Frankfort is a charming city (77.1% white and 16.5% African American) set amid the gentle hills of Kentucky but not a vibrant metropolis that attracts web-based business and venture capitalists. The conclusion that Carl Kramer drew from his study underlined this idea: "The community's rich historical and architectural heritage, combined with its scenic beauty and role as capital, offers opportunities for the tourist business which still are largely untapped. Frankfort's strategic location between Louisville and Lexington is advantageous for tourism and industrial development alike" (Kramer 1986, 396). He also stressed the importance of Kentucky State University. Although it only recently became a center for graduate studies, and the links between town and gown do not seem to be very close (personal communication), its economic impact was estimated at $100 million in 2010. But Kramer's analysis only emphasizes the fact that the capital depends heavily on outside corporations (like the automobile industry) and metropolises (tourists coming from the state's biggest MSAs). As for its "strategic location," Frankfort is too far from Louisville and Lexington to become an edge city, and still too close to attract the larger corporations (fig. 8.5). City leaders prefer to attract small businesses, so that the city remains diversified; they nevertheless participate in a new

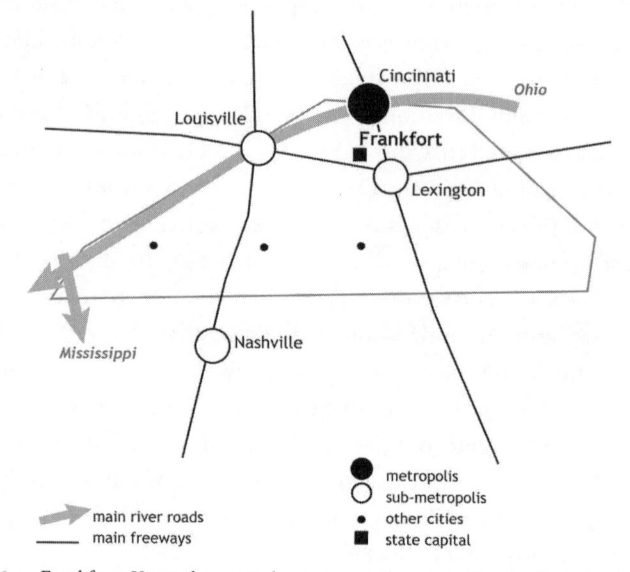

FIGURE 8.5 Frankfort, Kentucky, an in-between small capital. Credit: Christian Montès and M. L. Trémélo.

FIGURE 8.6 Frankfort, Kentucky, a charming city. Taken from the Downtown Frankfort Organization offices, this view shows the prominence of the statehouse in the landscape as well as the charm of a small, Victorian-style town: the family shoe store might have disappeared, but the city is busy trying to revive interest in its downtown. Photo by Christian Montès, June 2012.

regional partnership (BEAM: Bluegrass Economic Advancement Movement), under the aegis of Louisville and Lexington, that focuses on advanced manufacturing (June 2012, personal communication).

Anthony Trollope's description in *North America* (1862) still rings true: "Frankfort is the capital of Kentucky, and as quietly dull a little town as I ever entered. It is on the river Kentucky, and as the grounds about it on every side rise in wooded hills, it is a very pretty place. . . . Daniel Boone lies here." The contemporary version of this text (on the city's website) tries to lure visitors: "Frankfort, Kentucky's State Capital, is where history awaits you. It's not that time stands still here, but the charm and beauty of this historic town have been so well preserved throughout the years, a visit to Frankfort is almost like a visit back in time." (fig. 8.6)

These three cases bring to a close our study of American state capitals. At the beginning of the book, we asked if they were public company towns or mainly symbolic towns and wondered why they did not form an immediately understandable corpus owing to their various sizes and positions in the national urban system. The answer is evidently not a simple one. Each capital has its own history, which makes it unique on the map and in the hearts of (almost all) the citizens of its state, but the previous pages have shown that they are all integrated in the broader picture of the territorial construction of the United States. At first, from the seventeenth to the nineteenth century, they embodied the often delicate workings of democracy in an era of uncertainty in which compromises and spatial changes were interwoven to explain the selection and subsequent evolution of definitive capitals. They then embodied the constant readjustment between political and economic forces in a rapidly growing nation, and they perhaps testify to the failure of politicians to control the economy, which now largely dictates the evolution of state capitals, except in their plats and public monuments. But before putting the last period to this study, a look at the fate of cities that have lost capital status might shed a new light on the subject.

9

Losing Status

The Place of Former Capitals in Today's America

Losing capital status was a bitter pill to swallow for most former capital cities. Even the biggest commercial towns resisted losing such prestige. New York, for instance, tried desperately to recover the capital in 1865, almost one century after having lost it, when the Senate appointed a committee of three "to ascertain by correspondence with various municipalities of the State, on what terms the grounds and buildings necessary for a new Capitol and public offices could be obtained" (Beach 1879, 406). The Big Apple offered the committee a share of any public park in the city, including Central Park, to build a new capitol, as well as a magnificent governor's mansion on Fifth Avenue.

This episode also shows that even if many towns vied for such a high honor (among them Yonkers, Saratoga Springs, Athens on the Hudson, Whitestown, Argyle, Sing Sing, Fulton, and Margaretville), there were at least as many villages and cities (e.g., Buffalo, Oswego, and Utica) that wanted nothing to do with it. The president of the village of Sandy Hill, Washington County, was the most vehement, writing that "if the time has come when our capitol is to go to the highest bidder, like almost everything that has any connection with our present legislation, then I would plainly and frankly say that our people are not the ones to offer large bribes or inducements for the purpose of building up their place or people to the detriment or inconvenience of all the rest of the people of the State" (Beach 1879, 406). The infuriated president of Sandy Hill may have been driven by partisan politics as much as high morality.

Sometimes the loss of capital status was received with great equanimity. This was the case for the citizens of Newport, Rhode Island, when the General Assembly decided in 1900 that Providence was to become sole capital.

Newport had been the principal co-capital for more than two centuries. One of the socialites who adorned the town at the turn of the century, Mrs. John King Van Rensselaer, commented in her usual haughty and deprecating manner: "When the State House at Newport was abandoned the fantastic title was curtailed, and Providence became the reigning business capital of the smallest State in the Union, while Newport is self-elected the social capital not only of her own island and miniature State, but also maintains a claim to be that of the whole Union" (King Van Rensselaer 1905, 123–24). So much for Providence! But she was right: in 1905, Newport was indeed one of the social capitals of the Union, especially in the summer—the season of regattas and horseracing.

The objective here is to estimate the effects of losing the seat of government and to assess the place of former capitals in today's urban and rural America. In fact, most former capitals participate fully in the process of re-actualization of memories described by Bernard Debarbieux (1995, 106). Such a re-actualization was desirable, because former capitals are generally in a worse situation than current ones, at least from a demographic point of view (see table 9.1).

Certain trends are revealing. First, out of the eighty-seven former capitals here studied, seven (8%) have disappeared from the charts. The second trend is that twenty-seven of them (34%) lost population between 1990 and 2010. The third trend is that the other former capitals have mostly remained small towns or cities. Six have fewer than 1,000 inhabitants; ten have from 1,000 to 10,000; twelve, from 10 to 20,000; eighteen, from 20 to 50,000; ten, from 50 to 100,000; twelve, from 100 to 250,000; five, from 250 to 500,000; three, from 500,000 to 1 million; and four have more than 1 million (New York, Los Angeles, Philadelphia, and Houston). More than half of them (57.5%) have fewer than 50,000 inhabitants; that figure is 24% for current capitals. Seventy percent have fewer than 100,000 inhabitants; for current capitals, that figure is 38%. Only 15% of former capitals have more than 250,000 inhabitants vs. 26% for current capitals.

Only a small minority of former capitals are among America's greatest conurbations. The diversity of these six metropolises reflects the multinational history of the founding of the United States, for they have been settled (according to historical order) by the Dutch (New York), British (Philadelphia), French (St. Louis and New Orleans), Spanish (Los Angeles), and by independent Texans (Houston). We need not study them here; the American metropolization process has already been analyzed by scores of

Table 9.1 Population of former capitals, 1990–2010

State	Former capital	2010	1990	2010/ 1990 (%)
Alabama	St. Stephens	—	—	—
	Huntsville	180,105	159,880	+12.7
	Cahawba	—	—	—
	Tuscaloosa	90,468	77,866	+16.2
Alaska	Sitka	8,881	8,588	+3.4
Arizona	Prescott	39,843	26,592	+49.8
	Tucson	520,116	415,444	+25.2
Arkansas	Arkansas Post	—	—	—
California	Monterey	27,810	31,954	−3.0
	Los Angeles	3,792,621	3,485,557	+8.8
	San Jose	945,942	782,224	+20.9
	Vallejo	115,942	109,199	+6.2
	Benicia	26,997	24,431	+10.5
Colorado	Colorado City (CDP)	2,193	1,149	+90.9
	Golden	18,867	13,127	+43.7
Connecticut	New Haven	129,799	130,476	−0.5
Delaware	New Castle	5,285	4,837	+9.3
Florida	St. Augustine	12,975	11,695	+10.9
	Pensacola	51,923	59,198	−12.3
Georgia	Savannah	136,286	137,812	−1.1
	Augusta–Richmond City	195,744	186,616	+4.9
	Louisville	2,493	2,486	+0.3
	Milledgeville	17,715	17,727	−0.1
Idaho	Lewiston	31,894	28,082	+13.6
Illinois	Kaskaskia	14	32	−56.2
	Vandalia	7,042	6,114	+15.2
Indiana	Vincennes	18,423	19,867	−7.3
	Corydon	3,122	2,661	+17.3
Iowa	Burlington	25,663	27,208	−5.7
	Iowa City	67,862	59,735	+13.6
Kansas	Shawnee (Mission) City	62,209	37,962	+63.9
	(Fort) Leavenworth	35,251	38,495	−8.4
	Pawnee Rock City	252	367	−31.3
	Lawrence	87,643	65,608	+33.6
	Lecompton City	625	619	+1.0
Kentucky	Lexington-Fayette	295,803	225,366	+31.3
Louisiana	Mobile (Alabama today)	195,111	199,973	−2.4
	New Orleans	343,829	496,938	−30.8
	Donaldsonville	7,436	7,949	−6.5
	Shreveport	199,311	198,518	+0.4
Maine	Portland	66,194	64,157	+3.2
Maryland	St. Mary's	—	—	—
Massachusetts	Beverly	39,502	38,195	+3.4
Michigan	Detroit	713,777	1,027,974	−30.6
Mississippi	Columbia	6,582	6,815	−3.4
	Natchez	15,792	19,460	−18.8
	Washington	—	—	—
Missouri	St. Louis	319,294	396,685	−19.5
	St. Charles	65,794	54,555	+20.6
Montana	Virginia City town	190	142	+33.8
Nebraska	Bellevue	50,137	39,240	+27.8
	Omaha	408,958	344,463	+18.7

New Hampshire	Portsmouth	20,779	25,925	−19.8
	Exeter town	14,306	12,481	+14.6
New Jersey	Elizabeth	124,969	110,002	+13.6
	Burlington	9,920	9,835	+0.9
	Perth Amboy	50,814	41,967	+21.1
New York	New York	8,175,133	7,322,564	+11.6
	Kingston	23,893	23,095	+3.5
	Poughkeepsie	32,736	28,844	+13.5
North Carolina	New Bern	29,524	20,728	+42.4
	Fayetteville	200,564	75,850	+164.4
North and South Dakota	Yankton (SD)	14,454	12,703	+13.8
Ohio	Marietta	14,085	15,026	−6.3
	Cincinnati	296,943	364,114	−18.4
	Chillicothe	21,901	21,923	−0.1
	Zanesville	25,487	26,778	−4.8
Oklahoma	Guthrie	10,191	10,440	−2.4
Oregon	Oregon City	31,859	14,698	+116.8
Pennsylvania	Philadelphia	1,526,006	1,585,577	−3.8
	Lancaster	59,322	55,551	+6.8
Rhode Island	Newport	24,672	28,227	−12.6
	Bristol	22,954	21,625	+6.1
	South Kingstown CDP	30,639	24,631	+24.4
	East Greenwich	13,146	11,865	+10.8
South Carolina	Charleston	120,083	88,256	+36.1
Tennessee	Knoxville	178,874	169,761	+5.4
	Murfreesboro	108,755	44,922	+142.1
Texas	San Felipe de Austin	747	618	+20.9
	Washington / Brazos	—	—	—
	Houston	2,099,451	1,654,348	+26.9
Utah	Fillmore city	2,435	1,956	+24.5
Vermont	Rutland	16,495	18,230	−9.5
Virginia	Jamestown	—	—	—
	Williamsburg	14,068	11,409	+23.3
West Virginia	Wheeling	28,486	34,882	−18.3
Wisconsin	Belmont (village)	986	823	+19.8

Source: US Census Bureau, 1990 and 2010.
Note: Hawaii, Minnesota, Nevada, New Mexico, Washington, and Wyoming had only one capital. The dash (—) means that the city no longer exists.

social scientists. One point is worth mentioning with regard to this study: St. Louis's wish to become capital of the United States, which shows that the ambitions of large economic centers went far beyond the boundaries of their states. St. Louis did not really fight to remain as Missouri's state capital in 1820; on its way to becoming the alleged "fourth city in the nation" (with the help of some census tricks), however, the city did fight to become the national capital. Newspapers had already put the idea forward during the 1850s, but the movement did not begin seriously until the arrival in 1866 of Logan U. Reavis in St. Louis. Coming from Illinois, he launched a broad news campaign in favor of St. Louis. The Midwest had numerous arguments in its favor: it was the home of the Republican Party, the main contributor to the Union army, as well as the geographic and demographic center of the country (Primm 1998, 273). A meeting was held on the subject of removal to St. Louis in October 1869, with delegates and governors from twenty-one states and territories, but it led to nowhere. First, because in 1870, President Grant proved to be opposed to the move on legal grounds, stating that even if it were constitutional, it would be similar to amending the Constitution (Bowling 1988, 178). Second, because the House of Representatives decided to finance new federal buildings in Washington, definitively settling the matter. Since that time, St. Louis has regularly declined in American rankings. Although it is still a second or third level metropolis, the city is experiencing great difficulties in coping with its economic problems as well as a continuous decline in its population.

On the other end of the spectrum, we find dwindling villages returning to oblivion. Such a situation is naturally not unheard of in the United States, where abandoned settlements abound. They are testimonies to the roughness of territorial construction, where a grand history was (until the twentieth century at least) not enough to resist economic downfalls or geographical misfortunes. Kaskaskia, Illinois, is our first example. Founded as Notre Dame de Cascasquias by French Jesuits who wanted to compete with Sainte Famille de Kaoquias, which had been founded by another congregation, Kaskaskia was well situated on the Mississippi River fur trade routes and became Illinois's most important town during the eighteenth century. Then called "the Paris of the West," the following description emphasizes its early "splendor": "The principal buildings are the church and Jesuits house, which has a small chapel adjoining to it. These, as well as some other houses in the village, are built of stone, and, considering this part of the world, make a very good appearance. . . . Sixty-five families reside in the village, besides merchants, other casual people, and slaves."[1] In 1787, Kaskaskia had 191 inhabitants and was a "handsome little village" according to General Josiah Harmar. First capital of the new territory and then of the state, from 1809 to 1820, it disappeared because of the Mississippi, although the river had created its prosperity. It was frequently flooded and disappeared during the great flood of 1881, which changed the course of the river. It is now Kaskaskia Island, the only part of Illinois west of the Mississippi (Carrier 1993, 11). In 2010, only fourteen people lived in what remained of the capital, a loss of eighteen since 1990.

Louisville, Georgia's capital from 1796 to 1807, fared only slightly better after losing the capital to Milledgeville.[2] The town tried to revive its declining economy by making the Ogeechee River navigable to the city. This plan having failed, the economy remained based on the cotton that grew in its fertile soil. Shortly after losing the capital, in 1810, the town's population was only 524. It never succeeded in becoming a "real" town. In 1870 the census reported 356 inhabitants. The cotton weevil took its toll, and one century later, in 1970, there were still only 2,691 residents. Since that period, the village has vegetated; it had only 2,493 people in 2010.

Aside from these two extreme categories, a pattern of evolution emerges, stemming from the growth of heritage awareness among Americans, which helps to provide a central motif, even if it does not fit every former capital. Especially since the bicentennial celebration in 1976, this awareness has fostered a renaissance in most of these towns, which are part of the growing tourist and heritage economy. Iowa City provides us with a telling example. In 1970, the city completed the construction of its capitol, begun in 1840 and left unfinished when the capital was transferred to Des Moines in 1857 (fig. 9.1). It was dedicated on July 3, 1976, the day before the bicentennial (Keys 1986).

FIGURE 9.1 The former state capitol, Iowa City, Iowa. Today the heart of the University of Iowa campus, this building shows that the classical domed tradition endures, and that citizens carefully protect such testimonies to their history. The dome was only finished in 1976. Photo by Christian Montès, August 2011.

Former Capitals Re-created in the Name of Heritage and Tourism

Williamsburg is the seminal instance of the process of historic recreation of a former capital that would otherwise have perished in anonymity,[3] as indicated in the following quotation: "The past constitutes Williamsburg's livelihood, its present, and its future" (WWP Virginia 1940, 314). Later examples are numerous, and include New Castle (Delaware), St. Augustine (Florida), Guthrie (Oklahoma), and even Charleston (Hamer 1998, 2–7). Such a process was possible because the loss of capital status ossified these cities and preserved them from most urban and industrial misdeeds.

Tourism may not play a particularly great role in present state capitals, but heritage and political tourism often prove to be the bases for the renewal of the identity and economy of former capitals. They count on their heritage more than present capitals do. The paradox is only apparent: they have more to gain. When a city is a capital, it can count on the numerous jobs offered by the administration (more than half of the total employment for the smallest), but cities that have lost this honor must find other sources of revenue. This process is quite banal: tourism replaces politics, as it might replace agriculture in other cities. The fact that most former capital cities have remained small is an asset, because their monuments have been better preserved than those in other cities. Besides, being the capital often led to the construction of some prestigious mansions, which can also be seen as an advantage.[4] Former capitals also benefit from the over-valorization of small towns in the American collective mentality (cf. Francaviglia 1996). Their shops offer souvenirs of well-known people that make the process of re-actualizing memories easier. Tourism has also been stimulated by the surge of historic awareness among Americans during the 1976 bicentennial celebrations, which led to the multiplication of local heritage associations and to the passing in 1981 of the Economic Recovery Tax Act, which provided for a 25% federal tax credit for rehabilitation (reduced in 1986), inducing a boom in preservation. Because former capitals are small, however, they have developed a leisure economy more than a broad tourist economy. Their lodging capacities are modest, and their beauties can be visited in one day at most, often by way of a pedestrian tour—an exception in a country devoted to the automobile. All are far from having been transformed into major tourist locations, but most former capitals emphasize the remnants of their former status with pride on their websites.[5] Three categories can be distinguished.

The Re-creation of the Past through Archaeological Parks

We begin with Texas, because heritage preservation began very early there, earlier even than for Williamsburg. One possible explanation is the pride with which Texans recall the 1836–1845 period, when Texas was an independent state. Hollywood has largely dwelt on this period (e.g., in the movie *Alamo*, 1960, with John Wayne and Richard Widmark), but more "serious" heritage celebration was already under way: Washington on the Brazos became a historic park in 1915. Like another former capital, San Felipe de Austin, Washington on the Brazos had not fared well since the late nineteenth century, due more to being bypassed by the railroad than to the loss of the capital. For present capitals, being bypassed by the railroad meant slow growth; for former capitals, it meant stagnation or, even worse, regression.

San Felipe de Austin, founded in 1824 as headquarters of the first colony of American pioneers, led by Stephen F. Austin, was the first Anglo-American settlement in Texas. It thus became the unofficial capital of American settlement. The site was chosen for its central location, on a high bluff overlooking fertile lands and the Brazos River, where a ferry was already in operation (see the Handbook of Texas Online website, http://www.tshaonline.org/handbook/online). It had six hundred inhabitants in 1835 and was an important Texas trading center, second only to San Antonio, when it served as the capital of the provisional government. But its citizens fled during the 1836 battles against the Mexicans, and the government had to find new quarters. San Felipe de Austin never recovered from that blow, despite having been made the seat of Austin County in 1837, an honor it lost to Bellville in 1846. In the 1870s, the town refused the railroad, and nearby Sealy grew instead. Population declined to 177 by 1890, and only 313 inhabitants lived there in 1940, when the WPA writers described it as a "shabby, weed-grown community," an "almost a deserted village of dusty, narrow thoroughfares" (WWP Texas 1940, 598). Although San Felipe did survive the move of the capital, it was still only a small village of 747 inhabitants in 2010. To honor its more glorious past and attract visitors, the town created the 4,200-acre Stephen F. Austin Memorial Park in 1928, including a replica of Austin's log cabin, and donated it to the state in 1940.

Washington on the Brazos (the latter part of the name being added after the Civil War) could have had a great destiny. In addition to its grand name, it has been dubbed the Texan Philadelphia.[6] Indeed, it was here that, on March 2, 1836, the Texas Declaration of Independence was drawn up, a national constitution was written, and an ad interim government was adopted. It also became a major political and commercial center. Like San Felipe de Austin, it was created on the site of a ferry crossing (in service since 1822) at the upper limit of Brazos River navigation, when the Washington Town Company was established in 1835. In December of the same year, it became General Houston's headquarters, and the town's hundred citizens succeeded in attracting the 1836 convention by offering a free assembly hall. It became the county seat and was incorporated in 1837. It lost the county seat in 1841, but became the capital from 1842 to 1845, when Mexicans captured San Antonio and threatened Austin. President Houston tried to remove the archives from Austin to Washington, but the citizens of Austin resisted, and the archives remained at Austin; this bloodless skirmish has been called the Archives War (Roberts 1898, 120). Steamboat navigation and cotton brought some prosperity, and the population grew to 750 in 1856. But in 1858, the town's refusal to pay the railroad to come to it signaled the beginning of a rapid decline. Thirty years later, the town was almost empty. Although the legislature transformed it into a park in 1915 and held the annual Texas Independence Day Festival there on March 2, Washington had only one hundred residents in 1940, when the WPA writers described it as follows: "Sprawling over a reddish bluff above the yellow, muddy waters of the Brazos River, is a down-at-the-heels country village largely populated with Negroes of the old plantation type" (WWP Texas 1940, 644).

Today travelers can visit Washington State Park, which has a reproduction of an early capitol of the Republic of Texas; the home of the last president of Texas, Anson Jones; and the Star of the Republic Museum, which opened in 1970 and displays the findings made by archaeologists during the 1960s. In 1976 and 1996, the initial 293.1 acres of the park were augmented by new acquisitions from private owners. As in other parts of the United States, the real development of the park only began in the 1990s, with an expenditure of $6 million.

Re-creation can become a true resurrection. Former ghost towns, such as St. Mary's City, Maryland, and Arkansas Post, Arkansas can become archeological parks. Alabama presents good examples.[7] This state had little luck in the choice of its first two capitals, because both Saint Stephens and Cahawba are now ghost towns.[8] But their rediscovery happened later than that of Texas's former capitals. As elsewhere in the nation, the 1976 celebrations were a powerful stimulus and led local associations to begin the process of protection even before the state came in to help. The leaflet given to the tourist willing to explore the remains of Cahawba shows that to lose the capital is certainly a big blow for a city, but it is far from deadly. A poor site and the "casualties of war" were far better town killers.

An early Indian settlement, Cahawba became capital of the brand new state of Alabama in 1819, occupying a site donated by President James Monroe near the Alabama River. After a brief stay in Huntsville to allow time to build a statehouse and accommodations, the legislature moved to Cahawba in 1820. But the new capital's opponents dwelled on its reputation for flooding and unhealthy atmosphere to argue for another location. In 1826, the capital moved to Tuscaloosa, almost emptying the former "capital village" of three hundred inhabitants. The loss of the capital and the exaggeration of the flooding by the *Montgomery Republican* and other newspapers that favored competing towns did not bring the end of the city. On the contrary, Cahawba became an active port for shipping cotton to Mobile, and it was linked to the railroad in 1859. In 1860 it had between two thousand and three thousand inhabitants (according to different sources). But the destruction of the railroad during the Civil War and the 1865 flood led to the transfer of the county seat, along with most of the population, to nearby Selma in 1866. Only 431 people remained in 1870. Cahawba then became the "Mecca of the Radical Republican Party" and a rural center for seventy former slave families, but they had disappeared by 1900, and the city returned slowly to nature and was finally unincorporated in 1989. Thanks to the work of the Cahawba Concern organization, founded in 1979, Cahawba became an Archaeological Park under the supervision of the Alabama Historical Commission, which aims to create "a full-time interpretative park." The park's leaflet is bucolic: "Visitors are welcome at Old Cahawba. Enjoy the wildflowers. Take the time to roam the abandoned streets, view the moss-covered ruins, talk with an archaeologist, read the interpretative signs, and contemplate Cahawba's mysterious disappearance."

More than eighty thousand visitors contemplated this mystery in 1997. Walking through the moss-covered old Spanish oaks and the damp atmosphere of mid-April in 2000, amid the heavy scents of decaying vegetation, was indeed quite strange. All that remains are the streets bordering empty fields, wells, and a burial ground grown into a wild forest, where one finds here and there a surviving tomb.[9] Slave quarters, isolated columns, and one 1841 house are still standing. They have belonged to the same family since 1894 and were recently bought by the Historical Commission. Near the entrance gates, some antique mobile homes rust. They seem to be abandoned, but ageless people out of the *Grapes of Wrath* occupy them, barely surviving. This really is the deepest Old South.

The Valorization of Heritage in Small Towns That Rely Heavily on Tourism

New Castle, Delaware, paints another picture. As an older town, it could rely on an important heritage, as long as that of many European towns.[10] Founded in 1651 as Fort Casimir by Peter Stuyvesant for the Dutch,[11] it was Swedish from 1654 to 1655 (under the name of Fort Trefaldighet [Fort Trinity]). Back in Dutch hands, it was called New Amstel for nine years. It became English in 1664 and in 1665 was renamed New Castle (WWP Delaware 1937). It was the seat of colonial government from 1704, but it lost its early economic prominence to Philadelphia, which was better located, and lost the capital to Dover in 1777. Despite this, the town benefited from the growth of maritime trade in the 1780–1830 period, and became a stop for the north-south traffic between Philadelphia and Baltimore and also a stop on the inland route between Washington and the north (Munroe 1993, 107). A turnpike linked New Castle to Frenchtown, and then the railroad came in 1831, after the completion of the Chesapeake & Delaware Canal (built 1824–29). But in 1838, the new Baltimore-Wilmington-Philadelphia Railroad bypassed New Castle, which lost its importance as a transit point. In 1881, Wilmington, the leading city (42,478 inhabitants in 1880) became the county seat of New Castle County because New Castle was too inconveniently located, according to the county's citizens. New Castle soon became a distant suburb of Wilmington, both cities being linked by electric trolley cars as early as 1900. In 1930, it was a small town of 4,131 inhabitants. Forty years later, the number was almost unchanged at 4,814, and remained nearly the same in 2010 (5,285). Although only around 1,000 people were living in New Castle in 1776, its position as capital enabled it to benefit from numerous "pleasant and dignified" eighteenth-century buildings, such as the Old Court House (1732). That explains the early creation of the New Castle Historical Society in 1934 and the later designation of the town as a National Historic Landmark Area, which gives it federal funding and protection.[12] The city's official website insists on the quality of life in a town where present and past are intertwined: "Unlike many historic communities, New Castle is a residential town where people live and work. Each house reflects the individuality of its past and present owners." This echoes earlier comments in praise of the town's atmosphere of permanence: "The town's enchantment is compounded of seemliness and taste, of completeness and a sense of authentic continuity with the past without loss of its living and livable quality" (WWP Delaware 1937, 19).

But heritage is not reserved to old eastern seaboard towns. Guthrie, Oklahoma's capital from 1890 to 1910, proves that the West can also boast a rich, though sad, history: "The story of Guthrie is one of youthful hope, defensive aspiration, and tragic defeat" (Forbes 1938, 7). But since these remarks were made in the 1930s, Guthrie has been reborn, thanks to its heritage (Leider 1990–91). As elsewhere in the nation, the movement toward preservation began in the 1970s and was enhanced by the 1981 Economic Recovery Tax Act. For Guthrie, the oil boom also helped. Faced with the threat of having to demolish historic buildings because of modernization pressures, local preservationists united. They succeeded in 1974 in classifying all the land included in the original 1907 incorporation of Guthrie as a historic district (the first commercial and residential historic district in the state). It is still the largest contiguous historic district in the nation. It covers four hundred city blocks on fourteen hundred acres, with 2,169 buildings.[13] Downtown Guthrie was recently designated a National Historic Landmark because of its outstanding collection of late nineteenth and early twentieth century commercial architecture, called Victorian (for the red brick façade trimmed in stone topped with metal cornices or masonry designs), although the main architect of the town was a Belgian immigrant, Joseph Foucart (the city hall he designed was torn down in 1955). It is also "the only continental U.S. territorial capital that is substantially the same as it was during the 1890s."[14]

Such a halt in historic development came from the loss of the capital to Oklahoma City in 1910. When oil and politics brought the new capital prosperity, Guthrie, being too close (twenty-six miles), was somehow petrified, its population stagnating at around 10,000 people. In 1970, it had 9,575 inhabitants—fewer than the 12,000 of the day of its foundation on April 22, 1889 or the 11,654 of 1910. Did such an extensive heritage policy, strongly pushed by enterprising local leaders, bring visible benefits? Heritage did make it a point of interest, with about a quarter of a million annual visitors. Its architectural treasures have been renovated, and seven museums have opened.[15] Despite its nicknames, "Queen of the Prairie" and the "Williamsburg of the West," Guthrie was unable to retain its population: after growing slightly in the 1970s and 1980s (10,312 in 1980, and 10,518 in

1990), the population dropped again to 10,191 in 2010. The county's population experienced a rise from 19,645 in 1970 to 41,848 in 2010, and the main reason was certainly Oklahoma City's suburban sprawl. Visitors stop at Guthrie, but only for a limited time: besides eighteen B&Bs,[16] there are only three motels and inns offering 173 rooms.[17] Clearly, as for New Castle, heritage does not suffice.

Nor did it suffice to maintain the population of Vincennes, a former territorial capital of Indiana (1800–1813), which dropped from 19,867 in 1990 to 18,423 in 2010 (fig. 9.2). Founded in 1732 by the Sieur de Vincennes, the town presents numerous attractions and is mentioned in the French *Guide Bleu* (Collective authors 1988, 427).[18] This mention may be due to the fact that it was a French creation, and that French town planning is still visible today (Conzen 2002). The text of Vincennes's website is entirely devoted to history: "If you are looking for the place where Indiana was born, where Indian Chief Tecumseh walked, where the first governor of Indiana Territory and 9th President of the U.S. lived, worked and strolled, where troops

FIGURE 9.2 Vincennes, Indiana, Indiana Territory capitol historic site. This historic site testifies to the humble beginnings of law and order during territorial times. The small building in the middle of the photo was Indiana's territorial capitol from 1810 to 1813. Photo by Christian Montès, June 2012.

gathered for the battle of Tippecanoe, where laws have been passed that still stand for the residents of Indiana, and where Indiana's first free newspaper was born, then you should plan to visit the State Historic Sites of Vincennes."[19] There are six state historic sites, including, of course, the former territorial capitol. However, Vincennes attracts most people more for an event than for its history as a capital. The Spirit of Vincennes Rendezvous is a gathering that has taken place for about thirty years. In a National Historic Park, 400–500 persons revive the capture in 1779 of a British fort for the thrill and pleasure of thirty-five thousand visitors.

Heritage Tourism as One Asset among Others

While history in a broad sense remains the basis of Vincennes' limited appeal, this is not the case in Monterey. Its economy relies almost entirely on tourism, but its status as a former capital is not the main attraction. A charming little city 115 miles south of San Francisco, in a quiet location on the southern coast near Carmel, Monterey had 27,810 inhabitants in 2010, slightly up from 22,618 in 1960. After a long spell as capital of the Spanish and Mexican province of Alta California, and a very short period as California's first capital (1849), its economy first turned to fishing (mostly whaling), which induced the creation of numerous canneries, only one of which remained in 1960. The atmosphere of the city was vividly described in John Steinbeck's novels *Cannery Row*, *Sweet Thursday*, and *Tortilla Flat* (Monterey hosts the National Steinbeck Center). The army also provided some work for the population, but tourism began to come to the Monterey Peninsula as early as the 1880s (the first golf course opened in 1897). It is today a major tourist destination, where heritage is second only to landscape. As the Chamber of Commerce proudly writes, "By all accounts, this area is one of the most beautiful land and coastal environments in the world." The well-known Monterey Bay Aquarium displays some of the living treasures of the bay and beyond. History is nevertheless important to the city: "Everywhere you look in Monterey there are remnants of other times and other lives. History is part of our daily routine: we do business and work in historic buildings and stroll along historic walkways" (City of Monterey website, http://www.monterey.org/). The buildings that were built when Monterey was the capital now attract more visitors to the city. Most of those who promote tourism and preservation are united under a common banner called Historic Monterey. Four million tourists come annually to the peninsula (tourism provides 12% of all jobs). The proximity of San Francisco is a major reason for this success.

On a far smaller scale, Golden, Colorado, includes its status as a former capital as part of a broader plan for its economy and tourism. Manufacturing and the natural surroundings are both at the heart of its economy, but, like Monterey, the city dwells on its "glorious" past as a capital to lure more tourists. The town was hit hard by the loss of the capital to nearby Denver (Abbott, Leonard, and McComb 1994, 300–301). The School of Mines was established in Golden in 1874 as a consolation prize, but although it remained the seat of Jefferson County and Adolph Coors Sr. opened his soon-to-be-famous brewery in 1873, the town lost population until World War II, having only around three thousand people. Denver had become the main center of growth, having captured the commercial flows by completing a more direct rail link to the first transcontinental railroad at Cheyenne. The Golden-Cheyenne link, named the Colorado Central Railroad, opened too late, in 1870. Built by William A. H. Loveland,[20] the associate of Golden's founder, George West, it moved its headquarters to Denver in 1880.

Golden's growth after the 1940s comes from two trends: the spreading of Denver's metropolitan growth and the renewal of interest in history. Golden was fortunate enough to be near a rapidly growing metropolis, but it was also to be protected from Denver's sprawl by the abrupt Front Range. The town's website dwells on that idea, noting that the city is "only fifteen minutes from Denver but a world apart." Golden's population quadrupled from 1940 to 1980 to 12,237, and increased by another 6,000 by 2010, to reach 18,867 inhabitants. Still a small town, it now rules over the 535,734 inhabitants of Jefferson County, where three municipalities have more than 100,000 inhabitants (Lakewood, Arvada, and Westminster). This county had in 2010 the eighth highest median household income in Colorado. Golden is a significant center of employment: 1,900 people were employed at MillerCoors Breweries in 2011, 1,200 more at Coors Tek, which manufactures ceramic components, and 2,775 at the National Renewable Energy Laboratory (Jefferson County website).

Despite these assets, which are partly linked to the proximity of Denver and to the presence of the School of Mines, Golden describes itself with the phrase "small town, big fun." To preserve its Old West look, it passed a preservation ordinance in 1983, with the support of the private Golden Landmarks Board, founded during the 1970s. No less than nine museums have opened, including the Colorado Railroad Museum, the Rocky Mountain Quilt Museum, and the Golden Pioneer Museum. Twelfth Street offers the visitor a historic walk. The city's website insists on its importance in the state: "Golden was crucial to the formation of the great state of Colorado; it played a key role in opening the West to settlement, and it helped build this nation's mineral industry." But tourists first come to Golden for kayaking, hiking, or biking around this "active sports mecca." They also come for the free MillerCoors Brewery tours and tastings (three hundred thousand each year), and then to see Buffalo Bill's Museum and his grave on nearby Lookout Mountain. Out of the estimated 2.5 million annual tourists, "only" 0.62 million visit Heritage Square. The city offers more than two thousand rooms to accommodate these tourists. Golden has thus to be put in a different category than the previous examples, where history was foremost. Although local history is not forgotten, the accent is on the interchangeable historical stereotypes that numerous towns and cities have experienced. Such a process can be compared to the booster movement, although it looks toward the past instead of toward the future. David Hamer rightly wrote, "There has been a growing and powerful commercial emphasis on what Edward Relph has called 'ambience,' a vague concept signifying 'environmental style and atmosphere,' which is fostered in old or old-seeming places and is a key element in the postmodern urban landscape" (Hamer 1998, xii).

Former capitals offer leisure activities more than traditional tourist attractions. They attract a mostly regional population, sometimes numerous, as for Cahawba. They will never attract masses of visitors and will have difficulties in experiencing real economic growth linked to heritage, as is the case for most of the examples here studied. Indeed, the idea that the United States has no history compared to Old Europe is leading Americans as well as foreigners toward the natural wonders that abound in the country or, if they go to town, toward the major museums, the most impressive skyscrapers, or the most comfortable places. The remains of a real past are nevertheless present, and numerous small towns bear testimony to a rich history. Among those, former capitals are popular because of their grander history and civic monuments. One could argue that tourism plays a role in these small cities (fewer than twenty thousand inhabitants) that is somewhat similar to the role of the capitol in smaller state capitals, which is linked to the building of the state, a secure (though small) basis of employment, and the regular coming of out-of-town people during the "season." But to attract numerous tourists and to retain or gain population, a broader economic or tourist basis is needed, as the examples of Monterey and Golden demonstrate. Another way to succeed is to add a small-town atmosphere to a beautiful and well-preserved heritage that induces people to settle and not only visit, before or after retirement.

Prescott, Arizona: A Retirement Community

Prescott, Arizona, "the model frontier settlement of the United States" (according to Benjamin C. Truman, U.S. postal inspector), has become one of Arizona's favorite retirement communities besides Phoenix's famed Sun City. When John Fremont, the "father of California," became governor of the Arizona Territory in 1878, he settled with his family in Prescott, which had just regained the seat of government from Tucson. His daughter Lily (1842–1919) left scores of letters that have been recently published. In a letter to Ella Haskell Browne, on December 9, 1878, she described Prescott with great fondness.

> To the northwest and west we have a really beautiful view ranging from the rugged grandeur of Granite Mount along the line of intervening pine covered hills to Thumb Butte, which rises a sheer columnar mass of rock from the top of an already tall mountain—right behind it the sun sets and when there are clouds "the effects" are wonderfully beautiful. In the immediate foreground one block below us the Plaza begins and it is the business center of the town; we overlook one side of it and one of the streets which leads off towards a busy part of the Territory—the Camp Verde & Santa Fe road—so that we see all sorts of teams from the "prairie schooner" with its twenty, or more, mules down through many gradations to the solitary miner going out on a prospecting tour and driving before him his one burro with its heavy pack . . . We have a right good market here, quite a decent lot of vegetables and in meats besides the usual muttons and beef lots of good fat venison and wild turkey. (Spence 1997, 221–23)

A small town of 2,165 inhabitants in 1890, when Phoenix became the capital, Prescott did not experience any real growth until 1950, when its population reached only 6,764 people. But it obtained later some crumbs from the economic and population growth experienced by both Arizona and the whole West. From 12,861 in 1960, Prescott's population soared to 19,865 in 1980, 26,592 in 1990, and 39,843 in 2010. The growth occurred not only in the municipality itself, but also in planned new communities around the town, such as Prescott Valley, a 1960s new town of individual homes. *Money Magazine* rated the town in 1994 as "the best place to retire in America," and Prescott's active Chamber of Commerce praises the city's numerous assets: "It's the home of the world's oldest rodeo and the place known as Arizona's Christmas City. It was in Prescott, on the steps of the County courthouse, where Barry Goldwater announced his candidacy for the Senate and office of President of The United States. Many of the structures that were erected in the early days have been preserved, and Prescott claims 525 buildings on the National Registry of Historic Places. Prescott's mild four seasons make it a lovely place to visit and enjoy at any time of the year" (*Prescott's Hometown Map and Guide* [Tempe, AZ: Hometown Map and Guide, c. 1995]).

Today, there are 637 listed buildings, and the city is proud to assert that Prescott "is internationally-known as 'Everybody's Hometown.'" If one wants to settle in Prescott, housing is varied, and the city is really charming.[21] Capital memories and buildings are only one more asset in a valley that is slowly becoming a "virtually gated community" in a pleasant and healthy location. Such charm is not always present when the former capital has become entangled in the suburban sprawl of a nearby booming metropolis.

"Growth" Due to Inclusion in a Metropolis

With the tremendous contemporary metropolitan sprawl, cities that were once rivals for prominence now belong to the same MSA. This explains Oregon City's important growth between 1990 and 2010, from 14,698 to 31,859 inhabitants; it is now located in the MSA of Portland, thirteen miles south of its core.

California offers the best example of this trend. Three out of four of its former capitals (excluding Los Angeles, which was only half a capital) are now included in San Francisco's MSA: San Jose, Benicia, and Vallejo. Their fates were different, but all have lost almost all traces of their terms as capitals, which were very brief indeed. Benicia was the less fortunate from the demographic point of view (26,997 inhabitants in 2010) as well as from the scenic one, its principal employer being a refinery. But it is also a residential suburb. Vallejo did not suffer from the loss of the capital, as shown by its 2010 population of 115,942 inhabitants. An excellent position on San Francisco Bay enabled the city to become an important shipping and naval center. It can also boast one of the first nationally registered heritage districts west of the Mississippi River, the Vallejo Heritage District. Its style is "Working Man's Victorian," from the 1860s to the 1890s, after the loss of the capitol. Visitors can also go to St. Vincent's Hill Historic District (an area of thirty-three square blocks), to the Vallejo Naval and Historical Museum, or to a concert by the city's symphony orchestra, founded in 1931. San Jose was perhaps the most fortunate, experiencing tremendous growth after 1945. Its population soared from 95,280 in 1950 to 445,779 in 1970, and 945,942 in 2010. Santa Clara County, still known in the early 1960s as the producer of one-third of the world's prunes, has become the heart of the world's most well known technopolis, Silicon Valley, which is included in the San Francisco CSA and employs more than a million people.

Ohio offers another facet of this movement. One of its former capitals, Chillicothe, is now on the southern border of the MSA of the state's capital, Columbus—thus regaining part of the status it had to cede almost two centuries ago. But the city still finds it difficult to reap the benefits of being near a metropolis. It is still much like the sleeping beauty it was during the first half of nineteenth century (its early nickname was "The old Town Beautiful"). It had been one of Ohio's biggest cities, and lodged seventy-one hundred people in 1850. It was home to the oldest newspaper west of the Alleghenies, the *Scioto Gazette*, which was founded in 1800 by Nathaniel Willis and was still active in the 1930s. But growth almost halted in Chillicothe because it did not become an industrial center. The impression that time had come to a standstill for Chillicothe was well put in the city pageant produced in 1896: "'Slow' and 'unprogressive' have been the epithets hurled at her by more pushing sister cities as they scurried by. But serene and undisturbed is she, remembering her children of glorious manhood in every walk of life and in almost every [clime]–the statesmen she has nourished for national fame [General Nathaniel Massie, who founded the city in 1796], the noble sons she has reared and sent forth to fight" (Renick 1896, 11–12).

Almost half a century later, the enduring memory of its glorious past was still the main appeal of the town, as shown in the small book written about Chillicothe in 1938 by the Ross County Northwest Territory committee of the Works Progress Administration. In 1937, the city had twenty-five thousand inhabitants (18,340 in 1930) and was proud to see its landscape figuring on the great seal of the state. As the state capital, it had wide shaded streets as well as numerous Greek Revival buildings that "[gave] the town the appearance of great dignity, peace and quiet" (WWP Ohio 1938, 48). The town remained mostly a rural center in a rural county with some manufacturing (paper, since 1812, and shoe manufacturing), a US Reformatory, and US Veteran's Hospital.

> The glory and prestige of the early days are still vivid in the minds of many Chillicothe citizens who revere the traditions of their town. Today, Chillicothe is a Saturday night town, like many Ohio towns situated in farming communities. The word Chillicothe is a part of the American language; like "Main Street" and "Babbitt," it has come to have definite meaning, particularly to persons who have never been to Chillicothe. It is a funny word and has been used as a sure[-fire] laugh in some of the plays produced by George Tyler, a Chillicothe boy. Actually it is an approximation of an Indian term meaning "town." People in Chillicothe do not see anything funny about its sound. (WWP Ohio 1938, 9)

But George Tyler has been forgotten since the 1930s,[22] and the city proved unable to retain its population: the 2010 Census registered only 21,901 persons, about the same as 1990's 21,923. The city is part of a growing county, Ross County (40,307 inhabitants in 1880, 41,556 in 1920, 61,211 in 1970, and 78,064 in 2010). Located forty-five miles south of the state capital, the county is now part of the outer fringes of the Columbus conurbation. The county's major employer is Adena Health System. Chillicothe also hosts a branch of Ohio University, with 2,305 students (fall 2011). Chillicothe is the site of one of Columbus's new activity zones, which are transforming the old economy into a new one and creating difficulties for the "old" workers left behind (employment declined by 8.6% between 2000 and 2010). Thus, although Chillicothe has attracted some of Columbus's growth, the small town is far from having recovered its lost status. It is only one of the numerous satellites of the state's star city, torn between its pride in an interesting heritage and its will to take an active part in the modern economy (fig. 9.3).

Present state capitals seem better off than most former ones, from the demographic and economic points of view. But is this still true when assessing the quality of life they embody, the beauty of their streets, the atmosphere that pervades them? Most former state capitals are still small, and present themselves as perfect examples of the small towns that almost every American dreams of.[23] This is naturally part publicity (see Tracy Kidder's *Home Town*), but it also underlines the fact that history is now deeply ingrained in the American mind and that many American towns go back more than mere decades (St. Augustine was founded in 1565) and

FIGURE 9.3 Chillicothe, Ohio, memorializes its capital days. A historical marker, a celebration, and a proud motto ("Ohio's first capital") show that, even if the state capitol has been replaced by the grand courthouse on the left (a replica has been built nearby), Chillicothe's citizens are still proud of their past and of the small-town atmosphere of their city, now in the outer reaches of Columbus's MSA. Photo by Christian Montès, June 2012.

offer inhabitants and visitors alike a glimpse of another world, long past. These towns, especially former capitals, are living testimonies to the aging of the United States and to the "Europeanization" of the way the country perceives its past, influenced by the growing forces of policies aimed at the preservation of heritage and expressions of identity (naturally with all their shortcomings). Pioneers are now far more often seen among the wax figures displayed in museums (as in Golden or Prescott) than in the flesh and on the trail.

10

State Capitals Today

Symbols of American Democracy

Long associated with mediocrity and forgotten by social scientists, who prefer to study metropolises and national capitals because information about them is easier to gather and their influence is easier to delineate, state capitals are today cities with long and rich histories and are deeply ingrained in the American mind. Their status in the United States is ambiguous. Economists, mainly concerned with quantitative returns and monetary relationships, insist on the redistribution of rents by governments to the primary city in which they are based, although to a limited extent in the case of federal countries (Glaeser 1999, 4). Sociologists underline the dichotomy between the legislative world and the citizens' world, visible in the capital's architecture as well as in their respective relations to the capital city. Historians emphasize the complex and shifting political framework of decision making that influenced the selection of capitals and subsequent evolution of power in the fifty legislatures. Geographers scrutinize spatial patterns at various scales in terms of the overlapping economic, social, and political logics. We need to put all these approaches together to begin to understand the workings of the fifty state capitals of the United States and to assess their position in the national urban hierarchy.

On the one hand, the development of the American urban hierarchy clearly dissociated economic and political capitals, but on the other hand, the Founding Fathers never wanted to separate economics and politics. What Federalists wanted was "to separate social authority from political authority," in order to avoid the growing influence of uneducated "New Men," for they longed for a "natural elite" (Wood 1991, 548)—but this desire nevertheless contributed to the separation of economics and politics. Numerous factors were at work that explain this apparent paradox. First, capitals were selected according to different rules; some have been capitals since colonial times; and others, since territorial times. Congress did not adopt a unique way for territories to choose capitals, and the Supreme Court also had a say in the process. States were frequently carved out of disparate areas, creating sectional factions and parochialism that favored a period of temporary or rotating capitals.

Whatever the rules, town citizens shared the belief that their future prosperity depended on government and favored centrally located cities, and they usually stormed against the aristocracy, wealth, and the corruption of cities. Territorial and state politicians fully participated in the creation of capitals as speculative ventures, along with the better-known town boosters—local capitalists, absentee speculators, or railroad executives (Schultz 1989, 12). The processes of capital choice took place in a period when the urban system was still fragmented. This allowed numerous small towns to compete fiercely for capital status and nurtured the practice of the spoils system, which dispatched the state institutions all over the states, often depriving the capital of a university. These processes resulted in an apparently incoherent corpus of capitals, confirmed by the typology presented in chapter 7.

However, a capital primarily produces laws rather than goods, and it thrives upon the gathering and sorting out of the state's interests and communities. Capitals rule the states, not as an economic metropolis would, but in a more immaterial way. Thus, two modes of territorial organization exist in the United States. The first relies upon the GDP and shows that the country lacks a unified urban system, no city being able to control the whole national economy. The local influence of New York or Los Angeles ends at the outermost suburbs, and both cities have as many (or even more) links with foreign metropolises as with their own country. The influence of state capitals is very different and closer to the classic representation of urban influence. First, their network covers the whole country, one in each of the fifty states. Second, they all have similar power: federalism grants exactly the same power to each individual state, whatever its size and population. Third, the power they exert is zonal. It extends to the very boundaries of each state (except in Native American reservations), under the umbrella of Washington—a wide umbrella indeed in states such as Utah or Nevada, where federal lands amount to more than half of the state's area.

State capitals therefore form rather uniform links in the chain constituted by state boundaries that have long been superseded by the economic network, in which capitals participate in almost the same way as other cities do ("almost," because state capitals somehow boost existing qualities). Such links actively reflect the memory of the ideal political order envisioned by the Founding Fathers in the democratic unity they form: Montpelier and Boston have the same political power, even though the second has eight hundred times the population of the first.[1] The memory is still alive, because democracy is still at work on a regional basis, while the economy functions on a network basis, both being interdependent. Wisconsin's case has shown, however, that to make democracy work, internal factionalism must be at its uppermost, divisions at all levels at their peak, and mediocrity dominant—a high price to pay. Wherever any group, lobby, corporation, municipality, or region leads the state—alone or through alliances—the Jeffersonian model goes awry, and the division between politics and economics disappears.[2] Are American local (and national) politics so fragile as to work only in a perfect balance of powers?

State Capitals as "Normal" Cities

The contemporary evolution of capitals toward "normality" works at different levels, the norm here being the way the other American cities are functioning and evolving. At the national and state level, the economic and political systems often overlap. This is the case when the state capital is also the economic capital. Since the 1950s, the spatial convergence between economic and political leaderships has grown. The factual convergence was indeed present from the beginning—from the power exerted by the Robber Barons and their local avatars to the current might of lobbyists. The spatial convergence can be traced to the postindustrial turn of the American economy. Most capitals were linked to water and land transportation patterns and were bypassed by the railroads and the industrial revolution experienced during the second half of the nineteenth century (with notable exceptions, such as Atlanta). The immaterial revolution brought to the fore cities with amenities, with air transportation facilities, with good universities, and with political leveling that attracted the lucrative military-industrial contracts or the knowledge economy. State capitals offer most of these assets: what is better than good old pork-barrel policy to obtain them? This explains the tremendous growth of some capitals (Phoenix, Austin, and, on a lesser scale, Raleigh and Nashville), along with the more general economic catch-up processes undergone by Sunbelt states.

But every capital did not take part in that growth. Three main factors can be put forward: the size of the capitals in 1950 (too small; they lacked the necessary basis to sustain large growth: Montpelier, Augusta, and Pierre are examples); their location relative to a metropolis (too close; they were unable to create autonomous growth: Olympia is an example); and their location with regard to the commercial and immaterial flows of the economy (too isolated; they were unable to capture these flows: Helena and Cheyenne are examples). Tourism is often the only means left to broaden their economy beyond state and basic urban jobs, because their symbolic status adds real value to a physical fabric that differentiates them most of the time from other cities. But tourism is not based on well-paid jobs and, until recently, has attracted more local than international visitors. However, every capital benefits from the "social safety net" created by their numerous state jobs, which give them a more balanced economy than other cities and some respite during economic downturns, even though state employment does not fully immunize them against crises.

At the municipal level, even if *capital* and *capitol* sound the same, they have to be dissociated. The capitol is often on an isolated hill, and citizens look with scorn as much as envy upon the magnificent buildings where politicians and lobbyists gather to determine state policies. Being a capital is seen as a hindrance as well as an asset. Most legislators commute to the capital, and most state employees live in the suburbs or nearby metropolises, where they pay almost no taxes, leaving the inner city to crumble. Mayors feel overshadowed by powerful legislators, the municipal level being traditionally weak in America. Local citizens, especially when they belong to minorities, often feel remote from political celebrities and magnificent buildings. In smaller capitals, however, capitols are integral parts of the town, and everybody can rub elbows at the local restaurants and cafés. Capitols and capitol grounds are open to everybody free of charge, and many retired volunteers serve as guides who provide visitors with useful and amusing information about the buildings and the history of their states. Capitals therefore retain their unique qualities in the urban world of the United States; they might be economically more integrated, but they are still socially special.

Being a capital is clearly a mixed blessing. The fifty state capitals are definitely not fifty Jerusalems established through Godly advice or might on American soil. Nor are they the "new purified communities" that William White described when analyzing early western county seats (White 1939, 26–27). Despite their diversity and their shortcomings (roughly the same as in every American city), however, they have not become the Babylons that some early legislators feared they would. Although they have very different physical and economic characteristics, they are all part of the foundation of American democracy and powerful symbols of the states they represent. Therefore, the cultural approach, more than the demographic or economic ones, is the best way to assess their accomplishments, which are indeed significant.

Acknowledgments

This book would not have been possible without Professor Michael Conzen, of the University of Chicago, who persuaded me that my project could be of interest to American readers and helped to persuade others that it was worthy of publication. I am grateful also for the encouragement of Professor James Morone, of Brown University, and Research Professor Deborah Stone, of Dartmouth College. Many thanks for their careful reading are due to Romain Garcier and especially to Isabelle Lefort, a true friend.

This book is also heavily indebted to the collections of the Rockefeller Library at Brown University, where I was a research fellow. I wish to thank every author that has lifted a corner of the veil that rested upon the history of state capitals. My gratitude also goes to the numerous people who have helped to locate information during that research: the staff at university libraries, at the Alabama State Archives, the Louisiana State Archives, the New Hampshire State Archives, the New Jersey State Archives, the New York Historical Society. Special thanks go to the wonderful people at the Vermont Historical Society and to everyone who answered my web inquiry or agreed to be interviewed about the history and workings of "their" state capitals, in Olympia, St. Paul, Helena, Salem, Madison, Des Moines, Springfield, Columbus, and especially in Pierre and Frankfort.

The illustrations were produced by Marie-Laure Tremelo, cartographer at Lyon University, who has my deepest gratitude. Many thanks also to Darius Kratzer, who corrected the first version of this text, and to the anonymous referees, for their enthusiasm and good advice.

This book would not exist without the University of Chicago Press. I was impressed by the quality of the editing process and would like to thank Abby Collier, who managed the entire process, Ruth Goring, project editor, and Micah Fehrenbacher, from the marketing department, as well as my manuscript editor, Nick Murray, for his very careful and precise reading and editing of the text.

I would also like to thank my American friends, without whom I would never have begun to be interested in American state capitals, and who were always there when needed: Scot, Alix, David, and Wei Li, thank you! Last, but not least, special thanks to Manuel for accompanying me through this long adventure and every day.

Appendix 1
Demographic and Historical Tables

Appendixes 1 and 2 are meant to complement the narrative part of this book by providing background information on all fifty state capitals. Appendix 1 (tables A.1–A.5) provides statistical tables, which enable us to assess the demographic and economic evolution of each capital as well as to compare them. Information in the tables is presented at the municipal level and at metropolitan or micropolitan levels when they are appropriate.

Because each one of the fifty American states has a specific history, the aim of appendix 2 is to provide the reader with a brief chronological framework for each state, with the date of its foundation as a colony, a territory, and a state, and the dates of the selection of each successive capital.

Tables A.1–A.3 are interesting in giving the capitals' historic demographic trends. However, caution is required in interpreting them, because the boundaries of these cities in 2010 are usually far different from the ones they had when founded. This is the result of annexation policies that have been very active in some cities and less so in others. From 1860 to 1920, annexations caused Boston to grow from 5 to 40 square miles; Richmond, from 2 for 24; St. Paul, from 5 to 52; and Providence, from 5 to 18 (Weil 1992, 80). The movement continued during the twentieth century. For instance, more than 90% of the large increase in Madison's population between 1940 (67,447 inhabitants), and 1970 (173,258) is explained by annexation. Phoenix is another telling example (Konig 1982, 28–29). In 1940, the incorporated municipality covered only 10 square miles. In 1960 the figure was 187.3 square miles, which largely explains the quadrupling of the city's population. In 2010, its area had soared to 516.7 square miles. Tallahassee grew from 1 square mile in 1824 (given by Congress to establish to capital) to 2 in 1912, 55 in 1986, and 100.2 in 2010 (Ellis and Rogers 1986, 19). Even for capitals that seem to have had a very simple evolution, the story can be complicated. Frankfort, Kentucky, provides an example (Kramer 1986). Between 1810 and 1820, its population grew from 628 people to 2,000, but we must take into account the secession of South Frankfort in 1812 (even if the numbers involved remained small). Both towns reunited in 1850, thus partly explaining the increase in population from 1,917 in 1840 to 3,308. Likewise, the 1890 population takes into account Frankfort's first annexation, which doubled its area and added 450 inhabitants, explaining half the 1880–90 growth. The population drop from 1930 to 1940, from 11,626 to 11,492, is explained by the closure of the state reformatory in 1937, which resulted in the departure of its 2,273 inmates. Between 1940 and 1970, a new wave of annexations more than tripled the city's area. There were still more annexations during the 1970s. Between 1990 and 2010, Frankfort gained some more land, increasing its area to 28.06 square miles. Although the city's area is not huge compared to that of other capitals (e.g., Phoenix), it is larger than the area of the third most populated municipality in France, Lyon (17.4 square miles for 485,000 inhabitants in 2010).

Table A.4 gives a broad sketch of the economic evolution of state capitals to show the extent to which they have been able to broaden their economic bases beyond government.

Table A.5 provides the per capita income in state capitals in 1999 compared to per capita income for their states and to that of some selected cities in those states to show whether capital status is an "income booster" for residents of capital cities.

Table A.1 Evolution of the population of state capitals, 1795–1880

Capital	1795	1810	1820	1830	1840	1850	1860	1870	1880
Montgomery, AL	—	—	< 100	695	2,179	8,728	8,843	10,588	16,713
Juneau, AK	—	—	—	—	—	—	—	—	Founded
Phoenix, AZ	—	—	—	—	—	—	—	240?	1,708
Little Rock, AR	—	—	< 100	< 500	1,000	2,167	3,727	12,380	13,138
Sacramento, CA	—	—	—	—	—	6,820	13,785	16,283	21,420
Denver, CO	—	—	—	—	—	—	4,400	4,759	35,629
Hartford, CT	2,682	3,955	4,226	7,076	9,468	17,996	29,162	37,180	42,015
Dover, DE	500	600	500	?	600	2,000	1,289	1,906	2,811
Tallahassee, FL	—	—	—	800	1,616	2,000	1,932	2,023	2,494
Atlanta, GA	—	—	—	—	—	2,572	9,554	21,789	37,409
Honolulu, HI				13,000	9,000	11,000	14,310	14,000	15,000
Boise City, ID	—	—	—	—	—	—	—	995	1,889
Springfield, IL	—	—	—	c. 800	2,579	4,533	9,320	17,364	19,743
Indianapolis, IN	—	—	—	1,200	2,692	8,090	18,611	48,244	75,056
Des Moines, IA	—	—	—	—		502	3,965	12,035	22,408
Topeka, KS	—	—	—	—	—	—	759	5,790	15,452
Frankfort, KY	400	1,099	2,000	1,682	1,917	3,308	3,702	5,396	6,958
Baton Rouge, LA	?	469	?	1,000	2,269	3,905	5,428	6,498	7,197
Augusta, ME	?	1,000	2,475	3,980	5,314	8,231	7,609	7,808	8,665
Annapolis, MD	1,600	1,600	2,260	2,623	2,792	3,011	4,529	5,477	6,642
Boston, MA	18,038	33,250	43,298	61,391	93,383	136,881	177,840	250,526	362,839
Lansing, MI	—	—	—	—	—	1,229	3,074	5,241	8,319
St. Paul, MN	—	—	—	—	—	1,112	10,401	20,030	41,473
Jackson, MS	—	—		645	2,100	3,000	3,199	4,234	5,204
Jefferson City, MO	—	—		1,200	1,064	2,301	3,082	4,420	5,271
Helena, MT	—	—	—	—	—	—	—	3,106	3,624
Lincoln, NE	—	—	—	—	—	—	—	2,500	13,003
Carson City, NV	—	—	—	—	—	—	701	3,042	4,229
Concord, NH	1,747	2,393	c.3,000	3,702	4,903	8,584	10,896	12,241	13,843
Trenton, NJ	1,000	3,002	3,003	3,925	4,035	6,460	17,228	22,874	29,910
Santa Fe, NM	?	c. 3,500				4,846	4,635	4,765	6,635
Albany, NY	5,500	9,356	12,630	24,238	33,721	50,763	62,367	69,422	90,753
Raleigh, NC	150	1,000	2,674	1,700	2,244	4,518	4,780	7,790	9,265
Bismarck, ND	—	—	—	—	—	—	—	—	1,758
Columbus, OH	—	—	1,450	2,435	6,048	17,882	18,554	31,274	51,647
Salem, OR	—	—	—	—	Founded	?	?	1,139	2,538
Harrisburg, PA	1,500	2,287	2,990	4,312	5,980	7,834	13,405	23,104	30,762
Providence, RI	4,850	10,071	11,767	16,832	23,171	41,512	50,666	68,904	104,857
Columbia, SC	350	1,000	3,000	3,500	3,500	6,060	8,052	9,288	10,036
Pierre, SD	—	—	—	—	—	—	—	—	< 100
Nashville, TN	375	1,200	3,076	5,566	6,929	10,165	16,988	25,865	43,350
Austin, TX	—	—	—	—	856	854	3,494	4,428	11,013
Salt Lake City UT	—	—	—	—	—	—	8,236	12,854	20,768
Montpelier, VT		500	900	1,792	3,725*	2,310	2,411	3,023	3,219
Richmond, VA	4,000	9,785	12,067	16,060	20,153	27,570	37,910	51,038	63,600
Olympia, WA	—	—	—	—	—	Founded	?	1,203	1,232
Charleston, WV	100	200	?	?	900	1,050	1,520	3,162	4,192
Madison, WI	—	—	—	376	1,525	3,500	6,611	9,176	10,324
Cheyenne, WY	—	—	—	—	—	—	—	1,450	3,456

Sources: Moffat 1992. Corrections and additions from Bolton 1998, 37; Luckingham 1989, 37; Mawn 1977, 214; Hansen 1967, 252; Johnson 1991, 12, 381; Goodstein 1989, 205; Condon 1977, 16; Bouton 1856, 766; WWP Nebraska 1939, 181; WWP Texas 1940, 170; The Handbook of Texas online, Austin, 1; Zauner 1984, 9; Abbott et al. 1994, 60.

Note: The dash (—) means not founded. Oklahoma City, founded in 1889, is not shown. Figures for Honolulu have a limited reliability.

* The decrease between 1840 and 1850 is explained by the secession of East Montpelier in 1848.

Table A.2 Evolution of the population of state capitals, 1890–1970

Capital	1890	1900	1910	1920	1930	1940	1950	1960	1970
Montgomery, AL	21,183	30,346	38,136	43,464	66,079	78,084	106,525	134,393	133,386
Juneau, AK	1,251	1,864	1,644	3,058	4,043	5,729	5,956	6,797	6,050
Phoenix, AZ	3,152	5,544	11,134	29,053	48,118	65,414	106,818	439,170	581,562
Little Rock, AR	25,874	38,307	45,941	65,142	81,679	88,039	102,213	107,813	132,488
Sacramento, CA	26,386	29,282	44,696	65,908	93,750	105,958	137,572	191,667	254,413
Denver, CO	106,713	133,859	213,381	256,941	287,861	322,412	415,786	493,887	514,678
Hartford, CT	53,230	79,850	98,915	138,036	164,072	166,267	177,397	162,178	158,017
Dover, DE	3,061	3,329	3,720	4,042	4,800	5,517	6,223	7,250	17,488
Tallahassee, FL	2,934	2,981	5,018	5,637	10,700	16,240	27,237	48,174	71,897
Atlanta, GA	65,533	89,872	154,839	200,161	270,366	302,288	331,314	487,455	496,973
Honolulu, HI	22,907	39,306	52,183	81,820	137,582	179,326	248,034	294,194	324,871
Boise City, ID	2,311	5,957	17,358	21,393	21,544	26,130	34,313	34,481	74,990
Springfield, IL	24,963	34,159	51,678	59,183	71,864	75,503	81,628	83,271	91,753
Indianapolis, IN	105,436	169,164	233,650	314,194	364,161	386,972	427,173	476,258	744,624
Des Moines, IA	50,093	62,139	86,368	126,468	142,559	159,819	177,965	208,982	200,587
Topeka, KS	31,007	33,608	43,684	50,022	64,120	67,833	78,791	119,484	125,011
Frankfort, KY	7,892	9,487	10,465	9,805	11,626	11,492	11,916	18,365	21,356
Baton Rouge, LA	10,478	11,269	14,897	21,782	30,729	34,719	125,629	152,419	165,963
Augusta, ME	10,527	11,683	13,211	14,114	17,198	19,360	20,913	21,680	21,945
Annapolis, MD	7,604	7,657	8,262	8,518	9,803	9,812	10,047	23,385	29,592
Boston, MA	448,477	560,892	670,585	748,060	781,188	770,816	801,444	697,197	641,071
Lansing, MI	13,102	16,485	31,229	57,327	78,397	78,753	92,129	107,807	131,546
St. Paul, MN	133,156	163,065	214,744	234,698	271,606	287,736	311,349	313,411	309,980
Jackson, MS	5920	7816	21,262	22,817	48,282	62,107	98,271	144,422	153,968
Jefferson City, MO	6,742	9,664	11,850	14,490	21,596	24,268	25,099	28,228	32,407
Helena, MT	13,834	10,770	12,515	12,057	11,803	15,056	17,581	20,227	22,730
Lincoln, NE	55,154	40,169	43,973	54,948	75,933	81,984	98,884	128,521	149,518
Carson City, NV	3,950	2,100	2,466	1,685	1,596	2,478	3,082	5,163	15,468
Concord, NH	17,004	19,632	21,497	22,167	25,228	27,171	27,988	28,991	30,022
Trenton, NJ	57,458	73,307	96,815	119,289	123,356	124,697	128,009	114,167	104,638
Santa Fe, NM	6,185	5,603	5,072	7,236	11,176	20,325	27,998	33,394	41,167
Albany, NY	94,923	94,141	100,253	113,344	127,412	130,577	134,995	129,726	115,781
Raleigh, NC	12,678	13,643	19,218	24,418	37,379	46,897	65,679	93,931	121,577
Bismarck, ND	2,186	3,319	5,443	7,122	11,090	15,496	18,640	27,670	34,703
Columbus, OH	88,150	125,560	181,511	237,031	290,564	306,087	375,901	471,316	539,677
Oklahoma City, OK	5,086	10,037	64,205	91,295	185,389	204,424	243,504	324,253	366,481
Salem, OR	3,836	4,258	14,094	17,679	26,266	30,908	43,140	49,142	68,296
Harrisburg, PA	39,385	50,167	64,186	75,917	80,339	83,893	89,544	79,697	68,061
Providence, RI	132,146	175,597	224,326	237,595	252,981	253,504	248,674	207,498	179,213
Pierre, SD	3,235	2,306	3,656	3,209	3,659	4,322	5,715	10,088	9,699
Columbia, SC	15,353	21,108	26,319	37,524	51,581	62,396	86,914	97,433	113,542
Nashville, TN	76,168	80,865	110,364	118,342	153,866	167,402	174,307	170,874	448,003
Austin, TX	14,575	22,258	29,860	34,876	53,120	87,930	132,459	186,545	251,808
Salt Lake City, UT	44,843	53,531	92,777	118,110	140,267	149,934	182,121	189,454	175,885
Montpelier, VT	4,160	6,266	7,856	7,125	7,837	8,006	8,599	8,782	8,609
Richmond, VA	81,388	85,050	127,628	171,667	182,929	193,042	230,310	219,958	249,621
Olympia, WA	4,698	3,863	6,996	7,795	11,733	13,254	15,819	18,273	23,111
Charleston, WV	6,742	11,099	22,996	39,608	60,408	67,914	73,501	85,796	71,505
Madison, WI	13,426	19,164	25,531	38,378	57,889	67,447	96,056	126,706	173,258
Cheyenne, WY	11,690	14,087	11,320	13,829	17,361	22,474	31,935	43,505	40,914

Sources: Luckingham 1989, 37; Fischer 1975, 4; WWP South Dakota 1952; Naske and Slotnick 1987, 70; http://www.montpelier-vt.org; http://pierre.org.

Table A.3 Area and population of state capitals, 1980–2010

Capital	City area (sq. miles) 2010	City 1980	City 2000	City 2010	(C)MSA 1990	*μSA*/MSA or CSA 2010
Montgomery, AL	159.6	177,857	201,568	205,771	290,000	427,691
Juneau, AK	2701.9	19,528	30,711	31,275	—	—
Phoenix, AZ	516.7	789,704	1,321,045	1,445,656	2,238,000	4,192,887
Little Rock, AR	119.2	159,151	183,133	193,524	513,000	877,091
Sacramento, CA	97.2	275,741	407,018	466,488	1,481,000	2,461,780
Denver, CO	153.0	492,686	554,636	600,158	1,980,000	3,090,874
Hartford, CT	17.4	136,392	121,578	124,775	1,158,000	1,330,809
Dover, DE	23.1	23,507	32,135	36,047	111,000	162,310
Tallahassee, FL	100.2	81,548	150,624	181,376	234,000	367,413
Atlanta, GA	133.1	425,022	416,474	420,003	2,960,000	5,618,431
Honolulu, HI	60.5	365,048	371,657	337,256	836,000	953,207
Boise City, ID	79.4	102,249	184,727	205,671	296,000	616,561
Springfield, IL	59.5	100,054	111,454	116,250	190,000	210,170
Indianapolis, IN	361.4	700,807	791,926	820,445	1,380,000	2,080,782
Des Moines, IA	80.9	191,003	198,682	203,433	393,000	639,784
Topeka, KS	60.2	118,690	122,327	127,473	161,000	230,824
Frankfort, KY	14.3	25,973	27,741	25,527	—	*70,758*
Baton Rouge, LA	76.9	220,394	227,818	229,493	528,000	825,905
Augusta, ME	55.1	21,819	18,560	19,136	—	*122,151*
Annapolis, MD	7.2	31,740	35,838	38,394	—	—
Boston, MA	48.3	563,000	589,141	617,594	5,455,000	7,559,060
Lansing, MI	36.0	130,414	119,128	114,297	433,000	534,684
St. Paul, MN	52.0	270,230	287,151	285,068	2,539,000	3,615,902
Jackson, MS	111.0	202,895	184,256	173,514	395,000	539,057
Jefferson City, MO	35.9	33,619	30,711	43,079	—	149,807
Helena, MT	16.3	23,938	25,780	28,190	—	*74,801*
Lincoln, NE	89.1	171,932	225,581	258,379	214,000	302,157
Carson City, NV	144.7	32,022	52,457	55,274	—	55,274
Concord, NH	64.2	30,400	40,687	42,695	—	*146,445*
Trenton, NJ	7.6	92,124	85,403	84,913	326,000	366,513
Santa Fe, NM	46.0	49,160	62,203	67,947	117,000	184,416
Albany, NY	21.4	101,727	95,658	97,856	862,000	1,168,485
Raleigh, NC	142.9	150,255	276,093	403,892	858,000	1,749,525
Bismarck, ND	30.8	44,485	55,532	61,272	84,000	108,779
Columbus, OH	217.2	565,021	711,470	787,033	1,345,000	2,071,052
Oklahoma City, OK	606.4	404,014	506,132	579,999	959,000	1,322,429
Salem, OR	47.9	89,091	136,924	154,637	278,000	390,738
Harrisburg, PA	8.1	53,264	48,950	49,528	588,000	683,043
Providence, RI	18.4	156,804	173,618	178,042	1,134,000	1,630,956
Pierre, SD	13.1	11,973	13,876	13,646	—	*19,988*
Columbia, SC	132.2	101,229	116,278	129,272	454,000	805,106
Nashville, TN	475.1	455,651	569,891	626,681	985,000	1,582,264
Austin, TX	297.9	345,890	656,562	790,390	846,000	1,716,000
Salt Lake City, UT	111.1	163,034	181,743	186,440	1,072,000	1,744,886
Montpelier, VT	10.1	8,241	8,035	7,855	—	—
Richmond, VA	59.8	219,214	197,790	204,214	866,000	1,258,251
Olympia, WA	17.2	27,447	42,514	46,478	161,000	252,264
Charleston, WV	31.5	63,968	53,421	51,400	250,000	304,282
Madison, WI	76.8	170,616	208,054	233,209	367,000	630,569
Cheyenne, WY	24.5	47,283	53,011	59,466	73,000	91,738

Sources: US Census Bureau, 1980, 2000, and 2010.

Note: The dash (—) means the capital has no MSA. Boundary and definition changes blur the 1990–2010 comparison.

Table A.4 Economic evolution of state capitals from 1850 to the early 2000s

Capital	Economy 1850–1900*	Economy 1900–1950	Economy 2000s
Montgomery	Trade and government	Trade and government	Trade, cultural center of the state. Government (25%), army, some high-tech
Juneau	Mining	Mines and government	Government (57%), tourism, mining, fishing, logging
Phoenix	Service station of central Arizona; railroad hub	Service station of central Arizona; railroad hub	Metropolis, high tech, tourism
Little Rock	Center of the state	Center of the state	Center of the state, hub, army, health, government (21%)
Sacramento	Agriculture, railroads (Central Pacific), trade	Agriculture, trade	Trade (hub), two air bases, air and space industry, state univ., metropolization
Denver	Metropolis of the state	Metropolis of the state and beyond; trade, tourism, banks, manufacture	Metropolis (energy), federal jobs, high-tech, huge airport
Hartford	Insurance and manufacturing (Colt, machines, air) Richest city in America	Major insurance center, manufacturing	US major insurance center General crisis: 30% of pop. are poor
Dover	Small center	1899 incorporation act and HQ	Business HQ (attractive fiscal laws)
Tallahassee	Center of Florida's life, but not for long	Comparative decline (rise of tourism, WWII elsewhere)	Government (38.1%); two universities
Atlanta	Distribution center for northern products, banking and insurance center of the state	Uncontested center of the state	Regional metropolis; (Coca-Cola, CNN, manufacturing, regional HQ, etc.)
Honolulu	—	Major center	Has 75% of population, univ., tourism, navy, government (22%)
Boise City	Trade center near mines	Center, WWII air base	Cultural, economic center (high-tech: HP, Micron), political capital of Idaho
Springfield	Mostly government	Mostly government	Government , health services, tourism, software
Indianapolis	Railroad hub , diversified manufacture	Hub and diversified manufacture (cars until 1937)	Hub, trade, manufacture (medical), univ.
Des Moines	Insurance Overall center of the state	Insurance Overall center of the state	Great Plains hub / insurance Overall center of the state

Table A.4 (*continued*)

Capital	Economy 1850–1900*	Economy 1900–1950	Economy 2000s
Topeka	Boom and bust agricultural center; government	Agribusiness, (but 3rd rank), air base, government	Secondary hub, government, psychiatric asylum
Frankfort	Government Tobacco and bourbon	Government Tobacco and bourbon	Government (53%) Tobacco, bourbon, auto parts
Baton Rouge	Secondary town	Secondary town Oil	Petrochemicals, port, univ., government
Augusta	Trade (roads, inland port)	Decline	Government (40%) Local service center
Annapolis	Naval Academy and government	Naval Academy	Naval Academy Government, heritage tourism
Boston	Metropolis	Metropolis	Metropolis
Lansing	Secondary center	Automobile (Oldsmobile, then GM) Michigan State University	Government, manufacturing (GM), some FIRE Michigan State University
St. Paul	Railroads, trade, and banks, linked with government	Twin Cities, overall metropolis of north-central US (transition from agriculture-based economy)	Twin Cities (CMSA) Overall metropolis (from banks to high-tech and 3M)
Jackson	Secondary center Government	Became first center	Main center, hub (oil, gas) manufacture
Jefferson City	Government (printing and shoe industry)	Government	Government Some manufacturing
Helena	Government, local banks Redistribution center for mines	Government Local service center	Government, tourism Local service center
Lincoln	Government universities, agricultural trade center	University and government Trade (agricultural center)	Government Universities, agricultural center
Carson City	Trading center near mines government	Government Economy pales (Reno and Las Vegas…)	Government, tourism
Concord	Government Local center	Government Local center	Government, tourism Local center
Trenton	Trade, manufacture	Manufacture	In New York's CMSA Government, trade
Santa Fe	Government, trade (but decline of Santa Fe Trail)	Decline (Albuquerque expansion: univ., manufacturing, military)	Government Cultural and heritage tourism
Albany	Hub, manufacturing	Relative decline	Hub, trade, GE, univ. Government (39% in county)
Raleigh	Government	Government	Center of the state (CMSA) Triangle Research Park
Bismarck	Railroad town in agricultural state	Railroad town in agricultural state (crisis)	Hub (rail), trade, oil, government

(*continued*)

Table A.4 (*continued*)

Capital	Economy 1850–1900*	Economy 1900–1950	Economy 2000s
Oklahoma City	Sooners	Oil, government, service center, air base	Center of the state, hub, army, manufacturing, oil and gas
Columbus	Balanced economy	Balanced economy in an industrial state	Balanced economy
Salem	Government University (1842), food processing, lumber	Government Univ., food processing, lumber	Government, food processing lumber, univ., high-tech In Portland's CMSA
Harrisburg	Railroad, industry, trade	Railroad, industry (metal), trade, convention, government	Railroad, government, manufacturing (electric)
Providence	Manufacturing (large growth in textile and metal), finance, university	Port and manufacturing (but crisis since 1930s), finance, university	Government, banking, university, heritage
Columbia	Government, inland trade center, manufacture (1880s)	Government, textile, U. of South Carolina (still small)	Government, army, trade, univ. (25,140 students)
Pierre	Government (railroad) Small agricultural center	Government (railroad) Small agricultural center	Government Small agricultural center
Nashville	Financial and political center of the state, trade (railroad)	Distribution point, finance, DuPont, university	Government, finances, trade, manufactures (auto), univ., music
Austin	Government and trade (modest), university (1881)	Government, modest trade, university, but pales compared to other cities	Metropolization High-tech, univ., government (air base until 1993)
Salt Lake City	Main center of Utah; Mormon center	Main center of a poor state	Metropolis
Montpelier	Government, small trade center	Government, small trade center, insurance, tourism center	Government (35%) Insurance (FIRE) (18%), tourism
Richmond	Trade, manufacturing, (tobacco), government	Manufacture (tobacco), banks, government	In megalopolis; trade, finance, law, three universities
Olympia	Government, small local center	Government, small local center	In Seattle's CMSA Government (39%), small active port, lumber
Charleston	Government, small manufacturing center	Government, small manufacturing center	Economic center of Kanawah Valley Chemicals, government
Madison	Government and college city	Government and university	Government, University of Wisconsin
Cheyenne	Railroad, small center of an agricultural and mining region, army, government	Small center of an agricultural and mining region, tourism, army, government	Government Small center of an agricultural and tourist region

Note: Government is indicated only if it has a major importance in the capital's economy.

* Some capitals were founded after 1850, but almost all were founded before 1900. Juneau became capital in 1900, and Oklahoma City in 1910.

Table A.5 Per capita income (dollars) in American states, state capitals and selected cities, 1999

State	1. Per cap. income	Capital	2. Per cap. income	Col. 2/ Col.1 (%)
Alabama	18,189	Montgomery	19,385	106.6
Alaska	22,660	Juneau	26,719	117.9
		Anchorage	25,287	111.6
Arizona	20,275	Phoenix	19,833	97.8
Arkansas	16,904	Little Rock	23,209	137.3
California	22,711	Sacramento	18,721	82.4
		San Francisco	34,556	152.2
Colorado	24,099	Denver	24,101	100.2
Connecticut	28,766	Hartford	13,428	46.7
Delaware	23,305	Dover	19,445	83.3
		Wilmington	20,236	86.8
Florida	21,557	Tallahassee	18,981	88.1
Georgia	21,154	Atlanta	25,772	121.8
Hawaii	21,525	Honolulu	24,191	112.4
Idaho	17,841	Boise City	22,696	127.2
		Butte	11,889	66.6
Illinois	23,104	Springfield	23,324	101.0
Indiana	20,397	Indianapolis	21,640	106.9
Iowa	19,674	Des Moines	19,467	98.9
		Cedar Rapids	22,589	114.8
		Iowa City	20,269	103.2
Kansas	20,506	Topeka	19,555	95.4
Kentucky	18,093	Frankfort	20,512	113.4
		Louisville	18,193	100.6
		Lexington-Fayette	23,109	127.7
Louisiana	16,912	Baton Rouge	18,512	109.5
		New Orleans	17,258	102.5
Maine	19,533	Augusta	19,145	98.0
Maryland	25,614	Annapolis	27,180	106.1
Massachusetts	25,952	Boston	23,353	90.0
Michigan	22,168	Lansing	17,924	80.9
Minnesota	23,198	St. Paul	20,216	87.1
		Minneapolis	22,685	97.8
Mississippi	15,853	Jackson	17,116	108.0
Missouri	19,936	Jefferson City	21,268	106.7
Montana	17,151	Helena	20,020	116.7
		Billings	19,207	112.0
Nebraska	19,613	Lincoln	20,984	107.0
		Omaha	21,756	110.9
Nevada	21,989	Carson City	20,943	95.2
New Hampshire	23,844	Concord	21,976	92.2
New Jersey	27,006	Trenton	14,621	54.1
New Mexico	17,261	Santa Fe	25,454	147.5
		Albuquerque	20,884	121.0
		Los Alamos	34,240	198.4
New York	23,389	Albany	18,281	78.2
North Carolina	20,307	Raleigh	25,113	123.7
		Durham	22,526	110.9
		Chapel Hill	24,133	118.8
North Dakota	17,769	Bismarck	20,789	117.0
		Fargo	21,101	118.8
Ohio	21,003	Columbus	20,450	97.4
		Cincinnati	19,962	95.0
		Cleveland	14,294	68.1

(*continued*)

Table A.5 (*continued*)

State	1. Per cap. income	Capital	2. Per cap. income	Col. 2/ Col.1 (%)
Oklahoma	17,646	Oklahoma City	19,098	108.2
		Tulsa	21,534	122.0
Oregon	20,940	Salem	19,141	91.4
		Eugene	21,315	101.8
Pennsylvania	20,880	Harrisburg	15,787	76.3
Rhode Island	21,688	Providence	15,525	71.6
South Carolina	18,795	Columbia	18,853	100.3
		Charleston	22,414	119.3
South Dakota	17,562	Pierre	20,462	116.5
		Sioux Falls	21,374	121.7
Tennessee	19,393	Nashville-Davidson	22,018	113.5
		Memphis	17,838	92.0
Texas	19,617	Austin	24,163	123.2
		Dallas	22,183	113.1
		Houston	20,101	102.5
Utah	18,185	Salt Lake City	20,752	114.1
Vermont	20,625	Montpelier	22,599	109.6
		Burlington	19,011	92.2
		Barre	18,724	90.8
Virginia	23,975	Richmond	20,337	84.8
Washington	22,973	Olympia	22,590	98.3
West Virginia	16,477	Charleston	26,017	157.9
		Wheeling	17,923	108.8
Wisconsin	21,271	Madison	23,498	110.5
		Milwaukee	16,181	76.1
Wyoming	19,134	Cheyenne	19,809	103.5
		Casper	19,409	101.4
Washington, DC	—	—	28,659	—
USA	—	—	21,587	—

Source: US Census Bureau, 2000.

Appendix 2
A Brief Chronology of Colonial, Territorial, and State Capitals

Appendix 2 has to be seen as a work in progress. Although it would seem easy to provide the date of the creation of a colony or a territory and to reconstruct the precise sequence of capitals in each state, numerous difficulties remain. The precise status of some lands and cities is ambiguous: was a territory "created" when it was claimed by a country or when a "real" settlement and a government had been established? Does a settlement become a "capital" when it is founded, when it has been granted official existence, or when the first governor arrives? The fact that the existing sources (books as well as Internet resources) provide conflicting dates is evidence of such ambiguities. Appendix 2 owes its existence to the seminal work begun in the 1940s by Frank Douglas Halverson. His "Chart Showing Changes in the Location of State Capitols," typewritten on a few sheets of paper, is to be found today in a few American libraries (in several editions). I have revised the chart according to the sources used for this book, which include the *Book of the States, 1998–99* (Council of State Governments 1998, 448–49), but there are certainly still some errors or omissions.

ALABAMA

The eastern half of Mississippi Territory was organized as the Alabama Territory in 1817. The original capital of the territory was St. Stephens (1817–1818). Alabama became a state in 1819, and Huntsville, which hosted the state constitutional convention and the first legislature, was the state capital from 1819 to 1820. Other capitals were Cahawba (1820–26) and Tuscaloosa (1826–46).

The present capital, Montgomery, was chosen in 1847.

ALASKA

Alaska was owned by Russia until its purchase by the United States in 1867. Under the Russians, its original capital was Paul's Harbor (1792). Sitka, created as New Archangelsk, became the capital in 1799, and was designated the permanent capital in 1804.

Under US ownership, Alaska was governed by the US Navy until a civil government (a district, without a legislature) was established in 1884, with Sitka remaining as the capital. The courts alternated between Sitka and Wrangell.

Juneau, the present capital, was chosen in 1900 but the move was effective only in 1906, and the Alaska Territory was created in 1912. Alaska became the forty-ninth state in 1959.

ARIZONA

Arizona was established as a territory in 1863 on lands ceded by Mexico in 1848. Prescott was selected as its original capital in 1864, after a brief spell at Fort Whipple. The capital was removed to Tucson from 1867 to 1877, but Tucson was considered by Prescott to be too close to Mexico. Political fights between Prescott and Tucson resulted in Prescott again becoming the capital from 1877 to 1889.

The present capital, Phoenix, was chosen in 1889, and Arizona became a state in 1912.

ARKANSAS

Arkansas was part of the Louisiana purchase (1803). The District of Arkansas was organized in 1806 and became part of the Missouri Territory (1812–1819). Arkansas was established as a territory in 1819. The original capital of the territory, Arkansas Post (1819), had been the residence of early French and Spanish governors of Louisiana.

Little Rock, the present capital, was chosen in 1820, and Arkansas became a state in 1836.

CALIFORNIA

California was established as a Spanish colony in 1768. The original capital of the colony was Loreto (1768–77). Monterey became the second capital in 1777 and remained as the capital until 1847, even after California became Mexican in 1821. Los Angeles vied with Monterey until 1847 as the seat of government for the Mexican Province of California. The American consul resided at Monterey from 1847 to 1849, and the Constitutional Convention took place there in September 1849. California became a state in 1850. The successive capitals were San Jose (1849–51); Vallejo (1851–January 1852); Sacramento (1852); Vallejo (January 3, 1853); and Benicia (1853– March 1854).

The present capital, Sacramento, was chosen in 1854.

COLORADO

Colorado was part of the Louisiana purchase (1803). In 1819, the Territory of Eastern Colorado was established, and in 1848 it became part of the Kansas Territory. Colorado was established as a territory in 1861. The original capital of the territory was Colorado City (1861, during five days). Denver briefly became the capital (1862), and then Golden was selected (1862–68).

The present capital, Denver, was chosen in 1868. Colorado became a state in 1876.

CONNECTICUT

Connecticut was established as a British colony in 1638. The original capital of the colony was Hartford (1636–1701); New Haven was selected as the capital of the colony of New Haven (1643–65). During the Confederation of the United Colonies (1643–1680s), New Haven and Hartford were two of the four capitals (along with Boston and Plimouth). Hartford and New Haven were designated as co-capitals from 1701 to 1873.

Connecticut became a state in 1788. The present capital, Hartford, was chosen in 1875 as the sole capital.

DELAWARE

Delaware was established as a Swedish and a British colony in 1638. It belonged to Pennsylvania from 1682 to 1704, the year it was established as a separate colony. New Castle was selected as its original capital in 1704. Until 1776, it was called Lower Counties. Until 1776, the governor resided in Philadelphia, and Pennsylvania remained powerful. Between 1777 and

1781, the capital wandered between Dover, Wilmington, Lewes, and New Castle.

The present capital, Dover, was chosen in 1781. Delaware became a state in 1787.

FLORIDA

Florida was established as a Spanish colony in 1513. St Augustine was selected as its original capital in 1565. Pensacola became joint capital in 1763, when Florida was divided into eastern and western provinces by England and then by Spain. The American purchase took place in 1819 and came into effect in 1821. Florida was established as a territory in 1822.

The present capital, Tallahassee, was chosen in 1824. Florida became a state in 1845.

GEORGIA

Georgia was established as a British colony in 1732 and became a Royal Colony in 1752. Savannah, founded in 1733 by Oglethorpe, was selected as its original capital (1733–78); Savannah was under British control from 1778 to 1782. Augusta became capital (1778–80) and then Ebenezer (1782).

Between 1782 and 1786, the legislature alternated between Savannah and Augusta, and the latter became sole capital from 1786 to 1795. Georgia became a state in 1788. The capital was removed first to Louisville (1795–1807) and then to Milledgeville (1807–68). Macon served as temporary state capital in 1864.

The present capital, Atlanta, was chosen in 1868.

HAWAII

Hawaii was an independent kingdom. The court came to Waikiki in 1804 and to central Honolulu in 1809. Honolulu was the royal capital from 1850 to 1893 and then served as the capital of the Republic of Hawaii (1893–1898). Hawaii was annexed to the United States in 1898 and became a territory in 1900.

The present capital, Honolulu, was chosen in 1900. Hawaii became a state in 1959.

IDAHO

Occupying land ceded by the 1846 treaty with Britain, Idaho was first included in the Oregon Territory and later in the Washington Territory. Idaho was established as a separate territory in 1863. The Idaho Territory once embraced Wyoming, Montana, and part of the Dakotas. Lewiston was selected as its original capital in 1863.

The present capital, Boise City, was chosen in 1864. Idaho became a state in 1890.

ILLINOIS

Illinois was part of the Northwest Territory (1787) and included in 1800 with the Indiana Territory. In 1809, Illinois was established as a separate territory, which included most of Wisconsin, a large part of Michigan, and eastern Minnesota. Kaskaskia was designated as its original capital in 1809.

Illinois became a state in 1818 and the capital was removed to Vandalia from 1820 to 1836. The present capital, Springfield, was chosen in 1837.

INDIANA

Part of the Northwest Territory (1787) at first, Indiana was established as a territory in 1800. The Indiana Territory embraced also Illinois, Wisconsin, and Northeast Minnesota. Vincennes was designated as its original capital (1800–13), and the capital was later removed to Corydon (1813–25).

Indiana became a state in 1816. In 1821, the site of the present capital, Indianapolis, was selected for the new state capital, although this change was not made effective until 1825.

IOWA

Part of the Louisiana purchase (1803), Iowa was successively part of the District of Louisiana (1804–5), the Louisiana Territory (1805–12), the Missouri Territory (1812–21), unorganized territory (1821–34), the Michigan Territory (1834–36), and the Wisconsin Territory (1836–38). Iowa was established as a separate territory in 1838.

Burlington was designated as its original capital (1838–41). The capital was removed to Iowa City from 1841 to 1857. Iowa became a state in 1846, and the present capital, Des Moines, was chosen in 1857.

KANSAS

Part of the Louisiana purchase (1803), Kansas became part of the Missouri Territory in 1812, then part of "Indian Country" (1834–54). Kansas was established as a territory in 1854. Fort Leavenworth was designated as its original capital in 1854, for fifty days. Other capitals were Shawnee Mission (1854–55), Pawnee (1855), and Lecompton (1856–61). The free-state government met at Topeka and Lawrence. The official meeting places were to be Minneola in 1858 and Lawrence in 1858–61, but Congress did not recognize the change.

Topeka, the present capital, was chosen in 1859 as temporary capital and gained permanent status in 1861, the year Kansas became a state.

KENTUCKY

Kentucky was part of Virginia, in which it became a county in 1776, until it was directly established as a separate state in 1792. Nine conventions were held at Danville (1784–90), seeking separation from Virginia. A first state Constitution was framed by a convention that assembled at Danville in April 1792. Lexington was designated as the original capital for a six-month period (1792–93).

The present capital, Frankfort, was chosen in 1793. During the Civil War, besides Unionist Frankfort, a Confederate capital was established at Bowling Green.

LOUISIANA

Louisiana was established as a French colony in 1682. Mobile was designated as its original capital (1702–20). New Biloxi became the second capital (1720–22), and New Orleans became the third in 1723. Louisiana was established as a territory in 1804, after the Louisiana purchase (1803), and became a state in 1812. The capital was removed in 1830 to Donaldsonville, but came back to New Orleans in 1831.

In 1849, Baton Rouge was selected as the capital and held that status until 1862, after the capture of Baton Rouge by Confederate forces. Opelousas (May 1862–January 1863) and then Shreveport were the Confederate capitals. After 1868, New Orleans once again became the capital.

Baton Rouge, the present capital, was selected by the 1879 constitutional convention, and the transfer took place in 1882.

MAINE

Maine was under the jurisdiction of Massachusetts until it was established as a state in 1820. The original proprietor of Maine established his seat of government at York. The Court was first held at Saco. Portland was designated as its original capital (1820–32).

Augusta, the present capital, was chosen in 1827, but, as the statehouse was not then completed, Portland remained as the seat of government until 1832.

MARYLAND

Maryland was established as a colony in 1632. The first Assembly of Freemen met on January 26, 1635, at St. Mary's, the original capital.

The present capital, Annapolis, was chosen in 1694. Maryland became a state in 1788.

MASSACHUSETTS

Massachusetts was at first two British colonies: Plimouth (1620) was established by the pilgrims, and the Massachusetts Bay Colony (1629) was established by Puritans from England. Newtown (Cambridge today) was briefly the first capital of the Massachusetts Bay Colony, as was Beverly (1632). Boston became capital of the Massachusetts Bay Colony in 1632.

In 1692, a charter was granted that united the two colonies into the Province of Massachusetts, with Boston, the present capital, as its capital. Massachusetts became a state in 1788.

MICHIGAN

Michigan was part of the Northwest Territory (1787) and of the Indiana Territory (1802). Michigan was established as a separate territory in 1805 and had jurisdiction over Minnesota, Iowa, and Wisconsin. Detroit was designated as its original capital in 1805.

When Michigan became a state in 1837, Detroit remained capital, until 1847. The present capital, Lansing, was chosen in 1847.

MINNESOTA

Part of the Northwest Territory (1787) at first, Minnesota was part of numerous territories before it was finally established as a separate territory in 1849. It had jurisdiction over parts of the Dakotas. Minnesota became a state in 1858.

St. Paul, the present capital, was chosen in 1849, and became the permanent capital in 1893.

MISSISSIPPI

Mississippi was established as a territory in 1798, and Natchez was designated as its original capital (1798–1802). Other capitals were Washington (1802–1820) and Columbia, which was selected as temporary capital in 1817, when Mississippi became a state, but the move only became effective in 1821.

Jackson, the present capital, was chosen in 1821 and officially became the capital in 1822.

MISSOURI

In 1765, Missouri was part of Spanish Upper Louisiana, with St. Louis as its capital. After the Louisiana purchase (1803), Missouri became in 1804 part of the District of Louisiana, and part of the Louisiana Territory in 1805. In 1812, Missouri was established as a separate territory that embraced a much larger area than the present state. St. Louis was designated as the capital in 1805. When Missouri became a state in 1821, the capital was removed to St. Charles (1821–26).

The present capital, Jefferson City, was chosen in 1826. Missouri's secessionist shadow government had two seats: Neosho (1861–1863) and Marshall, Texas (1863–1865).

MONTANA

Montana was part of the Louisiana purchase (1803) and was later included with several nearby territories, the last being the Idaho Territory (1863–64). Montana was established as a territory in 1864. Bannack City was designated as its original capital in 1864. The capital was removed to Virginia City in 1865, until 1875.

Helena, the present capital, was chosen in 1874, but became the permanent capital only in 1894, five years after Montana had become a state in 1889.

NEBRASKA

Part of the Louisiana purchase (1803), Nebraska was established as a separate territory in 1854. The Nebraska Territory was reduced to its present boundaries between 1861 and 1863. Bellevue briefly hosted the first governor. Omaha was designated as its original capital in 1855.

Nebraska became a state in 1867, and Lincoln, formerly known as Lancaster, was chosen as the present capital in that same year.

NEVADA

Part of the Louisiana purchase (1803), Nevada was under jurisdiction of the Territory of Utah before it was established as a separate territory in 1861. Genoa, founded in 1850 by Mormons, was the site of pioneer government before the creation of the territory.

The present capital, Carson City, was chosen in 1861. Nevada became a state in 1864.

NEW HAMPSHIRE

New Hampshire was established as a British colony in 1622 and 1629. In 1643, it was joined to Massachusetts. In 1679, New Hampshire became a separate Royal Province, but until 1740 it had the same governor as Massachusetts. Portsmouth was designated as its original capital in 1679 and remained so until 1775. Exeter was the second capital (1776–84), and then there was a period of wandering capitals (eight cities) until 1805. In 1781–82, the state Constitutional Convention met at Concord. New Hampshire became a state in 1788.

Concord, the present capital, was chosen in 1805 and became the permanent capital in 1814.

NEW JERSEY

New Jersey was established as a Dutch settlement in 1618 and became a British colony in 1664. Elizabethtown was designated as its original capital in 1665. In 1674, New Jersey was divided into East Jersey (with Perth Amboy as the capital from 1684), and West Jersey (with Burlington as the capital from 1677). East and West Jersey united in 1702, and alternate legislative sessions were held in both cities. Until 1738, the governor of New York was also governor of New Jersey. The first Provincial Congress of New Jersey met on July 21, 1774, at New Brunswick. The 1776–1788 period was one of wandering capitals. Princeton in 1783, and Trenton in 1784, were briefly Federal capitals.

New Jersey became a state in 1787. Perth Amboy and Burlington became co-capitals for two years (1788–1790), and the present capital, Trenton, was chosen in 1790.

NEW MEXICO

New Mexico was established as a Spanish colony in 1598. San Juan de los Caballeros, the first settlement, was created in 1598. The Territory of New Mexico originally extended to the borders of California from the 103rd meridian on the east. New Mexico became Mexican in 1821. After the Mexican cession of 1848, New Mexico became a territory in 1850 (larger than today's state) and a state in 1912.

The present capital, Santa Fe, was chosen in 1610. It was the first official capital. El Paso was the capital between 1681 and 1692, after an Indian uprising left Santa Fe almost deserted.

NEW YORK

New York was established as a Dutch settlement in 1623 and became a British colony in 1664. New York City was designated as its original capital in 1626 by the Dutch and in 1664 by the British. Between 1777 and 1797, the capital wandered between Kingston, Poughkeepsie, Hurley, New York City, and Albany. New York City was the seat of the Federal Government from 1785 to 1790.

New York became a state in 1788, and the present capital, Albany, was chosen in 1797.

NORTH CAROLINA

Carolina was established as a British colony in 1663. South and North Carolina originally formed the Province of Carolina. But, as the first settlements were far apart, it was divided, except for a period from 1691 to 1712. Edenton was designated as its original capital (1722–37, but also 1740–41, and 1743). Then followed a period of wandering capital (Wilmington, Halifax, Bath, Hillsborough, Fayetteville, etc.), during which New Bern was the most "active" (it was designated as capital in 1765–78, 1780–81, 1784–85, 1791–93, and 1794).

North Carolina became a state in 1789. The present capital, Raleigh, was chosen in 1792, after almost half a century of debates. The first legislative meeting was held in Raleigh in 1794.

NORTH DAKOTA

Part of the Louisiana purchase (1803), Dakota was established as a territory in 1861, and separated into North and South Dakota only in 1889. The original capital of the territory was Yankton (1861–83).

Bismarck, the present capital, was chosen as territorial capital in 1883, and as state capital in 1889, the year North Dakota became a state.

OHIO

Part of the Northwest Territory (1787), Ohio was established as a territory in 1800. Marietta was designated in 1788 as the capital of the first county formed in the territory. Cincinnati then became capital of the second county in 1790. When Ohio became a state in 1803, the capital was first removed to Chillicothe (1803–10) and then to Zanesville (1810–12), before it came back to Chillicothe (1812–16).

The present capital, Columbus, was chosen in 1816.

OKLAHOMA

Part of the Louisiana purchase (1803), Oklahoma was opened to homesteaders in 1889 and established as a territory in 1890. Guthrie was designated as its original capital in 1890. Muskogee was the seat of government for Indian Territory. The state of Oklahoma, established in 1907, is a consolidation of Indian and Oklahoma Territories.

The present capital, Oklahoma City, was chosen in 1910.

OREGON

Oregon was established as a territory in 1848 on land ceded after the 1846 Treaty with Britain. The Oregon Territory embraced Washington and part of Montana.

Oregon City was designated as its original capital in 1849.

The present capital, Salem, was chosen in 1852. The legislature commenced meeting at Salem in 1852 (the removal act was validated by Congress that year, after a legal battle). Corvallis was briefly the capital (1855), but Congress disagreed, and the removal act was repealed.

Oregon became a state in 1859, and Salem became the permanent capital after a popular vote in 1864.

PENNSYLVANIA

Pennsylvania was established as a grant to William Penn in 1681. He created Philadelphia as its original capital in 1682. From 1790 to 1800, Philadelphia was the seat of the Federal Government. Pennsylvania became a state in 1787. The capital was removed to Lancaster from 1799 to 1811.

The present Capital, Harrisburg, was chosen in 1812.

RHODE ISLAND

Rhode Island was established as a British colony in 1663. Between 1663 and 1681 Newport was the only capital. Rhode Island then experienced a system of rotating capitals—in an irregular fashion. Until 1900, the capital moved between Providence, Newport, East Greenwich, South Kingston, and Bristol. Rhode Island became a state in 1790. From 1854 until 1900, Providence and Newport served as dual capitals.

Providence, the present capital, was chosen as the sole capital in 1900.

SOUTH CAROLINA

Carolina was established as a British colony in 1663, but settlement began only in 1670. South and North Carolina originally formed the Province of Carolina. But, as the first settlements were far apart, it was divided, except from 1691 to 1712. Charles Town was designated as its original capital in 1670. The town was moved in 1680 to its present location, and its name was changed to Charleston in 1783.

South Carolina became a state in 1788. The present capital, Columbia, was chosen in 1790.

SOUTH DAKOTA

Part of the Louisiana purchase (1803), Dakota was established as a territory in 1861 and separated into North and South Dakota only in 1889. The original capital of the territory was Yankton (1861–82). Bismarck became the second territorial capital (1883–89).

The present capital, Pierre, was chosen in 1889, when South Dakota became a state.

TENNESSEE

Part of North Carolina until 1789, Tennessee was established as a territory in 1790. "Rocky Mount" (a mansion, also called Rogersville) was selected as its original capital (1790–92). The capital was removed to Knoxville in 1792. Tennessee became a state in 1796.

In 1807, the legislative assembly met at Kingston, but adjourned to Knoxville after the first day and never met there again. Other capitals were Nashville (1813–16), Knoxville (1817–19), and Murfreesboro (1819–26).

The present capital, Nashville, was chosen in 1826. Between 1826 and 1843, it was the de facto capital, and from 1843, the de jure capital.

TEXAS

Texas was officially claimed as a Spanish colony in 1519, but was formally established only at the beginning of the eighteenth century. Los Adaes (today in Louisiana) was selected as its original capital (1729–73). Other capitals were San Antonio (1773–1824), and, under Mexican rule, Saltillo (1824–33), Monclova (1833–35), and Saltillo again (1835).

In 1836, Texas became an independent state. Because of the war with Mexico, the capitals of Texas frequently changed: capitals during that time were San Felipe de Austin (1836), Harrisburg (1836), Houston (1837–39 and 1842), Austin (1839–42), and Washington on the Brazos (1842–45). During the early history of Texas, emergency governmental headquarters were established at Galveston, Velasco, and often at other points for brief intervals. These places were not capitals in the present sense of the term.

The present capital, Austin, was chosen in 1845, when Texas chose to be admitted to the United States of America. The choice of Austin as capital was ratified by election in 1850.

UTAH

In 1848, the Mormons created the unofficial State of Deseret on land ceded by Mexico, of which Salt Lake City was the unofficial capital. The Territory of Utah was established in 1850 (it was smaller than Deseret). Fillmore was designated as the territorial capital (1851–56).

The present capital, Salt Lake City, was chosen in 1856. During the "Utah War" against the Federal Government, Parowan was briefly capital in 1858. Utah became a state in 1896.

VERMONT

Both New Hampshire and New York originally claimed the area of Vermont. In 1777, a state Constitution was drafted in a convention held at Windsor. Vermont was established as a state in 1791. Its original capitals wandered between fourteen cities.

Montpelier, the present capital, was chosen in 1805 as one of the wandering capitals, and it was chosen as the sole capital in 1808.

VIRGINIA

The Virginia company charter was granted by James I in 1609. Jamestown was selected as its original seat in 1609. Virginia became a Royal Colony in 1624, with Jamestown as capital. Williamsburg—once known as Middle Plantation—became the second capital in 1698.

The present capital, Richmond, was chosen in 1779 and remained capital when Virginia became a state in 1788. From 1861 to 1865, Richmond was the Confederate capital.

WASHINGTON

Washington was part of the Oregon Territory in 1848. It was established as a territory in 1853 (Idaho was then part of Washington).

The present capital, Olympia, was chosen in 1853. Washington became a state in 1889.

WEST VIRGINIA

West Virginia was part of Virginia until 1863. Because it had refused to secede, it was established as a state in 1863. Wheeling was designated as its original capital in 1863. The capital was then removed to Charleston (1870–75) and then returned to Wheeling (1875–85).

The present capital, Charleston, was chosen in 1885.

WISCONSIN

Wisconsin was part of the Northwest Territory in 1787, the Indiana Territory in 1800, the Michigan Territory in 1805, the Illinois Territory in 1809, and the Michigan Territory again in 1818. In 1836, Wisconsin was established as a territory that included the present states of Minnesota, Iowa, and part of the Dakotas. Belmont (1836) and Burlington (now in Iowa; 1837–38) were designated as the temporary capitals.

The site of the present capital, Madison, was selected in 1836. In 1838, the legislature first met at Madison. In 1839, with the completion of the government buildings, Madison became permanent capital. Wisconsin became a state in 1848.

WYOMING

Part of the Louisiana purchase (1803), Wyoming was part of the Dakota, Oregon, Utah, and Idaho Territories. Wyoming was established as a separate territory in 1868.

The present capital, Cheyenne, was chosen in 1869. Wyoming became a state in 1890.

Notes

CHAPTER ONE

1. Searching the Library of Congress database yielded only three titles: Delia Goetz (1971) provides 159 pages of juvenile literature; Paul W. Pollock (1960) has 208 pages of popular literature (3 pages per capital); and Tracy Maurer (1999) presents 48 pages of vulgarization. There was also a survey of American capitals on four videotapes.

2. An MSA is a metropolitan statistical area, as defined by the US Bureau of the Census.

3. Yet J. F. Rishel wrote in 1992 that "much remains to be done, for the small city is not merely a microcosm of the metropolis" (1992, 5). Chapter 4 of his book deals with small towns (Maureen Ogle, "Beyond the Great City: Finding and Defining the Small City in Nineteenth Century America").

4. Clay Burnette, quoted in Moore 1993 (463). In 2010, the municipality of Columbia had 129,272 inhabitants, and its CSA included 805,106.

5. James Bryce, *The American Commonwealth* (New York: Macmillan, 1891), 2:855, quoted in Meinig 1998, 319. Bryce wanted to prove that the United States had no "real" capital at the national level. He did not even bother to study the state level.

6. Glaeser underlines the close links between major (Glaeser uses the word *primate*) cities and national capitals. The US case proves that the links are far more distended when it comes to the federated states.

7. James (1877–1969) was executive secretary of the American Civic Association and then of the American Planning and Civic Association from 1921 to 1958. She was very active in zoning and parks matters (http://www.aapra.org/Pugsley/JamesHarlean.html).

8. For instance, although the American Constitution created a national citizenship, it subordinated it to state citizenship (article IV, section 2, 1).

9. For an interesting overview of contemporary trends in American urban history—as well as of the impossibility of attaining a "general urban history"—see Gilfoyle 1999.

10. The definition of the *lieux de mémoire* given by Nora is "all significant unity of material or ideal order that through the will of mankind or the work of the time, was made a symbolic element of the memorial heritage of any community" (Nora 1984, 20).

11. This research was completed with the significant help of web resources, including the capitals' websites and the treasures offered online by the Library of Congress, the US Bureau of the Census, and other databases.

CHAPTER TWO

1. The quest for the "most American city" is an old one. For nineteenth-century candidates, see Strauss 1961, 104–5.

2. To achieve perpetuation—the French word is *perennialité*—places must embody a history, as do the three avatars of the small town put forward by Donald Meinig (1979): the New England village, the main streets of the Middle West, and the Pacific Coast suburbs. They are close to myth.

3. Such an assessment can seem strange in a nation where citizens' distrust of Washington was (and still is) overwhelming, and where classic studies oppose federal and state conceptions. It is one more instance that proves the utility of going beyond analyses that rest only upon one academic slot—here, political analysis.

4. This totem marked the boundary between two Indian tribes, the Bayougoulas and the Houmas. The 1721 French settlement soon disappeared (Gleason 1991, vii).

5. Salt Lake City was first called Great Lake City of the Great Basin, and then Great Salt Lake City until 1868, when the territorial legislature dropped the "Great."

6. Its complete name was La Villa Real de la Santa Fe de San Francisco.

7. The name was selected, after some difficulties, during a meeting held by the town's founders in 1854.

8. Arthur St. Clair (1734–1818) had been a general in the Continental Army during the Revolutionary War. He was then a member and president of the Continental Congress (1787), and later the governor of the Northwest Territory from 1789 to 1802 (http://politicalgraveyard.com).

9. The United States was not alone in using Greek names: the capital of the Brazilian state of Santa Catarina is Florianopolis. Brazil also has an Anapolis.

10. The Duke of York was the brother of King Charles II, who gave him the province of New York in 1664, after seizing it from the Dutch.

11. The trading post at Le Fleur's Bluff was renamed Jackson in 1821 to honor Major General Andrew Jackson, who had not yet become president.

12. Utah's first territorial capital (1851–1856), Fillmore, bears the name of the president at the time, Millard Fillmore, in honor of his courage in naming Brigham Young, leader of the Mormons, as Utah's first governor. Although it was the official capital, the capitol was never completed, and only one legislative session was held there (1855–56).

13. From 1855 to 1867, the first capital of the territory was Omaha, north of the Platte River.

14. Otto, Prince von Bismarck (1815–98), was prime minister of Prussia (1862–90) and chancellor of the German Empire (1871–90).

15. The exact reason remains unknown: "The name 'Columbia' was suggested by Senator Gervais: that of 'Washington' by Senator Barnwell. 'Columbia' secured a majority in the senate, and the bill was so sent to the house. There seems no reason for the name 'Columbia,' except that it was at that time popular" (Green 1932, 147–48).

16. Pierre Dorion Sr. (1740–1810) was a Frenchman from Quebec. He came to work as a fur trader along the Missouri River around 1780 and married a Yankton Sioux squaw. In 1804, he was hired by Lewis and Clark to work with the Indians again (http://www.sd4history.com). Fort Pierre, located across the Missouri River from the city of Pierre, draws its name from another Pierre, Pierre Chouteau Jr. (1789–1865), head of the American Fur Company's Western Department, who had the fort built in 1831. It was abandoned in 1857 (Schuler 1990, 30).

17. Milledge (1757–1818), a Republican, was governor from 1802 to 1806, and later became representative and senator.

18. The name was given in 1837 by the Western and Atlantic Railroad Company. Atlanta probably also hints at the name of the company. Railroad towns or streets were often named after a railroad employee. Some companies, like the St. Joseph and Denver City Railroad, were "naming the supply stations at the end of each completed section in alphabetical sequence" (Lingeman 1980, 247).

19. According to the census, a place is a city, a village, a town, a borough, or a Census Designated Place.

20. The open form of the city also allowed continual growth and was later perfectly adapted to railroads (Caves 1995, 59).

21. At the end of the eighteenth century, neo-Palladian influences had replaced baroque ones, as is evident in Jefferson's architectural works.

22. As isomorphism does not always mean influence, only proven influences are here indicated.

23. On June 11, 1805, less than one week after the Michigan Territory was created, the capital, Detroit, was destroyed by fire.

24. The idea that Versailles inspired the plan of Washington is found in most studies of the federal capital. For an interesting critical view, see Corboz 2003, 61–99. Washington was certainly more modeled on the palace's park than on the city of Versailles.

25. The reason was that, as a circle has no back, "no governor's wife would tolerate the fact that the family wash would have to be hung out to dry in plain view of all the town" (Meinig 1993, 443). Meinig describes the plan of Indianapolis as "the model geography of its day in many features" (ibid.). Besides the squares provided for government, there were squares for commerce (market) and religion, the three bases of American civic order.

26. The capitol, begun in 1866, had to wait until 1903 to be completed, at a cost of $3.2 million.

27. Although it acted illegally, the Seminole party was led by notables: William Couch, leader of the Boomers, and General James B. Weaver, former Greenback party presidential candidate.

28. The rivalry was not only land-based. Eventually, the Kickapoos gained political control, and the Seminoles retained economic control. Berlin B. Chapman (1960) provides a thorough study of Oklahoma City's first year of existence.

29. The American Legion is headquartered in Indianapolis.

30. The prestigious firm of McKim, Mead, and White was selected, and it built the capitol from 1896 to 1904. Made entirely of white marble, it cost $3 million.

31. Frederick Law Olmsted, who co-designed Central Park in New York City, worked for Providence.

32. The name was given to the most famous group of buildings of the World's Columbian Exposition, held in 1893.

33. Some states use the name *statehouse* (e.g., Indiana, Maryland, and Rhode Island). Some use other names: Delaware uses *legislative hall*; Washington and Nevada use *legislative building.*

34. "In democratic communities the imagination is compressed when men consider themselves; it expands indefinitely when they think of the state. Hence it is that the same men who live on a small scale in cramped dwellings frequently aspire to gigantic splendor in the erection of their public monuments" (Tocqueville 2000, chap. 12).

35. Topeka *Weekly Leader*, October 18, 1866, quoted in Richmond 1972, 251–52.

36. The Maryland statehouse is today the oldest functioning capitol in the United States.

37. Capitols therefore differ from the domestic, commercial, and religious architecture in cities and towns that imitated European buildings.

38. The design of South Dakota's new capitol (1910) was also inspired by Boston's capitol.

39. After 1776, however, architects drew their inspiration from democratic Greece rather than from imperial Rome, which was politically too close to imperial Britain.

40. The marble and limestone used for the lower façade were quarried on Prince of Wales Island in southeastern Alaska (www.legis.state.ak.us/students/capitol.htm).

41. Here, as in West Virginia, gilding was controversial, because of its cost and because it symbolizes wealth.

42. In 1976, "the state's copper industry gave 15 tons of copper to cover the dome" (Arizona State Capitol Museum leaflet).

43. The current capitol replaced the neoclassical "Anglo" territorial capitol built in 1900.

44. Tennessee offers yet another example. Nashville's capitol is compared to a "grand marble wedding cake" (Goodstein 1989, 120).

45. Thirty capitols offer all six elements; thirteen offer four or five of them (Goodsell 2001, 15, 29).

46. An assemblyman wrote the text in 1863, when Albany felt the need to replace its too-small capitol.

47. In Boston, as in Richmond, the capitol serves as the zero mileage point for the state's roads. In France, this point is situated just in front of Notre Dame Cathedral in Paris.

48. Privatization comes from the outside—that is, the physical urban fabric—and also from the inside, through the growing use and power of media. For an in-depth study of the relationships between television, geography, and community in the United States, see Adams 2002, chap. 10.

49. This raises the question of the state's changing role in the production and regulation of public space (the question was posed by Trevor Griffey, University of Washington, on the web list of H-URBAN, July 11, 2003).

50. See the works of Erwin Goffman, Yi-Fu Tuan, or Ulf Hannertz.

51. Capitol sites having often been selected for their eminence, the capitol is often essentially out of the city's core, out of the city's "normal" life (cf. Atlanta or Providence). However, we must remember that state business is further discussed after the legislative day in downtown restaurants and bars.

52. "Official" histories were ordered by some states. In Florida, the history of the state capital was written "in order to educate state visitors and Florida citizens about the history of Tallahassee, the capital city, in furtherance of F.S. 266.116 (9) (18) & (19)" (Ellis and Rogers 1986, 77).

53. Penelope Lemov describes the capitals' downtowns as "blighted, boarded-up civic embarrassments" (1993, 47). The third most common grievance—that the state obstructs the capital's economic development plans—is discussed in chapter 7, as it deals with a different theme.

54. South Frankfort seceded from Frankfort in 1812, but both towns reunited in 1850.

55. This makes Montpelier the fourth smallest state capital, behind Annapolis, Trenton, and Harrisburg.

56. For instance, East-siders laughed at the western "literary fellows" by spelling the river's name "Demoine."

57. "Des Moines—Iowa State Capital." *New York Daily Graphic*, 1878, September 17; reproduced in Petersen 1970, 226–27.

58. The federal government paid three-fourths of the costs.

59. The Capitalize Albany Corporation tried in 1994 to heal some of the 1960s functionalist mistakes by trying to reconnect the city with its waterfront, from which it was separated by a motorway (http://capitalizealbany.com).

60. The successful completion of such a grand design in an underpopulated state—fewer than one million inhabitants at the time—is partly explained by the

wealth of the legislature. In fact, Congress granted the new state 132,000 acres of land in 1889, the income from which was only to be used to finance buildings at the state capital. Mostly consisting of rich timber areas, these lands are still contributing to the state's finances.

61. Established in 1909, Honolulu is the largest single municipally governed area in the nation.

62. Before 1776, mayors were quite weak, being appointed by governors or proprietors. After independence, "the legislatures quickly assumed control over municipal corporations," undermining the power of governors over municipalities. By 1840, the popular vote—first restricted to white men—was the almost universal way to elect mayors, who soon became political leaders (Adrian 1999, 53–54). Although there are four structures of municipal government in the United States, that fact has no perceptible influence on the processes discussed here (Kemp 1999, 63–68). The only form of municipal government bearing the name of a state capital is the commission form, which is also called the Des Moines Plan (Brigham 1911; Rice, 1976). By 1987, only 3% of cities with more than five thousand inhabitants used that form (Topeka, Kansas, for instance, abandoned it).

63. "Almost all governors agreed that experience in the state legislature just prior to election as governor is invaluable. . . . Mayoral offices and most specialized statewide elective offices below the governorship do not afford experience with the intricate array of state problems" (Sabato 1983, 33).

64. From other points of view, state capital mayors in 2003 did reflect the evolution of American society: seven were women, five were African American, three were Hispanic, and one was overtly gay.

65. After 1987, Fulton established a private firm dealing in governmental relations, a subject he knew well.

66. We should note earlier cases, such as that of Thomas Taggart (1856–1929), who served as both mayor of Indianapolis (1895–1901) and Democratic national chairman (1900–8) during his long political career, from the 1880s until his death (Bodenhamer and Barrows 1994, 165).

67. His ancestor, the first Erastus Corning, was mayor of Albany from 1834 to 1837.

68. Charles Adrian contends that administrators have replaced political parties in the coordination of functions and activities among the three levels of intergovernmental relations. A large majority of local elections were indeed nonpartisan in 1986: 60% in mayor-council cities, 74.6% in commission cities, and 81.9% in council-manager cities (Adrian 1999, 57–58).

69. Historians have shown that such awareness of the rise of non-places had already occurred during the nineteenth-century growth of industrial cities. Non-places are therefore more a recurrent topos than a novelty described by postmodernists.

70. The words *public* and *private* should be discussed not only in terms of places but also in terms of activities.

71. "Jaywalking," however, is considered a crime in numerous cities.

72. The capitol in Pierre, South Dakota, is proud to announce that it is open 365 days a year from 8 a.m. to 10 p.m.

73. The September 11, 2001, attacks, however, certainly altered the "openness" of capitols.

74. They are open less frequently than the White House, however, owing to their far more modest scale. An exception is the mansion in Frankfort, which reproduces the Petit Trianon of Versailles.

75. The best-known case is naturally Springfield, Illinois, where great pride (and tourist income) results from promoting reminiscence about the days when future President Lincoln was a member of its assembly.

76. See *Renaissance on the Missouri: The South Dakota Capitol* (1983), the book produced for the touring exhibition sponsored by the Memorial Art Center in Brookings (cited in Schuler 1989, 49).

77. The oldest one is Capitol Hill United Neighborhood, Inc., founded in 1969, which organizes the annual Capitol Hill People's Fair (drawing more than two hundred and fifty thousand people). In 1991, the Center for the People of Capitol Hill was founded.

78. In 1750, Savannah "was still much more noteworthy for its spacious and elaborate plan than for its actual growth and substance" (Meinig 1986, 190).

CHAPTER THREE

1. Stevens was also superintendent of Indian Affairs and engineer in charge of surveying a northern route for a transcontinental railroad from the Great Lakes to Puget Sound (Dodds 1986, 94). Stevens was born in 1818 in Massachusetts.

2. Clark County, with 1,134 people, was slightly more populated than Thurston County, with 996 inhabitants.

3. Known today simply as Yakima, it is situated near the geographic center of the state.

4. The state's total population was 423,000, with only 2,000 people in the newly opened western third of the state.

5. In the Treaty of San Lorenzo, also called Pinckney's Treaty, Spain recognized the US borders at the Mississippi and the Thirty-First Parallel. It also granted the United States the right to deposit goods for transshipment at New Orleans, prompting the subsequent acquisition of Louisiana.

6. The pattern of settlement still roughly followed the French one: Vincennes, founded in 1732, was an outpost for French Louisiana on river fur trade routes.

7. In 1776, the most populated towns in Delaware were Wilmington (1,200 inhabitants) and New Castle (1,000 inhabitants).

8. Ravaged by raids during the Revolution and having lost much of its entrepreneurial talent, Newport never recovered its former prosperity (Meinig 1986, 364).

9. The two-thirds rule was not restricted to capital choice. It still rules the passage of the budget, creating great inconveniences.

10. I am using the nineteenth-century terms.

11. "Sacramento was a city born to floods" (Brienes 1979, 3). It took almost two decades to build the necessary levees against the powerful Sacramento and American rivers, and more floods occurred in 1861, in 1862–63, and in 1878. The fight against the floods enhanced the legend of the city and "confirmed Sacramento's moral legitimacy as a commercial and governmental center and as the major city of the valley" (18).

12. Ironically, Broderick was narrowly defeated in the senate election. He nevertheless succeeded in 1857.

13. These processes are therefore close to those experienced by the seemingly "stable" capitals during the nineteenth century.

14. For the time being, Olympia, Washington, can be discarded as an exception. Already badly hit by an earthquake on February 28, 2001 (after the big 1949 one), it is waiting for the "huge earthquake [that] will hit this community in the next 300 years. When that happens I expect the location of the capital to be discussed once again," according to the Honorable Stan Biles, mayor of Olympia (March 2, 2003, personal communication).

CHAPTER FOUR

1. Boorstin was one of the first historians to offer explicit discussion of the question of capital movements.

2. The actual number of capitals was higher, but when a state experienced a system of rotating capitals, this was considered a single event with one main factor of explanation, generally centrality or expression of sectionalism.

3. For Kansas, for instance, this meant looking at Cutler 1883; Adams 1903–4; Zornow 1957; Richmond 1972; and Socolofsky and Self 1988.

4. A major cause of early urban growth was the influx of new settlers. This was more important in the Northeast, which profited from regular arrivals of Puritans, than in the Southeast, which followed Anglicanism, the established religion.

5. Later accounts that Mormons had miraculously transformed a barren wilderness into a new Garden of Eden belong largely to a mythical reconstruction of the past.

6. The broader question of the role of fires in the history of American towns is addressed by Robert M. Preston in chapter 3 of Rishel 1992. The argument is that fires were not always catastrophes.

7. Today, a dam at about this location has created a large lake, Jordan Lake, southwest of the Raleigh-Durham-Cary CSA.

8. Such a vision was only made possible by obliterating the people that already lived there—the Native Americans, indeed, but also the Spanish, Dutch, Swedes, and French.

9. Commentators at the time compared the federal form of the American republic to the Hebrew political order and the Constitution to a second Decalogue.

10. Zagarri's book only studies the thirteen original states, but it is the first to discuss capital removals thoroughly.

11. An "All-American City" seems to be a city with good intergovernmental relations and a good school system. Both are necessary bases for a healthy democracy, according to Jefferson and numerous others since.

12. On the general question of morality in America, see Boyer 1978, especially chapter 9, "American Protestantism and the Moral Challenge of the Industrial City."

13. Philadelphia was at the time America's second city. In 1790 New York City had 33,131 inhabitants, and Philadelphia had 28,522. In 1800, the figures were respectively 60,515 and 41,220.

14. During the Civil War, a Missourian secessionist shadow government conducted by the governor was set up that settled from 1863 to 1865 in Marshall, Texas (Geise 1962–63, 193–207). The "capitol" and the governor's mansion were razed in 1950.

15. "Une Constitution basée sur la haine et l'injustice ne saurait durer" (*L'Abeille de la Nouvelle-Orléans*, March 7, 1845, 1).

16. Opelousas was briefly Louisiana's capital during the Civil War, from May 1862 to January 1863 (Bernard 1995, 475–80).

17. Such a sum was enormous for Manchester, which counted only 20,107 inhabitants in 1860, against 10,896 for Concord.

18. The story did not end with the victory of Concord. When Concord's capitol once again proved too small in 1909, and the legislature voted to enlarge it, Manchester offered $1 million dollars to attract it. In 1910, Manchester had grown to 70,063 inhabitants because of industrialization, while Concord had only 21,497 inhabitants. Concord nevertheless retained the coveted prize.

19. Summary in ABC Clio-Serials of Christopher S. Davies (1986, 443–554).

20. See Thomas Bender, who states that his "interpretation challenges the notion that American thinkers of that period were simply anti-urban" (1987, x).

21. Machor adds that this myth came from a specific group: white, predominantly middle-class, articulate Americans.

22. Schultz nevertheless forgets that the entire population could not benefit from all these services, as was proved, for example, by Jacob Riis in the famous study he published in 1890, *How the Other Half Lives: Studies among the Settlements of New York* (New York: Charles Scribner's Sons).

23. Agnew and Smith show that such urban forms, while celebrating lost ideals, also serve to maintain and exacerbate social divisions, and to undermine the city.

24. Fayetteville's supporters were wrong. Fayetteville is a medium-sized city, while Raleigh is a bustling center in the millionaire Raleigh-Durham-Cary CSA.

25. "Des Moines—Iowa State Capital," *New York Daily Graphic*, September 17, 1878; reprinted in *The Palimpsest* 51 (1970): 225–26. The text does not specify what Mitchell promised to Warren County leaders to win their acceptance.

26. The situation had been different during colonial times, because establishing an English colony and founding a town were synonymous.

27. A *sutler* was a merchant who accompanied an army in order to sell provisions to the soldiers.

28. In the final ballot, Montgomery received 68 votes, Tuscaloosa 39, Selma 11, Wetumpka 9, and Mobile 3. The first ballot yielded the following results: Tuscaloosa 38, Wetumpka 33, and Montgomery 27.

29. Political pamphlets were not a new feature in American politics. Many "libels" had circulated under British rule, and independent American politicians did not end the tradition (Pessen 1985, 113).

30. The headquarters of the Afro-American Colonization Company were in Guthrie.

31. "Money was the talisman that would open the door to elevated social status, the key to political influence, the portal to cultural magnificence" (Billington 1974, 653).

32. For a complete history of Wisconsin's capitals, see Cravens 1983, 100–68. Sixty-nine pages were a necessity, for the capital issue "was not only one of the most controversial of the day, but would reemerge repeatedly over the course of the ensuing century" (103).

33. The selection of Phoenix was not primarily due to bribes, which were commonplace at the time. See chapter 5.

34. This folded paper is carefully kept at the Vermont Historical Society, Montpelier.

35. A livelier (and longer) account appears in Hayes and Cox 1889, 109–10.

36. Lancaster was soon renamed Lincoln, and Mr. Cadman soon overcame his defeat by becoming a businessman in Lincoln.

37. Ferraro gives two examples. First, William Cronon, *Nature's Metropolis: Chicago and the Great West* (New York: W. W. Norton, 1991). He explains Chicago's development as "an expanding metropolitan economy creating ever more elaborate and intimate linkages between city and country" (1998, xv).

Second, Mary P. Ryan, *Civic Wars: Democracy and Public Life in the American City during the 19th Century* (Berkeley and Los Angeles: University of California Press, 1997). She argues that, through open and volatile expression in public spaces, people in New York, New Orleans, and San Francisco created viable civic identities despite cultures that were "full of cultural differences" and "fractured by social and economic changes" (3).

38. For a broader perspective, see table 46 in Meinig 1986 (250–51) and table 33 in Meinig 1993 (270–71). Meinig distinguishes first the colonial port and then the mercantile frontier from the speculative frontier.

39. "Des Moines—Iowa State Capital," *New York Daily Graphic*, September 17, 1878; reproduced in *The Palimpsest* 51 (1970): 223.

40. The other "roads" were the east-west road from Marietta to Cincinnati and the Gallipolis-Springfield road.

41. Centers could shift, of course, when boundaries changed, as they frequently did during territorial times.

42. Capitals located within a twenty-five mile radius of the state's geographic center are considered geographically central. Washington, DC, was centrally located when it was selected, the national population center being situated twenty-three miles east of Baltimore, according to the 1790 census.

43. Journal of Dr. W. H. Simmons, commissioner, "The Selection of Tallahassee as the Capital," *Florida House Journal* (1903), reprinted in the *Florida Historical Quarterly* 1 (April 1908): 34.

44. Journal of John Lee Williams, the *Florida Historical Quarterly* 1 (July 1908): 19. Nevertheless, Williams did mention the "good rice and cotton" that such land would produce and that the stream was "sufficiently large to turn an overshot mill" (23).

45. Such a vision was already present in Aristotle's *Politics*, where he describes the capital as a "common center."

46. Support for removal was not universal, however, especially in cities that were to lose the capital. The fact that opponents never united under a common banner undermined their efforts to eliminate removals.

47. In 1807, New Hampshire's most distant county sent 79% of its legislators to the capital, while nearby counties sent more than 90% of theirs. This naturally influenced the result of votes.

48. Another cause was independence, which released the former colonies "from the external power that had kept capitals on the seacoast as beachheads of imperial authority" and allowed the "more or less free play of local geopolitical forces" (Meinig 1986, 366–67).

49. Since the 1850s, the debate was based more on individual equality and proportional representation than on spatial equality, which had soon shown its democratic limitations. Reapportionment replaced removal as the way to attain equality (Zagarri 1987, 33–34).

CHAPTER FIVE

1. A political study of Orange County, NY, showed that "politics was not dominated from Albany down, but that politics was a process in which pressure could come from either direction" (Tick 1981).

2. Individuals often belonged to a privileged class that was organized in groups of shared interests, such as the "Richmond Junto," which ruled Virginia in the first half of the nineteenth century, or the "Irish Aristocracy" of Albany.

3. Wyoming was the first state to grant suffrage to women. It did so in 1869, after the selection of the permanent capital, Cheyenne, in that year.

4. Montana's 1870 census showed that men outnumbered women three to one in Helena, and two-thirds of them were between twenty-one and forty-two years old.

5. W. Rostow has proved that "a significant proportion of the Protestant immigrants belonged to the wealthy middle class" (see Eysberg 1989, 186).

6. In 1877, Booker Taliaferro Washington (1856–1915) was not yet the educa-

tor and major African-American spokesman he would soon become. At twenty-one, he was a teacher in Tinkersville, WV, and already politically active at local level.

7. "Many pioneers seem to have come to Nebraska for the express purpose of carving political careers for themselves in the new territory, others to use politics as a means to financial gain" (Olson and Naugle 1997, 131).

8. This excluded slaves—and also Native Americans.

9. There were exceptions: Texas, Louisiana, New Mexico, and Utah entered the union already constituted under other logics. The question is further analyzed in chapter 6 from an economic point of view.

10. American Indians were living on these lands, of course, but they were regarded as "savages" who had to give way to the idea of "civilization" held by white, Christian men.

11. According to another source, St. Stephens had eight hundred inhabitants in 1818, fewer than Mobile's one thousand (Moffat 1992, 1–3).

12. For an example of the former, see the case of California described in chapter 3.

13. Another constitutional provision allowed the governor to convene or remove the seat of government elsewhere if from any cause it became impossible or dangerous to meet as designated.

14. This can be seen as the translation at the state level of "Calhounism" (Brege and Crouzatier 1991, 35). John Caldwell Calhoun (1782–1850), was a US representative and senator from South Carolina, and a vice president of the United States (1825–32). His political philosophy rested on the notion of sectional balance. Balance was obtained by giving to each section, through its own majority, a veto on the acts of the federal government (called "nullification").

15. In 2010, Stillwater had 45,688 inhabitants, owing to the presence of the main campus of Oklahoma State University; Edmond had 81,405, owing to the state's third largest university, the University of Central Oklahoma; and Norman, which had the main campus of the University of Oklahoma, had 110,925. The question of universities is discussed in chapter 7.

16. Today, Port Townsend is one of the only three Victorian seaboard cities in the nation and has been designated a national landmark. Like Olympia, it was bypassed in the 1890s by the railroad, which preferred Seattle, the future metropolis of the state (www.culture.com/PT-Home/).

17. Yankton, located in the southeastern corner of the vast Dakota Territory, had witnessed the first influx of pioneers in 1868 and the arrival of the railroads in the 1870s.

18. The North Dakota legislators' vote for Bismarck was secured by an article in the constitution assigning locations for thirteen institutions, one for almost every town of the state.

19. Ray Billington argued that Western states provided "manhood suffrage and a governmental structure that reflected complete faith in the people," with

regard to capital selection, but such faith was not really "complete" (Billington 1974, 652).

20. The WWP "forgot" to mention the help of the black leader Booker T. Washington.

21. The capital matter was only settled in the 1920s. Charleston had to wait for the completion of the capitol that replaced the one that had burned in 1921 to settle "for all time the question of Charleston as the permanent seat of state government" (Goodall 1968a, 132).

22. There were only two contenders instead of the four of 1889, since the state's largest cities, Fargo (28,619) and Grand Forks (17,112), on the eastern border of the state, were not central, and both already had state universities. Bismarck was geographically central, but Jamestown, located in the prairie between Bismarck and Fargo, could boast demographic centrality. In 1930, Jamestown had 8,187 inhabitants, and Bismarck had 11,090. In 2010, Jamestown—nicknamed Buffalo City—had 15,427 inhabitants, down from 15,571 in 1990.

23. Exeter was the first town of future New Hampshire, founded in 1638 by Reverend John Wheelwright, who had fled Boston's dogmatic leaders (he was a victim of the Ann Hutchinson religious controversy).

24. On March 8, 1847, the senators voted fifty-one times without reaching a decision. The next day they approved the House bill.

25. Lansing is 87 miles from Detroit, 86 from Monroe, 91 from Mt. Clemens, 89 from the mouth of the Grand River, and 87 from the mouth of the Kalamazoo River. When Lansing was proposed, a map was produced to show the legislators—with broad black lines that could be seen from every point in the House—how centrally located it was.

26. Corvallis is today the main center of Oregon State University; its population is 54,462.

27. Several Indian tribes had lived there, the Anasazi and the Navajo being the most famous, but Arizona became a territory in the Western sense of the word only through Spanish colonization.

28. We can add political causes, since Tucson had been left aside because of the presence of Confederate sympathizers, according to Madeline F. Parré (1965, 115).

29. Federal appropriation for the university was only $25,000, while it topped $100,000 for the insane asylum (Fauk 1970, 183).

30. *Tucson Arizona Weekly Citizen*, December 15, 1888 (Ehrlich 1981, 240).

31. Pierre took more than forty years to repay its debt.

CHAPTER SIX

1. However, although the founding fathers meant by "happiness" an efficient government, the sense of the word has since become far more materialistic (Maier 1997).

2. Such an objective was even more pronounced in Western cities, where

"economic development concerns have dominated Western thinking and policy making. One of the reasons for this emphasis . . . is the lack of economic diversification and Western reliance on outside interests" (Weatherby and Witt 1994, 7).

3. These are average figures. Employment remained mainly based on agriculture, forestry, and mining in southern and mountain states at the beginning of the twentieth century.

4. The movement continued: between the 1840s and 1930s, 2,205 towns, villages, and hamlets were abandoned in Iowa—an average of 22 a year (Lingeman 1980, 328).

5. François André Michaux, *Lowcountry and Upcountry South Carolina as Seen by a Famous Botanist, 1802–1803*, quoted in Zagarri 1987, 32.

6. Prince Achille Murat, nephew of Emperor Napoleon Bonaparte, and his wife Catherine, great grandniece of George Washington, were among the first settlers of central Florida.

7. The *State Journal Power Press Print* published the report in 1870, some pages of which were reprinted in Hayes and Cox 1889, 90–94.

8. Ebenezer Cook, *The Sot-Weed Factor* (1708), cited in Reps 1972, 136. Cook compared Annapolis—a town already sixty years old—with Southwark, which was an active trading place in the seventeenth century and the second largest urban area in England. A borough of contemporary London, south of the Thames, it gave America two prestigious sons, John Harvard and William Penn. Also situated in central London, Tottenham Court was an ancient manor, with a small "village" around it. Today, only the name remains, in Tottenham Court Road.

9. Simonin, *The Rocky Mountain West*, 63–34, cited in Reps 1981, 89.

10. A generation later, Tallahassee experienced the same arrival of wealthy Virginians. Among them was a grandson of Thomas Jefferson, Francis W. Eppes, who served three terms as mayor of Tallahassee and recreated the plantation system, based on a white patrician minority, in Florida's Leon County (Ellis and Rogers 1986, 6–9); see also note 6 above, on the Murats.

11. In 1923, Lincoln prided itself on being "The Athens of the West" (Lincoln Chamber of Commerce 1923, 157). There is at least one "real" Athens, in Georgia.

12. Hamer's book includes interesting photographs of Helena.

13. Annexations must be taken into account in the evolutionary pattern of capital cities–but this is true for every American city.

14. Chapter 2 has shown, however, that "grandness" was certainly present in most state capitals.

15. "Central places systems always nestled within the exogenous mercantile system" (Conzen 1975, 361).

16. Among Cheyenne's early leaders were a judge at the probate court, a banker, a restaurant owner, a saloonkeeper, a liquor dealer, a hotelkeeper, and a livery stable operator. Their headquarters were not the territorial capitol, but the Variety Theater. Territorial officials helped the movement but did not create it.

17. Springfield's population grew from 19,743 inhabitants in 1880 to 71,864 in 1930.

18. This was naturally the case when a city was already the leading place in its state before being selected as the capital, such as Oklahoma City, thanks to an enterprising citizenry (Meredith and Shirk 1977).

19. James H. Madison, "Economy," in Bodenhamer and Barrows 1994, 61; this chapter (61–71) is the main source of the paragraph.

20. George W. Geib, "Politics," in Bodenhamer and Barrows 1994, 161.

21. James Russell Lowell (1819–1891), a poet, editor, and satirist, is one of the Schoolroom Poets with Whittier, Holmes, and Longfellow. The poem is quoted in Harrison 1979, 334. Besides the "hornet's nest," of Harrison's title, other epithets have been used—the "Gilded Age" or the "great barbecue"—to describe the nauseating political atmosphere at the end of the nineteenth century.

22. Dakota was divided into North and South Dakota in 1889.

23. Abraham Lincoln, 1864, cited in "Washington, Babylon?" *Economist*, July 29, 2000, 85.

24. For instance, in 1925, Providence and West Greenwich were each entitled to one senator in Rhode Island's General Assembly. The first was a city of 267,918 inhabitants; the second, a village of 407 (Conley and Campbell 2012).

25. Federal expenditures passed state and local ones during World War II (Gray and Herbert 1996, 320).

26. Some states changed from annual to biennial meetings.

27. There were naturally exceptions, as in Albany, where the old Dutch aristocrats were replaced around the mid-nineteenth century by Irish ones. "Like the patricians of the Netherlands, they directed the orderly growth of their community in the best interests of all its citizens, and so created a unique continuity between Albany's historic heritage and its present position" (Kenny 1987, 173; see also Rowley 1971).

28. Contemporary historical research has shown that "party machines" owed as much to myth as to reality: town "bosses" were not as powerful as was claimed: strong conflicting factions challenged their grip, and most of the time they were overturned. They nevertheless existed.

29. This did not prevent Indianapolis from becoming Indiana's leading city. In 1870 the capital had 48,244 inhabitants, compared to Fort Wayne's 17,718 and Terre Haute's 16,103.

30. This study used factorial analysis to show the socioeconomic status of medium-sized cities. The classification relied on two indicators of community life: an input indicator drawn from aspects of structural variation and an output indicator related to public policy. The socioeconomic status dimension relies on the extent to which a city deviates from the average on several variables and on governmental activity.

31. As was the case for state capitals, such a situation seems strange. Besides the excellent summaries of Donald Meinig (1986, 1993, and 1998), which are the main sources of this paragraph, an official book on the question exists, written for

the bicentennial, but it is more a judicial narration than a thorough analysis (Van Zandt 1976). Several books concerned with one particular state are also available (Schwartz 1979; Hemperly and Jackson 1993). A recent book by a playwright provides a wealth of details for each state (Stein 2008).

32. This did not prevent disputes that often arose from "the application of different boundary criteria in the charters to adjacent lands" (Meinig 1986, 220).

33. Louisiana is a unique case, the addition of Florida parishes being due to "geopolitical expediency" (Meinig 1993, 436) and not to any discussion of boundaries.

34. The twelve are Salt Lake City, Denver, Minneapolis-St. Paul, Des Moines, Oklahoma City, Little Rock, Atlanta, Boston, Richmond, Columbus, Indianapolis, and Nashville.

35. James Vance's book is the main basis for the following analysis.

36. The rapidity was linked to a specific American process, based on easier and less expensive engineering and propulsion. Total rail mileage rose from 8,879 miles in 1850 to 267,000 in 1916.

37. This obstacle was the Continental Divide in the Montana Front Range at Mullan Pass, where the tracks had to rise 1,618 feet in twenty miles.

38. Prescott, the first territorial capital of Arizona Territory in 1864, had to wait until 1893 to be reached by a branch line of the Santa Fe Railroad, four years after the transfer of the capital to Phoenix.

39. As consolation prizes, Milledgeville received the Middle Georgia Military and Agricultural College in 1880 and the Georgia Normal and Industrial School (Georgia College) in 1889.

40. See *The Handbook of Texas Online: Austin* (Texas State Historical Association website: http://www.tshaonline.org/handbook/online).

41. The former period, 1800–1860, was the "trading city era."

42. In the 1930s, Richmond had three hundred factories, including "one of the largest fertilizer plants and one of the largest cigar factories in the world" (WWP Virginia 1940).

43. The city boasted to have the world's largest tool factory, file factory, engine factory, screw factory, and silverware factory. It was first in the nation in the production of woolens and jewelry, and third in machine and machinery tools (Conley and Campbell 2012).

44. Annapolis was for six months in 1783 the capital of the United States; see also Papenfuse 1975.

45. The twelve were Salt Lake City, Denver, Minneapolis-St. Paul, Des Moines, Oklahoma City, Little Rock, Atlanta, Boston, Richmond, Columbus, Indianapolis, and Nashville (see figure 6.1). The rise in the number of capitals from 1840 to 1910 was due to the rise in the number of states and cities far more than to the rise in the capitals' influence.

46. The analysis of banking relations in 1910 shows that, except for Boston, no state capital was able to control exchange: the western capitals that controlled physical movements, Denver and Salt Lake City, were under the umbrella of Chi-

cago's banks, which controlled the whole West, except for a small area controlled by St. Louis's banks. In the East, New York reigned almost supreme, along with Boston and Philadelphia (Conzen 2001, 341–43). The other capitals in that classification were Providence, Minneapolis–St. Paul, Indianapolis, Albany, Columbus, Nashville, Richmond, Atlanta, Springfield, and Helena.

CHAPTER SEVEN

1. The nickname refers mostly to the 1940s–60s, when the city's conservative leaders tried to maintain its "small town" atmosphere. Things have greatly changed since then. With 6.5 square miles, downtown Indianapolis has about twenty thousand inhabitants (about 1% of the MSA) but a hundred and twenty thousand jobs and thirty-six hotels. Under the aegis of Downtown Inc., created in 1992 to develop that part of the city, $9 billion has been invested since the 1990s. (www. indydt.com)

2. Carroll and Meyer used a linear panel regression model to study the population of state capitals to 1970.

3. D. P. Doyle and T. W. Harlte, "A Funny Thing Happened on the Way to New Federalism," *Washington Post* (national weekly edition), December 2, 1985, 23 (quoted in Burns, Pelatson, and Cronin 1987, 64).

4. The other three are Montana, Nevada, and North Dakota, but they can hold extraordinary sessions, as in Montana. In Nevada, the biennial sessions were limited to 120 days by a 1998 popular vote. Arkansas has a fiscal session between regular sessions, and Kentucky (in 2000) and Oregon introduced a short session between the biennial regular ones.

5. "The careerist orientation of legislators is having an enormous impact on legislative life. It is largely responsible for the increasing political nature of legislatures and partly responsible for their greater fragmentation as well" (Rosenthal 1989, 71). The overall percentage of avowed full-time legislators was about 14% in 1993 (Gray and Jacob 1996, 175). The New Hampshire House of Representatives is the only one retaining for its four hundred members the $200 salary per biennium without per diem (but with mileage) established in 1889. This perhaps explains the state's mostly gray-haired assembly.

6. In the nation as a whole, large cities' votes counted less than half of the open country's (David and Eisenberger 1961–62, 10).

7. The *Baker v. Carr* 1962 judgment was specified by the Supreme Court on June 15, 1964, in the *Reynolds v. Sims* case introduced by Alabamans. On April 1, 1968, the Supreme Court extended the "one person, one vote" doctrine to elected local governments.

8. The federal government also fought electoral fraud on a racial basis. Under a 1965 law, federal registrars went to several southern states where less than half of the citizens of voting age had voted in the previous presidential election (Dabney 1971, 555).

9. The influence of interest groups naturally varies from state to state, from being dominant in seven of them to being complementary/subordinate in five of

them, according to a 1994 study, the results of which appear in Gray and Jacob 1996, 122–158.

10. As usual, there are exceptions. Vincent Cianci was the mayor of Providence, a capital, for more than twenty years and was the main force behind its resurrection. During his first term as mayor, however, twenty-two persons associated with his administration were convicted of corruption, and in April 2001, Cianci was indicted on federal charges of racketeering, conspiracy, extortion, witness tampering, and mail fraud (Stanton 2003).

11. Ryan Mulcahy, assistant to the mayor of Madison (March 2, 2003, personal communication). But Todd Dieffenderfer, from Columbus, wrote that the idea of a "recession proof economy has been challenged in this most recent economic downturn" (March 2, 2003, personal communication).

12. At the capital county level, the percentages were 18.3% for Albany, 20.2% for Austin, 14% for Lansing, and 14.2% for Sacramento, c. 1987 (Bromley 1990, 12).

13. Dover's very low figure (56.9) is perhaps due to the weight of Wilmington (headquarters of the DuPont company) in Delaware, which produces 77.1% of the state's GP with 64.55% of the state's population.

14. However, more than thirty-five states "have provisions where the state may make a payment in lieu of taxes, but Kansans have never enacted such a law" (Donna Freel, Topeka Town Hall [March 3, 2003], personal communication).

15. "An estimated 70,000 people travel to Madison from outside the city each day to work, shop or recreate. That represents real wear and tear on city infrastructure, roads, and the like, for which we are not compensated" (Ryan Mulcahy, assistant to the mayor of Madison [March 2, 2003], personal communication). He nevertheless added that it was "one of those hidden costs of being the state capital that we accept graciously."

16. City regimes do not seem to influence the process, according to Lewis Randolph (1990). Local regimes all use nonmarket development policies that require public involvement. The capital status of Richmond and Columbus is not even mentioned in the abstract of the PhD dissertation.

17. Carroll and Meyer based their study on empirical analyses of the capital Standard Metropolitan Statistical Areas (thirty-five of the fifty state capitals). Small capitals are thus excluded from the restructuring of the urban hierarchy.

18. Stanback and Noyelle analyze the evolution of seven cities between 1959 and 1976. The fact that five of them are capitals emphasizes their overall process of growth.

19. Ross used a technique based on vertical integration through intraorganizational linkages between cities that analyze the flows of control from one urban place to another. However, only the 158 SMSAs were studied.

20. Bromley's goal was to develop effective public/private partnerships. "The project focus was on a broad participation and the mutual benefits to be gained from coordinated problem-solving and action" (Bromley 1990, 3). The five capitals were Albany, Tallahassee, Austin, Sacramento, and Lansing. All five were secondary MSAs in their states, with interior locations and major universities.

Columbus and Raleigh were said to belong to the same pattern. Therefore, we cannot extend the study's conclusions to all capitals.

21. Table 7.7 shows, however, that traffic at all three airports is not impressive.

22. However, this was not always the case; during his two terms as mayor of Oklahoma City (1939–47), Robert Heffner introduced city planning and fostered interest in economic development (Trafzer 1984).

23. Mayor Dennis Eisnach, *Pierre Chamber News*, March 2003, 5.

24. And Pierre is becoming a rural "creative city." It succeeded in attracting in 2008 Eagle Creek Software Services, a Minnesota company offering more than a hundred well-paid jobs.

25. Railroads still bring prosperity but mostly to the handful large railroad companies that control the remaining lines (173,000 miles of tracks vs. 430,000 in 1930) and no longer to the cities they serve. Five companies control about 80% of total market shares, and none of them are based in a state capital. The Union Pacific is headquartered in Omaha, Nebraska; the Burlington Northern Santa Fe, in Fort Worth, Texas; the Norfolk Southern in Norfolk, Virginia; the Canadian National in Montreal, Canada; and the CSX left Richmond, Virginia, for Jacksonville, Florida, in 2003.

26. Techno-scientific rankings were based upon information technology, airspace, and biopharmaceutical industries.

27. See GACC (Greater Austin Chamber of Commerce), *1996–97 Economic Review and Forecast* (Austin, TX: GACC, 1996); and S. E. Engelking, "Austin's Opportunity Economy: A Model for Collaborative Technology Development," *Annals of the New York Academy of Sciences* 798 (December 1996): 29–47.

28. High-tech research programs multiplied at the University of Texas, and the state created in the 1980s a state research and technology fund to supply matching money for federal grants.

29. This was the subtitle of Bromley 1990.

30. The numerous rankings that appear before each academic year naturally have shortcomings: their methodology is almost never indicated and the categories they offer are often too broad.

31. For the role of the media, see Adams 2002.

32. As chapter 2 showed, Springfield is a popular name (twenty-three occurrences), but not as popular as Salem or Madison. It is fortunate that the cartoon does not refer to Illinois's capital, because the motto of Springfield, NT's mayor is "Corruptus in extremis"!

33. An earlier version of this paragraph appeared in "Les illustres inconnues ou identité, patrimoine et tourisme dans les capitales d'Etat aux Etats-Unis," *Géocarrefour* 76 (2001):115–21.

34. See, for instance, http://www.office-tourisme-usa.com/.

35. Santa Fe is an exception. Its Plan of 1912, by introducing community-wide historic restoration, "helped broaden the scope of [the] American historic preservation movement," then still restricted to individual structures (Wilson 1997, 232).

36. There are two exceptions. The Historical Annapolis organization was created in 1952, and the joint City Planning Commission/Providence Preservation Society report in 1959 paved the way for the creation of historic districts. Providence's College Hill District was created by Rhode Island's legislature in 1961.

37. The development of historic districts is more connected to nineteenth-century American traditions of urban boosting than to any capital project. In Harrisburg, historic preservation has been used since 1982 as an urban renewal tool (Hamer 1998, 105).

38. President Lincoln's fame draws about one million visitors a year to Springfield. For Lincoln's importance to Springfield, see, for instance, Krohe 1976.

39. Field battles are easier to stage than battles of eloquence.

40. The leaflet's front page nevertheless shows an old photograph of the state capitol's cupola.

41. Hawaii had to wait for the democratization of air travel to realize its tourism potential. In 1953 there were only 66,296 tourists (Johnson 1991, 367). In 2011 the archipelago welcomed 7.3 million tourists.

42. It owes much as well to the open-air opera festival it organizes each summer and to its picturesque setting.

43. "Santa Fe has methodically transformed itself into a harmonious Pueblo-Spanish fantasy through speculative restoration, the removal of overt signs of Americanization, and historic design review for new buildings" (Wilson 1997, 232).

44. Potter's aim is to look behind the veil of gentility that pervades the traditional histories of Annapolis. He stresses, for instance, the importance of the Chesapeake Bay watermen community, which contrasts with the sophisticated image of Maryland's capital. See also Shakel 1993.

45. One can add the popular district of Alkali Flat (three historical districts in 1984), where railroad employees lived.

46. This derogatory nickname refers to the way the state legislature is supposed to work.

47. The "principal city capitals" are Boston, Atlanta, St. Paul, Denver, Phoenix, and Indianapolis. Among "classic government towns" are Harrisburg, Springfield, Trenton, and Baton Rouge. However, chapters 4 and 5 have shown that some of Bromley's arguments were too hasty, as when he stated that small-town capitals stemmed from a "neutrality between different regions and interest-groups within the state, and the deliberate separation of government from commerce and industry" (Bromley 1990, 7).

48. This regional geography of North America mentions only thirty state capitals (Bismarck is situated in South Dakota in the index), and capital status is only briefly referred to for six of them. This would point to their still secondary rank in the national urban hierarchy.

49. The modernization movement was still hardly noticeable in the 1960s, as indicated by a contemporary article (Lemmon 1966). The "New South" in question does not apply to the modernization of the southern economy after World

War II but to the 1876–95 period of reconciliation with the North and the improvement of cities in that period. The study concludes that Raleigh proved to be "a reasonably good example of the New South even though it was not in the mainstream of development" (Lemmon 1966, 285). Raleigh entered the "mainstream of development" in the 1960s, and its population exceeded one hundred thousand. By 2000, it was ranked sixty-third among the nation's most populated cities (with an increase of 83.8% between 1980 and 2000) and forty-first among MSAs.

50. In 1950, the population of metropolitan Maricopa County, of which Phoenix is the seat, represented 44% of Arizona's population. Another factor was the small size of the incorporated municipality of Phoenix, which covered less than ten square miles. In 2010, Phoenix's area was 516.7 square miles.

51. In the 1930s, Lincoln, Nebraska, home to forty insurance companies, was nicknamed "the Hartford of the West," a title today held by Des Moines (WWP Nebraska 1937, 16).

52. A research project ranks Indianapolis among the "minor regional-global centers," ranked 21st in the nation, just after Cleveland, and 114th in the world (Lang and Taylor, 2003).

53. In 2001, manufacturing accounted for 14% of Indianapolis's workforce, and government for 12.7% (Chamber of Commerce website).

54. Founded in 1801, the University of South Carolina is the state's major university and was ranked forty-third among public research universities in 2011.

55. J. R. Simplot Company, founded in the 1930s and headquartered in Boise, is an agribusiness firm with sales of $4.5 billion in 2011. It is one of the world's largest frozen potato processors; it also raises beef cattle and manufactures fertilizers. J. R. Simplot was an early investor in Micron Technology, a leader in microprocessors that was founded in Boise in 1978. This further shows the low importance of capital status for Boise City's economic growth.

56. FIRE employment amounted to 9.5% of the MSA's workforce in 2001. The state employed 40,500 people in 2003. In 2011, six Fortune 500 companies where headquartered there, the most important being Altria (the former Philip Morris; 4,387 jobs), a powerful remainder of Virginia's long-standing staple, tobacco.

57. The Naval Academy also is an important feature in the city's economy.

58. We must also consider the effects of shale gas production, which explains the 2.6% unemployment rate (July 2012) and the recent surge in population and construction.

59. *Travel and Leisure Magazine*, November 2002 and CNN, 2003. Rankings are naturally to be taken *cum grano salis*. Therefore, more than the accuracy of such rankings, this shows the almost frantic national quest for quality of life, which generates scores of often scientifically suspicious surveys.

60. The book describes contemporary life in Northampton, Massachusetts, and shows that, beneath the homely façade of a carefully restored New England small town, all the evils of modern America are lurking.

61. Montarry's essay offers only a first glimpse, since the persons interviewed

were mostly from the white, educated part of the city's population. Any generalization would therefore be hasty.

62. Baton Rouge nevertheless has a symphony orchestra, an opera, a ballet, a Gilbert and Sullivan Society, jazz and blues bars, and a yearly blues festival.

CHAPTER EIGHT

1. Vincennes became the capital of the western part of the Northwest Territory (now Indiana) when it was subdivided in 1800 during the second stage of territorial administration (until 1813). Detroit became in 1805 the first capital of Michigan (until 1847). Kaskaskia became in 1809 the first capital of Illinois (until 1820).

2. In 1849, cholera killed 162 people in the town and 116 in the penitentiary, and 225 more died during the 1850 epidemic.

3. Franklinton never recovered and was annexed by Columbus in 1864.

4. *Ohio State Journal*, August 15, 1848, quoted in Cole 2001, 52.

5. In 1855, when the state had been divided into two districts, the courts went to Cincinnati.

6. In fact, Columbus only had 2,435 inhabitants in 1830 and 6,048 in 1840, almost the same as Dayton and Cleveland, but far less than Cincinnati, which already had 46,338 inhabitants and was ranked sixth among American cities.

7. The Willoughby Medical College, which moved to Columbus in 1846 and became the Starling Medical College in 1848, also helped establish the future university.

8. About two-thirds of the population of the United States and Canada live within six hundred miles of Columbus. The landing of America's first air freight (from Dayton) in 1910, added air transportation to an already rich crossroads.

9. Despite a 22% loss in manufacturing employment between 1958 and 1998, Ohio remained the largest employer in the Great Lakes area. With 5.77% of all US manufacturing jobs in 1990, Ohio ranked third in the nation, before Pennsylvania. In 1980, it ranked fourth, after Pennsylvania but with 6.22% of national manufacturing employment (Thomas 2000, 99).

10. Gains in the service sector were 288% for Ohio, 329% for Cincinnati, and 308% for Akron and Dayton.

11. A location quotient greater than 1 means that the activity is more locally concentrated than nationally. If it is more than 2, the area has a strong specialization in that activity. For instance, Cleveland is strongly specialized in primary metals ($LQ = 8.6$) and fabricated metal products ($LQ = 2.3$) (Thomas 2000, 128–29).

12. A nonprofit organization, it was founded in 1929 in Columbus. It had 2,900 employees in 1982, and 2,629 (Columbus only) in 2011 (more than 20,000 worldwide).

13. Wexner created Victoria's Secret, and built the Easton Town Center shopping village. He is a trustee of OSU and was behind the Downtown Plan proposed

by the Columbus Downtown Development Corporation (CDDC). His wife Abigail is also influential with regard to the children's hospital.

14. Central renewal is not complete. Columbus still empties at night, and segregation remains high—more than 80% of all African Americans reside in the central city, amounting to 28% of its population.

15. Hunker nevertheless affirms that an "administrative" town may become a high-tech nexus based on information (2000, 60).

16. Created respectively in 1833 and 1836, these eastern Iowa cities had benefited from the first influx of pioneers.

17. In 1878, there were, for instance, five flour mills, two flax seed oil mills, four foundries and machine shops, four pork-packing houses, two barbed wire factories, four plough factories, five breweries, and eight brick yards (Petersen 1970, 244).

18. In 1938, Des Moines ranked sixth in the US publishing (WWP Iowa 1959). In 1970, Des Moines published *Look*, a magazine that ranked fourth in the nation, with 7,731,177 subscribers; *Better Homes & Gardens*, which ranked eighth, with a circulation of 7,055,967 in 1967; and *Successful Farming*, with a circulation of 1,330,991 (Petersen 1970, 222). In 2002, *Look* disappeared, but the others were still part of the successful and growing Meredith Corporation (see its website), founded in 1902 and still headquartered in Des Moines. *Better Homes & Gardens* is ranked fifth among American magazines, with a circulation of 7.6 million in 2010; *Successful Farming* has declined to 420,000. New titles came, such as the *Ladies' Home Journal* (3.8 million) and *American Baby* (2 million). August Home Publishing, founded in 1979, publishes five magazines that focus on woodworking and the home, with a global circulation of more than a million (according to its website).

19. 2012 fall enrollment at Drake was 3,200 full-time students in this independent and private university. This is quite small compared to the University of Iowa at Iowa City, the enrollment of which was 30,893 for the fall 2011 semester.

20. Hoyt's brother, Lampson Sherman, founded the future *Iowa State Journal* in 1850.

21. See "What's Next Downtown?," a planning project by the city and Polk County that was endorsed by the Downtown Community Alliance and approved in 2008.

22. Louisville was the 42nd MSA in the nation in 2010, with 1,283,566 inhabitants in the CSA (although its central city has been losing population since 1970, the CSA is growing), and Lexington-Fayette was the 106th MSA, with 472,099 people (687,173 in the CSA).

23. According to the city's website, Frankfort leaders offered the commission town lots, building materials, and the rents from a tobacco warehouse. Eight citizens offered to pay $3,000 if Frankfort was chosen as Kentucky's capital.

24. The only half-exception came during the Civil War, when Bowling Green became the Confederate capital of the state. Kentucky was officially against

secession, but thirty thousand Kentuckians fought under the Confederate banner and designated their own "capital." At the beginning of the twentieth century, Frankfort might have lost its status if the federal money given to build the new capitol had not been clearly earmarked for Frankfort (David Buchta, personal communication).

25. Bearing the name of the French Royal family, it is produced in Bourbon County (created in 1786) by distilling barley and rye and at least 51% of corn.

26. Renamed the Genesco Shoe Company, it had to close in 1977.

27. A convention bureau was established in 1926. A seventy-room hotel was completed in that year. With that and the one-hundred-room Capital Hotel (1923), Frankfort could accommodate visitors with comfort.

28. Other major manufacturing plants in Frankfort were the Buffalo Trace Distillery, with 318 employees and the Jim Beam Brands Co., with 265 (Kentucky Capital Development Corporation website: www.kycapitaldevelopment.com).

CHAPTER NINE

1. Philip Pittman, *The Illinois Villages, 1765–68* (quoted in Angle 1968, 42–3).

2. See Lee A. Caldwell, review of Holmes 1996, *Georgia Historical Quarterly* 81 (1997): 173–75.

3. *Seminal* refers to the process and scope of the renaissance and not to the date of the process, as some historic parks had already been set up by states before 1922 (cf. the case of Texas, studied later in the chapter).

4. Milledgeville was created to be the capital of Georgia in 1803, and was the capital from 1807 to 1868. With 17,715 inhabitants in 2010, the town contends that its slow economic growth after losing the capital "contributed to a wealth of well preserved Federal-style architecture, [and] enhanced the noteworthy Greek Revival, Victorian, and Classic Revival Houses, that can be found throughout the city today. Milledgeville is proud today to be a Main Street City and a center of attraction on Georgia's Antebellum Trail" (www.milledgevillega.com). It has a historic district and celebrated its bicentennial in 2003.

5. For some examples, see www.ci.benicia.ca.us, www.bellevue.net, www.belmontwi.com, www.visitnewbern.com, www.visitfayettevillenc.com, and www.birthplaceoftexas.com.

6. See www.tpwd.state.tx.us/state-parks/washington-on-the-brazos and www.birthplaceoftexas.com

7. Montana also offers ghost capitals: Bannack City (1864–65), a state park since 1954, and Virginia City (1865–1875), purchased by the Montana Historical Society in 1997. After having been the hub of the Northwest, it has become a tourist village where 130 people live.

8. St. Stephens vanished, although it had desperately tried to become a commercial node. For example, on Monday, December 7, 1818, a petition by St. Stephens merchants, traders, and inhabitants was presented by Mr. Williams, of Mississippi, to the US Senate, "praying to be made a port of entry." The petition was read and then "referred to the committee on commerce and manufacture,

to consider and report thereon" (*Journal of the Senate of the United States of America, 1789–1873* [Washington], 57; reproduced on the Library of Congress American Memory website: http://memory.oc.gov/ammem/amlaw/lwsj.html). Epidemics of yellow fever and the development of shallow-bottom boats allowing travel beyond the shoals resulted in the disappearance of St. Stephens within twenty years (Fairley 1998). The Old St. Stephens Historical Commission was established by the Alabama legislature in 1988 to study the site of the former capital, but work began only in 1996.

9. Called the "Negro Burial Ground," it was created in 1819, and the last burial was held in 1957. It is a testimony to Cahawba's rich African American past.

10. Many European cities are indeed far older than American ones, but most of their houses and monuments were built after the sixteenth century.

11. The name probably honors Count Ernest Casimir of Nassau, a member of the Dutch royal family (the Orange-Nassau).

12. A National Historic Landmark is "a district, site, building, structure, or object in public or private ownership, which the Secretary of the Interior judges to have national significance in American history, archaeology, architecture, engineering, or culture, and worthy of preservation as an illustration or commemoration of the history and prehistory of the United States. . . . Upon designation, NHL's are automatically included in the National Register of Historic Places; however, they are subject to a higher level of protection from potential damage or destruction by a Federal undertaking than are other (non-NHL) National Register properties" (www.atlantisforce.org).

13. Of these buildings, 1,408 were designated as contributing historically and 761 as noncontributing (Guthrie Chamber of Commerce website).

14. Sixty-nine percent of all buildings and 90% of the commercial ones were built before 1910.

15. The seven museums are the State Capital Publishing Museum (1975); the Oklahoma Territorial Museum (17,658 visitors in 1980; 33,695 in 1986); the Scottish Rite Masonic Temple, the largest in the world, built in 1929, and open to visitors since 1983; the Oklahoma Frontier Drugstore Museum; the National 4-String Banjo Hall of Fame Museum; the Oklahoma Sports Museum; and the Lighter Museum.

16. The B&Bs, offering fifty-one rooms located in beautiful historic homes, earned Guthrie the title of "B&B capital of Oklahoma."

17. In 1987, 71% of the tourists came from Oklahoma, only 5% from Europe, and only 16% stayed one day or more (Leider 1990–91, 415.)

18. The *Guide* grants Vincennes one star (out of a maximum of three), nine lines, and the epithet of "historic town."

19. See the websites of George Rogers Clark National Historic Park and Vincennes State Historic Sites.

20. Loveland's store served as a meeting hall for the legislature.

21. Prescott has interesting museums, beautiful nineteenth-century homes—and a remarkable microbrewery!

22. George Tyler (1841–1912) was one of Broadway's most famous producers.

23. Murfreesboro, former capital of Tennessee (1819–25), with 108,755 inhabitants in 2010, is boastful on its website: "Named the Most Livable Town in Tennessee, Murfreesboro is the fastest growing city in the state—and it's not hard to see why. Murfreesboro provides the quality of life that attracts growth." The city is participating in the National Main Street Program and has a historic district.

CHAPTER TEN

1. The contemporary multiplication of lobbying groups (e.g., the National Association of Cities) shows that capital cities are perhaps beginning to realize that they might have collective power.

2. Inside the state, geographic boundaries are more malleable. When the basis of representation changed from territory to population, around the 1850s, the concept of a district was introduced, the boundaries of which can be redrawn as often as is deemed necessary (Zagarri 1987, 148).

References

Abbott, Carl. 1978. "Indianapolis in the 1850s: Popular Economic Thought and Urban Growth." *Indiana Magazine of History* 74:293–315.

Abbott, Carl, Stephen J. Leonard, and David McComb. 1994. *Colorado: A History of the Centennial State.* Niwot: University Press of Colorado.

L'Abeille de la Nouvelle Orléans, Journal Officiel de l'Etat. February and March 1845. Microfilm. Baton Rouge: Louisiana State Archives.

Abler, Ronald, John S. Adams, and John R. Borchert. 1976. *The Twin Cities of St. Paul and Minneapolis.* Cambridge, MA: Ballinger Publishing.

Adams, Franklin G. 1903–4. "The Capitals of Kansas." *Transactions of the Kansas State Historical Society* 8:331–51.

Adams, Paul C. 2002. "Mediascapes." In Agnew and Smith 2002, 292–316.

Adrian, Charles R. 1999. "Forms of Local Government in American History." In *Forms of Local Government: A Handbook on City, County, and Regional Options,* edited by Roger L. Kemp. Jefferson, NC: McFarland.

Agnew, John A., and Jonathan M. Smith, eds. 2002. *American Space/American Place: Geographies of the Contemporary United States.* New York: Routledge.

The Alabama Journal. January 11 and December 13, 1843. Microfilm MN-895. Montgomery: Alabama State Archives.

Alexander, Thomas G., and James B. Allen. 1984. *Mormons and Gentiles: A History of Salt Lake City.* Vol. 5. Boulder, CO: Pruett.

Allen, William B. 1872. *A History of Kentucky.* Louisville: Bradley & Gilbert.

Alvergne, Chrystel, and Daniel Latouche. 2003. "Le système urbain nord-américain à l'heure de la 'nouvelle économie.'" *Mappemonde* 2:21–23.

Anderson, Leon W. 1981. *To This Day: The 300 Years of the New Hampshire Legislature.* Canaan, NH: Phoenix.

Anderson, Lowell E. 1974. *A Guide to the Illinois Old State Capitol.* Springfield, IL: Rudin.

Andriot, John L. 1980. *Population Abstract of the United States.* McLean, VA: Andriot Associates.

Angle, Paul M. 1950. *"Here I Have Lived": A History of Lincoln's Springfield, 1821–1865.* New Brunswick, NJ: Rutgers University Press.

———, ed. 1968. *Prairie State: Impressions of Illinois, 1673–1967, by Travelers and Other Observers*. Chicago: University of Chicago Press.

Annual Reports of the City Plan Commission, Providence, RI, for the Years 1915–1920. 1921. Providence, RI: Oxford Press, City Printers.

Annual Reports of the City Plan Commission, Providence, RI, for the Years 1929–1931. 1932. Providence, RI: Oxford Press, City Printers.

Arrington, Leonard J. 1994. *History of Idaho*. 2 vols. Moscow: University of Idaho Press.

Augé, Marc. 1995. *Non-Places: Introduction to an Anthropology of Supermodernity*. Translated by John Howe. New York: Verso.

Bailyn, Bernard. 1967. "The Origins of American Politics." In *Perspectives in American History*. Vol. 1. Cambridge: Warren Center for Studies in American History, Harvard University.

Baker, Nancy T. 1986. "Annapolis, Maryland, 1695–1730." *Maryland Historical Magazine* 81:191–209.

Ball, William W. 1932. *The State That Forgot: South Carolina's Surrender to Democracy*. Indianapolis: Bobbs-Merrill.

Beach, Allen C. 1879. *The Centennial Celebrations of the State of New York*. Albany, NY: Weed, Parsons.

Beardsley, Arthur S. 1941. "Early Efforts to Locate the Capital of Washington Territory." *Pacific Northwest Quarterly* 32:239–87.

Beck, Warren A. 1962. *New Mexico: A History of Four Centuries*. Norman: University of Oklahoma Press.

Bell, Charles G. 1986. "Legislature, Interest Groups and Lobbyists: The Link Beyond the District." *Journal of State Government* 59:14–15

Bender, Thomas. 1987. *Toward an Urban Vision: Ideas and Institutions in Nineteenth-Century America*. Baltimore, MD: Johns Hopkins University Press.

Bergeron, Paul H., Stephen V. Ash, and Jeanette Keith. 1999. *Tennesseans and Their History*. Knoxville: University of Tennessee Press.

Bernard, Shane K. 1995. "From Baton Rouge to Opelousas: An Eyewitness Account of the Wartime Removal of the State Capital." *Louisiana History* 36:475–80.

Bettinger, Julie S., and Heidi T. King. 1995. *Tallahassee: Tradition, Technology and Teamwork*. Montgomery, AL: Community Communications and Tallahassee Area Chamber of Commerce.

Billington, Ray Allen. 1974. *Westward Expansion: A History of the American Frontier*. New York: Macmillan.

Bird, Annie L. 1945. "A Footnote on the Capital Dispute in Idaho." *Pacific Northwest Quarterly* 36:341–46.

Blue, Mathew Powers. 1963. "The State Capitol in Montgomery." *Alabama Historical Quarterly* 25:245. (Orig. written between 1866 and 1878.)

Bodenhamer, David J., and Robert G. Barrows, eds. 1994. *The Encyclopedia of Indianapolis*. Bloomington: Indiana University Press.

Bolton, S. Charles. 1998. *Arkansas, 1800–1860*. Fayetteville: University of Arkansas Press.

Boorstin, Daniel J. 1965. *The Americans: The National Experience*. New York: Random House.

Bouton, Nathaniel. 1856. *The History of Concord, from its First Grant in 1725, to the Organization of the City Government in 1853, with a History of the Ancient Penacooks, the Whole Interspersed with Numerous Interesting Incidents and Anecdotes, down to the Present Period, 1855; Embellished with Maps; with Portraits of Distinguished Citizens, and Views of Ancient and Modern Residences*. Concord, MA: Benning W. Sanborn.

Bowling, Kenneth R. 1988. "Neither in a Wigwam nor the Wilderness: Competitors for the Federal Capital, 1787–1790." *Prologue* 20:163–79.

Boyer, P. 1978. *Urban Masses and Moral Order in America, 1820–1920*. Cambridge, MA: Harvard University Press.

Brege, A. M., and J. M. Crouzatier. 1991. *Les fondements de la démocratie américaine*. Toulouse: Presses de l'Institut d'Etudes Politiques de Toulouse.

Bridenbaugh, Carl. 1966. *Cities in the Wilderness: The First Century of Urban Life in America, 1625–1742*. Oxford: Oxford University Press.

Brienes, Marvin. 1979. "Sacramento Defies the Rivers, 1850–1878." *California History* 58:2–19.

Brigham, Johnson. 1911. *Des Moines: The Pioneer of Municipal Progress and Reform of the Middle West, Together with the History of Polk County, Iowa*. Vol. 1. Chicago: S. J. Clark.

Bromley, Ray. 1990. *Doing Business in a Capital City: Report of the Capital Cities Project*. Albany, NY: The University at Albany Foundation and Norstar Bank of Upstate New York.

Brown, Ralph H. 1948. *Historical Geography of the United States*. New York: Harcourt, Brace and World.

Browning, Clyde E. 1970. "The State Capitals: Meaningful Geographic Analysis vs. Memorization." *Journal of Geography: A Magazine for Schools* 69:40–44.

Brugger, Robert J. 1988. *Maryland: A Middle Temperament, 1634–1980*. Baltimore, MD: Johns Hopkins University Press.

Bureau of Business Research, The Ohio State University. 1966. *The Columbus Area Economy: Structure and Growth, 1950 to 1985*. Vol. 1. *Employment and Value Added by Manufacture*. Monograph No. 126. Columbus: The Ohio State University.

Burns, J. McG., J. W. Pelatson, and T. E. Cronin, eds. 1987. *Government by the People: Bicentennial Edition, 1987–1989*. Englewood Cliffs, NJ: Prentice Hall.

Burtschi, Joseph C. 1954. *Documentary History of Vandalia, Illinois: The State Capital of Illinois from 1819 to 1839*. Vandalia, IL: Chamber of Commerce of Vandalia.

Burtschi, Mary. 1963. *Vandalia: Wilderness Capital of Lincoln's Land*. Decatur, IL: Hudson-Patterson.

The Capital of Vermont: Journal of the Proceedings and Debates of the General Assembly of Vermont at Their Special Session, Feb. 1857. 1857. Montpelier, VT: E. P. Walton, Printer-Publisher.

Carmony, Donald F. 1998. *Indiana, 1816–1850: The Pioneer Era*. Indianapolis: Indianapolis Historical Bureau and Indianapolis Historical Society.

Carrier, Lois A. 1993. *Illinois: Crossroads of a Continent*. Urbana: University of Illinois Press.

Carroll, Glenn G., and John W. Meyer. 1982. "Capital Cities in the American Urban System: The Impact of State Expansion." *American Journal of Sociology* 88:565–78.

Caves, Roger W., ed. 1995. *Exploring Urban America*. Thousand Oaks, CA: Sage.

Cayton, Andrew R. L. 1996. *Frontier Indiana*. Bloomington: Indiana University Press.

Certeau, Michel de. 1980. *L'invention du quotidien. I Les arts de faire*. Paris: UGE.

———. 1988. *L'écriture de l'histoire*. Paris: Gallimard.

Chaffee, Eugene B. 1938. "The Political Clash between North and South Idaho over the Capital." *Pacific Northwest Quarterly* 29:255–67.

———. 1963. "Boise: The Founding of a City." *Idaho Yesterdays* 7:2–7.

Channing, Steven A. 1977. *Kentucky: A Bicentennial History*. New York: W. W. Norton; and Nashville, TN: American Association for State and Local History.

Chapman, Berlin B. 1960. *Oklahoma City, from Public Land to Private Property*. Oklahoma City: Oklahoma Historical Society.

Chartkoff, Kerry. 1990. "Michigan's 'Atomic Age' Capitol." *Michigan History* 74:28–31.

Chudacoff, Howard P. 1981. *The Evolution of American Urban Society*. Upper Saddle River, NJ: Prentice Hall.

City of Albany. 1924. *Albany's Tercentenary, 1624–1924, Historical Narrative Souvenir*. Albany, NY: City of Albany.

Clarke, Susan E. 1995. "Institutional Logics and Local Economic Development: A Comparative Analysis of Eight American Cities." *International Journal of Urban and Regional Research* 19:513–33.

Cleland, Robert G. 1959. *From Wilderness to Empire: A History of California*. New York: Alfred A. Knopf.

Cochran, Thomas C. 1978. *Pennsylvania: A Bicentennial History*. New York: W. W. Norton.

Cole, Charles C., Jr. 2001. *A Fragile Capital: Identity and the Early Years of Columbus, Ohio*. Columbus: Ohio State University Press.

Coleman, Kenneth, ed. 1991. *A History of Georgia*. Athens: University of Georgia Press.

Collective authors. 1988. *Etats-Unis. Côte Est. Guides Bleus*. Paris: Hachette.

———. 1989. *Les Etats-Unis. Centre et Ouest. Guides Bleus*. Paris: Hachette.

Columbus Board of Trade. 1904. *Columbus, Ohio*. Columbus: Columbus Board of Trade.

Condon, George E. 1977. *Yesterday's Columbus: A Pictorial History of Ohio's Capital*. Miami, FL: E. A. Seemann.

Conley, Patrick T., and Paul R. Campbell. 2012. *Providence. "375 Years at a Glance."* Available online at http://www.providenceri.com/archives/history/city-history.

Conley, Patrick T., Robert B. Jones, and William Ray Woodward. 1988. *The State Houses of Rhode Island: An Architectural and Historical legacy.* Providence: Rhode Island Historical Society Preservation Commission.

Conzen, Michael P. 1975. "A Transport Interpretation of the Growth of Urban Regions: An American Example." *Journal of Historical Geography* 1:361–82.

———. 1977. "The Maturing Urban System in the US, 1840–1910." *Annals of the Association of American Geographers* 67:88–108.

———. 2001. "The Impact of Industrialism and Modernity on American Cities, 1860–1930." In McIlwraith and Muller 2001, 333–55.

———. 2002. "L'héritage morphologique de l'urbanisme français aux Etats-Unis." *Géocarrefour* 77:161–73.

Corboz André. 2003. *Deux capitales françaises. Saint-Pétersbourg et Washington.* Gollion: Infolio.

The Council of State Governments. 1998. *Book of the States, 1998–99.* Vol. 32. Lexington, KY: The Council of State Governments.

———. 2002. *Book of the States, 2002.* Vol. 36. Lexington, KY: The Council of State Governments.

Cravens, Stanley H. 1983. "Capitals and Capitols in Early Wisconsin: From Belmont to Madison, the Colorful Story of the Selection of Sites and the Building of Capitols in the Territory and State." In *1983–1984 State of Wisconsin Blue Book,* edited by the State of Wisconsin, 100–68. Madison: The State of Wisconsin.

Crowder, James L., Jr. 1992. "'More Valuable than Oil': The Establishment and Development of Tinker Air Force Base, 1940–1949." *Chronicles of Oklahoma* 70:228–57.

Cumber, John T. 1989. *A Social History of Economic Decline: Business, Politics, and Work in Trenton.* New Brunswick, NJ: Rutgers University Press.

Cutler, William C. 1883. *History of the State of Kansas.* Chicago: A. T. Andreas. Available online at http://www.kancoll.org/books/cutler/.

Cuvillier, Elian. 2001. "Babylone et Jérusalem." *Réforme* 2952 (November): 8.

Dabney, Virginius. 1971. *Virginia: The New Dominion.* Garden City, NY: Doubleday.

Dalton, John E. 1944. "A History of the Location of the State Capitals in South Dakota." Master's thesis, University of South Dakota.

Daniels, Bruce C. 1986. "The Colonial Background of New England's Secondary Urban Centers." *Historical Journal of Massachusetts* 174:11–24.

David, Paul T., and Ralph Eisenberger. 1961–62. *Devaluation of the Urban and Suburban Vote: A Statistical Investigation of Long-Term Trends in State Legislative Representation.* Charlottesville: University of Virginia, Bureau of Public Administration.

Davies, Christopher S. 1986. "Life at the Edge: Urban and Industrial Evolution of Texas, Frontier Wilderness—Frontier Space, 1836–1986." *Southwestern Historical Quarterly* 89:443–554.

Davies, Richard O. 1998. *Main Street Blues: The Decline of Small-Town America.* Columbus: Ohio State University Press.

Davis, Edwin A. 1959. *Louisiana: The Pelican State.* Baton Rouge: Louisiana University Press.

Dean, William B. 1908. "A History of the Capitol Buildings of Minnesota. With some account of the struggles for their location." *Collections of the Minnesota Historical Society* 12:2–42.

DeArmond, Robert N. 1967. *The Founding of Juneau.* Juneau, AK: Gastineau Channel Centennial Association.

———. 1995. *From Sitka's Past.* Sitka, AK: Sitka Historical Society.

Debarbieux, Bernard. 1995. "Le lieu, le territoire et trois figures de rhétorique." *L'Espace géographique* 2:97–112.

Di Méo, Guy. 2002. "L'identité: une médiation essentielle du rapport espace/société." *Géocarrefour* 77:175–84.

Dodds, Gordon B. 1986. *The American Northwest: A History of Oregon and Washington.* Arlington Heights, IL: Forum Press.

Donahue, John D. 1997. *Disunited States.* New York: Basic Books.

Dunbar, Willis F., and George S. May. 1995. *Michigan: A History of the Wolverine State.* Grand Rapids, MI: W. E. Eerdmans.

Dunn, Jacob P. 1910. *Greater Indianapolis: The History, the Industries, the Institutions, and the People of a City of Homes.* Vol. 1. Chicago: Lewis.

Earle, Carville. 1992. *Geographical Inquiry and American Historical Problems.* Stanford, CA: Stanford University Press.

Edgar, Walter B. 1998. *South Carolina: A History.* Columbia: University of South Carolina Press.

Edgar, Walter B., and Deborah K. Wooley. 1986. *Columbia: Portrait of a City.* Norfolk, VA: Donning.

Eggert, Gerald G. 1993. *Harrisburg Industrializes: The Coming of Factories to an American Community.* University Park: Pennsylvania State University Press.

Ehrlich, Karen L. 1981. "Arizona's Territorial Capital Moves to Phoenix." *Arizona and the West* 23:231–42.

Elazar, Daniel J. 1970. *Cities of the Prairie: The Metropolitan Frontier and American Politics.* New York: Basic Books.

———. 1987. *Building Cities in America: Urbanization and Suburbanization in a Frontier Society.* Lanham, MD: Hamilton Press.

Ellis, Mary Louise, and William Warren Rogers. 1986. *Tallahassee, Leon County: A History and Bibliography.* Tallahassee: Florida Department of State, Historic Tallahassee Preservation Board.

Etulain, Richard W., ed. 1994. *Contemporary New Mexico, 1940–1990.* Albuquerque: University of New Mexico Press.

Eysberg, Cees D. 1989. "The Origins of the American Urban System: Historical Accident and Initial Advantage." *Journal of Urban History* 15:185–95.

Fairley, Nan. 1998. "The Lost Capitals." *Alabama Heritage* 48:18–31.

Ferejohn, John, and Barry R.Weingast, eds. 1997. *The New Federalism: Can the States Be Trusted?* Stanford, CA: Hoover Institution Press, Stanford University.

Ferraro, William M. 1998. "Representing a Layered Community: James, Lampson P., and Hoyt Sherman and the Development of Des Moines, 1850–1900." *Annals of Iowa* 57:240–73.

Fischer, LeRoy H. 1975. "Oklahoma Territory, 1890–1907." *Chronicles of Oklahoma* 53:3–8.

Fohlen, Claude. 1992. *Thomas Jefferson*. Nancy: Presses Universitaires de Nancy.

Forbes, Gerald. 1938. *Guthrie, Oklahoma's First Capital*. Norman: University of Oklahoma Press.

Foster, Lee. 1976. "Old Sacramento: Most Ambitious Historic Restoration in the West." *American West* 13:20–7.

Francaviglia, Richard V. 1996. *Main Street Revisited: Time, Space, and Image Building in Small-Town America*. Iowa City: University of Iowa Press.

Fries, Sylvia D. 1977. *The Urban Idea in Colonial America*. Philadelphia, PA: Temple University Press.

Gannon, Michael, ed. 1996. *The New History of Florida*. Gainesville: University Press of Florida.

Gaquin, Deirdre A., and Katherine A. DeBrandt, eds. 2002. *County and City Extra: Special Decennial Census Edition*. Lanham, MD: Bernan.

Geise, William R. 1962–63. "Missouri's Confederate Capital in Marshall, Texas." *Southwestern Historical Quarterly* 66:193–207.

Gibson, Arrell M. 1981. *Oklahoma: A History of Five Centuries*. Norman: University of Oklahoma Press.

Gilfoyle, Timothy J. 1999. *White City, Linguistic Turns, and Disneylands: The New Paradigms in Urban History*. Available online at http://www.luc.edu/history/fac_resources/gilfoyle/WHITECIT.HTM.

Glaeser, Edward L. 1999. *Urban Primacy and Politics*. Discussion Paper Series, No. 1874, Harvard Institute of Economic Research. Cambridge, MA: Harvard University.

Gleason, David K. 1991. *Baton Rouge*. Baton Rouge: Louisiana State University Press.

Goetz, Delia. 1971. *State Capital Cities*. New York: Morrow.

Goodall, Cecile R. 1968a. "Development of Municipal Government, Charleston, West Virginia, 1794–1936." *West Virginia History* 29:97–137.

Goodall, Elizabeth J. 1968b. "The Charleston Industrial Area Development, 1797–1937." *West Virginia History* 30:358–412.

Goodsell, Charles T. 2001. *The American Statehouse: Interpreting Democracy's Temples*. Lawrence: University Press of Kansas.

Goodstein, Anita S. 1989. *Nashville, 1780–1860: From Frontier to City*. Gainesville: University Press of Florida.

Gordon Winifred, and Douglas Gordon. 1972. "The Dome of the Annapolis State House." *Maryland Historical Magazine* 67:294–97.

Gottmann, Jean. 1983. *The Coming of the Transactional City*. College Park: University of Maryland Institute for Urban Studies.

Gray, Virginia, and Herbert Jacob, eds. 1996. *Politics in the American States: A Comparative Analysis*. Washington, DC: CQ Press.

Green, Edwin L. 1932. *A History of Richland County*. Vol. 1. *1732–1805*. Columbia. SC: R. L. Bryan.

Halverson, Frank Douglas. 1955. *Chart Showing Changes in the Location of State Capitals*. Salt Lake City: Halverson Research Bureau.

Hamer, David. 1990. *New Towns in the New World: Images and Perceptions of the Nineteenth-Century Urban Frontier*. New York: Columbia University Press.

———. 1998. *History in Urban Places: The Historic Districts of the United States*. Columbus: Ohio State University Press.

Hancock, John L. 1967. "Planners in the Changing American City, 1900–1940." *Journal of the American Institute of Planners* 33:290–304.

Hansen, Harry, ed. 1967. *California: A Guide to the Golden State*. New York: Hastings House.

Hanson, Joseph M. 1916. *The Book of the Pageant of Yankton, 1916*. Yankton, SD: Yankton Printing Company.

Hanson, Russell L. 1998. *Governing Partners: State-Local Relations in the United States*. Boulder, CO: Westview Press.

Harris, Chauncy D. 1943. "A Functional Classification of Cities in the United States." *Geographical Review* 33:86–99.

Harrison, Robert. 1979. "The Hornet's Nest at Harrisburg: A Study of the Pennsylvania Legislature in the Late 1870s." *Pennsylvania Magazine of History and Bibliography* 103:334–55.

Hayes, A. B., and Sam D. Cox. 1889. *History of the City of Lincoln, Nebraska, with Brief Historical Sketches of the State and of Lancaster County*. Lincoln, NE: State Journal Company.

Heffernan, Nancy C., and Ann P. Stecker. 1996. *New Hampshire: Crosscurrents in Its Development*. Hanover, NH: University Press of New England.

Hemperly, Marion R., and Edwin L. Jackson. 1993. *Georgia's Boundaries: The Shaping of a State*. Athens: Carl Vinson Institute of Government, University of Georgia.

Hill, Ellen C., and Marylin S. Blackwell. 1983. *Across the Onion: A History of East Montpelier, VT, 1781 to 1981*. Montpelier, VT: East Montpelier Historical Society.

Hitchcock, Henry R., and William Seale. 1976. *Temples of Democracy: The State Capitols of the U.S.A.* New York: Harcourt, Brace, Jovanovich.

Höbling, Walter W., and Karl Franzens. 1990. "From Main Street to Lake Wobegon and Half-Way Back: the Ambiguous Myth of the Small Town in Recent American Literature." In *Mythes ruraux et urbains dans la culture américaine. Actes du colloque des 2, 3 et 4 mars 1990, Aix en Provence*, 51–66. Aix-en Provence: Publications de l'Université de Provence.

Holdsworth, Deryck W. 2002. “Historical Geography: The Ancients and the Moderns— Generational Vitality.” *Progress in Human Geography* 26:671–78.

Hollingsworth, J. Rogers, and Ellen J. Hollingsworth. 1979. *Dimensions in Urban History: Historical and Social Science Perspectives on Middle-Sized American Cities*. Madison: University of Wisconsin Press.

Holmes, Yulssus Lynn. 1996. *Those Glorious Days: A History of Louisville as Georgia’s Capital, 1796–1807*. Macon, GA: Mercer University Press.

“The Homeless Governor.” 2003. *The Economist*, February 22, 50.

Hubler, Dave. 1968. “Des Moines River Navigation: Great Expectations Unfulfilled.” *Annals of Iowa* 39: 287–306.

Hull, Sarah, and Stephen Keeling. 1999. *The Rough Guide to New England*. London: Penguin.

Hunker, Henry L. 2000. *Columbus, Ohio: A Personal Geography*. Columbus: Ohio State University.

James, Harlean. 1925–26. “Seven Southern State Capitals.” *Social Forces* 4: 386–94.

Johnson, Donald D. 1991. *The City and County of Honolulu: A Governmental Chronicle*. Honolulu: University of Hawaii Press, City Council of the City and County of Honolulu.

Johnson, Norman J. 1988. *Washington’s Audacious State Capitol and Its Builders*. Seattle: University of Washington Press.

Johnston, R. J. 1982. *The American Urban System: A Geographical Perspective*. New York: St. Martin’s Press.

Journal of the Constitutional Convention of the State of Oregon, (Held at Salem, Commencing August 17, Together with the Constitution Adopted by the People, November 9, 1857). 1882. Salem, OR: W. H. Byers, State Printer.

Journals of the General Assembly of the State of Vermont at their Session Begun and Holden at Montpelier in the County of Jefferson on Thursday the 14th of October A.D. 1813. 1813. Rutland, VT: Fay and Davidson.

Journals of the Honorable Senate and House of Representatives, June Session, 1864. 1864. Concord: Amos Hadley, State Printer. Concord: New Hampshire State Archives LS 328.7 N533 C4.

Kemp, Roger L., ed. 1999. *Forms of Local Development: A Handbook on City, County, and Regional Options*. Jefferson, NC: McFarland.

Kennedy, William. 1983. *O Albany!* New York: Viking Press.

Kenny, Alice. 1987. “The Transformation of Albany Patricians, 1778–1860.” *New York History* 68:151–73.

Kestenbaum, Justin L. 1981. *Out of the Wilderness: An Illustrated History of Greater Lansing*. Woodland Hills, CA: Windsor.

———. 1986. “A Choice in the Wilderness.” *Michigan History* 70:45–51.

———, ed. 1990. *The Making of Michigan, 1820–1860: A Pioneer Anthology*. Detroit, MI: Wayne State University Press.

Keys, Margaret N. 1986. "Old Capitol Restored." *Palimpsest* 57:122–28.

Kidder, Tracy. 1999. *Home Town*. New York: Random House.

King Van Rensselaer, J. 1905. *Newport, Our Social Capital*. Philadelphia, PA: J. B. Lippincott.

Knepper, George W. 1989. *Ohio and Its People*. Kent, OH: Kent State University Press.

Knowles, Anne K. 2001. "Afterword: Historical Geography since 1987." In McIlwraith and Muller 2001, 465–75.

Konig, Michael F. 1982. "Phoenix in the 1950s: Urban Growth in the 'Sunbelt.'" *Arizona and the West* 24:19–38.

———. 1984. "Postwar Phoenix: Arizona Banking and Boosterism." *Journal of the West* 23:72–6.

Kramer, Carl E. 1986. *Capital on the Kentucky: A Two Hundred Year History of Frankfort and Franklin County*. Frankfort, KY: Historic Frankfort.

Krohe, James, Jr., ed. 1976. *A Springfield Reader: Historical Views of the Illinois Capital, 1818–1976*. Springfield, IL: Sangamon County Historical Society.

Lambert, Louis E. 1959. "The Indiana State Board of Accountants: The Law in Formation." *Indiana Magazine of History* 55:111–68.

Lang, Rob, and Peter Taylor. 2003. *The Global Connectivity of US Cities: A Study of Its Importance and Policy Relevance for 'Lower Level' Cities*. Available online at http://www.lboro.ac.uk/gawc/projects/projec32.html.

Lang, William L. 1987. "Spoils of Statehood: Montana Communities in Conflict, 1888–1894." *Montana* 37:34–45.

Langdon, William C. 1916. *The Pageant of Corydon: The Pioneer Capital of Indiana, 1816–1916*. New Albany, IN: Baker's Printing House.

Lass, William E. 1998. *Minnesota. A History*. New York: W. W. Norton.

Lees, Andrew. 1985. *Cities Perceived: Urban Society in European and American Thought, 1820–1940*. Manchester, UK: Manchester University Press.

Leider, Charles L. W. 1990–91. "Capitol Townsite Historic District, Guthrie, Oklahoma: A Case Study, 1980–1986." *Chronicles of Oklahoma* 68:396–423.

Lemmon, Sarah McCulloh. 1966. "Raleigh: An Example of the 'New South'?" *North Carolina Historical Review* 43:261–85.

Lemov, Penelope. 1993. "Governing in the Shadow of the Dome." *Governing* 6:46–9.

Leonard, Stephen J., and Thomas J. Noel. 1990. *Denver: Mining Camp to Metropolis*. Niwot: University Press of Colorado, 1990.

Lincoln Chamber of Commerce. 1923. *Lincoln, Nebraska's Capital City, 1867–1923*. Lincoln, NE: Woodruff Printing.

Lingeman, R. 1980. *Small Town America: A Narrative History, 1620–the Present*. New York: G. P. Putnam's Sons.

Littrell, William, Esq. 1806. *Political Transactions in and Concerning Kentucky, from the First Settlement Thereof, Until it Became an Independent State, in June 1792*. Frankfort, KY: William Hunter, Printer to the Commonwealth.

Logan, George B. 1925–26. "Guides to Periodical Readings." *Social Forces* 4:394–95.

Luckingham, Bradford. 1989. *Phoenix: The History of a Southwestern Metropolis*. Tucson: University of Arizona Press.

Lynch, Jean, and David R. Meyer. 1992. "Dynamics of the U.S. System of Cities, 1950 to 1980: The Impact of the Large Corporate Law Firm." *Urban Affairs Quarterly* 28:38–68.

Machor, James L. 1987. *Pastoral Cities: Urban Ideals and the Symbolic Landscape of America*. Madison: University of Wisconsin Press.

Mahoney, Nell S. 1945. "William Strickland and the Building of Tennessee's Capitol, 1845–1854." *Tennessee Historical Quarterly* 4:99–153.

Mahoney, Timothy R. 1990. *River Towns in the Great West: The Structure of Provincial Urbanization in the American Midwest, 1820–1870*. Cambridge: Cambridge University Press.

Maier, Pauline, 1997. *American Scripture: Making the Declaration of Independence*. New York: Alfred A. Knopf.

Malone, Michael P., Richard B. Roeder, and William L. Lang. 1991. *Montana: A History of Two Centuries*. Seattle: University of Washington Press.

Margadant, Ted W. 1992. *Urban Rivalries in the French Revolution*. Princeton, NJ: Princeton University Press.

Marienstras, Elise. 1991. *Les mythes fondateurs de la démocratie américaine. Essai sur le discours idéologique aux Etats-Unis à l'époque de l'indépendance, 1763–1800*. Paris: Editions Complexes.

Maurer, Tracy. 1999. *State Capitals*. Vero Beach, FL: Rourke Press.

Mawn, Geoffrey P. 1977. "Promoters, Speculators, and the Selection of the Phoenix Townsite." *Arizona and the West* 19:207–24.

McCormick, Richard P. 1981. *New Jersey from Colony to State 1609–1789*. Newark: New Jersey Historical Society.

McCready, Eric S. 1974. "The Nebraska Capitol: Its Design, Background and Influence." *Nebraska History* 55:325–461.

McEneny, John J. 1981. *Albany: Capital City on the Hudson. An Illustrated History*. Woodland Hills, CA: Windsor.

McIlwraith, Thomas F., and Edward K. Muller, eds. 2001. *North America: The Historical Geography of a Changing Continent*. Lanham, MD: Rowman & Littlefield.

McInerney, Suzanne. 1994. "A Capital Idea! A Brief and Bumpy History of Pennsylvania's Capitols." *Pennsylvania Heritage* 20:24–31.

McKee, D. Duff. 1992. "The People vs. Caleb Lyon and Others: The Capital Relocation Case Revisited." *Idaho Yesterdays* 36:2–18.

McKee, James L., and Arthur A. I. A. Duerschner. 1976. *Lincoln: A Photographic History*. Lincoln, NE: Salt Valley Press.

McKnight, Tom. 1992. *Regional Geography of the United States and Canada*. Englewood Cliffs, NJ: Prentice Hall.

McLemore, Richard A., ed. 1973. *A History of Mississippi*. 2 vols. Hattiesburg: University and College Press of Mississippi.

McLoughlin, William G. 1978. *Rhode Island: A Bicentennial History*. New York: W. W. Norton.

McRaven, William Henry. 1949. *Nashville, "Athens of the South."* Chapel Hill, NC: Scheer and Jervis.

Meinig, Donald W. 1986. *The Shaping of America: A Geographical Perspective on 500 Years of History*. Vol. 1. *Atlantic America, 1492–1800*. New Haven, CT: Yale University Press.

———. 1993. *The Shaping of America: A Geographical Perspective on 500 Years of History*. Vol. 2. *Continental America, 1800–1867*. New Haven, CT: Yale University Press.

———. 1998. *The Shaping of America: A Geographical Perspective on 500 Years of History*. Vol. 3. *Transcontinental America, 1850–1915*. New Haven, CT: Yale University Press.

Meredith, Howard L., and George H. Shirk. 1977. "Oklahoma City Growth and Reconstruction, 1889–1939." *Chronicles of Oklahoma* 55:293–308.

Minto, John. 1902. *Salem, Oregon: Past, Present, an Historical Sketch*. Salem, OR: Schaefer Printing.

Moffat, Riley. 1992. *Population History of Eastern U.S. Cities and Towns, 1790–1870*. Metuchen, NJ: Scarecrow Press.

Montarry, Anne-Laure. 1995. *Baton-Rouge, petite capitale d'Etat*. Master's thesis, Université Lumière-Lyon 2.

Moore, John H. 1993. *Columbia and Richland County. A South Carolina Community, 1740–1990*. Columbia: University of South Carolina Press.

Morone, James A. 1990. *The Democratic Wish: Popular Participation and the Limits of American Government*. New York: Basic Books.

Morrissey, Charles T. 1981. *Vermont: A Bicentennial History*. New York: W. W. Norton.

Moussalli, Stephanie D. 1997. "Choosing Capitals in Antebellum Southern Frontier Constitutions." *Southwestern Historical Quarterly* 101:58–75.

Munroe, John A. 1993. *History of Delaware*. Newark, DE: University of Delaware Press.

Murphy, Raymond E., and Marion Murphy. 1937. *Pennsylvania: A Regional Geography*. Harrisburg: Pennsylvania Book Service.

Nash, Gary B. 1987. "The Social Evolution of Preindustrial American Cities, 1700–1820." *Journal of Urban History* 13:15–45.

Nash, Gerald D. 1985. "Urban Development in the Southwest." *Journal of Urban History* 11:471–80.

Naske, Claus-M., and Herman E. Slotnick. 1987. *Alaska: A History of the 49th State*. Norman: University of Oklahoma Press.

Nesbit, Robert C. 1989. *Wisconsin: A History*. Second edition, revised and updated by William F. Thompson. Madison: University of Wisconsin Press.

Newby, Rick. 1987. "Helena's Social Supremacy: Political Sarcasm and the Capital Fight." *Montana* 37:68–72.

Newton, Robert E. 1977. "Relocating Alaska's State Capital." *State Government* 50:165–9.

Nicandri, David, and Dereck Valley. 1980. *Olympia Wins: Washington's Capital Controversies*. Olympia: Washington State Capitol Museum.

Nora, Pierre, ed. 1984–92. *Les lieux de mémoire*. Paris: Gallimard.

North, James W. 1981. *The History of Augusta, Maine*. Facsimile of the 1870 edition with a new foreword by Edwin A Churchill. Somersworth, NH: New England History Press.

Nouailhat, Yves-Henri. 1982. *Evolution économique des Etats-Unis du milieu du XIXème siècle à 1914*. Paris: SEDES-CDU.

Olson, James C., and Ronald C. Naugle. 1997. *History of Nebraska*. Lincoln: University of Nebraska Press.

Osborn, Norris G., ed. 1925. *History of Connecticut: In Monographic form*. Vol. 1. New York: The States History Company.

Owen, Mary B. 1949. *The Story of Alabama: History of a State*. Vol. 1. New York: Lewis History Publishing.

Oxford, June. 1995. *The Capital That Couldn't Stay Put: The Complete Book of California's Capitols*. Fairfield, CA: James Stevenson.

Papenfuse, Edward G. 1975. *In Pursuit of Profit: The Annapolis Merchants in the Era of the American Revolution*. Baltimore, MD: Johns Hopkins University Press.

Parré, Madeline F. 1965. *Arizona Pageant*. Phoenix: Arizona Historical Foundation.

Pessen, Edward. 1985. *Jacksonian America: Society, Personality, and Politics*. Urbana: University of Illinois Press.

Petersen, William J., ed. 1970. "Des Moines, Iowa: State Capital." *Palimpsest* 51:217–49.

Phillips, Paul C., ed. 1937. *Historical Reprints. Sources of Northwest History, No. 24*. Missoula: Montana State University.

Pollock, Paul W. 1960. *The Capital Cities of the United States*. Phoenix: P. W. Pollock, PO Box 1352.

Potter, Merle. 1932. "The North Dakota Capital Fight." *North Dakota Historical Quarterly* 7:25–36.

Potter, Parker B., Jr. 1994. *Public Archeology in Annapolis: A Critical Approach to History in Maryland's Ancient City*. Washington, DC: Smithsonian Institution Press.

Potts, James B. 1988. "The Nebraska Capital Controversy, 1854–1859." *Great Plains Quarterly* 8:172–82.

Powell, William S. 1989. *North Carolina Through Four Centuries*. Chapel Hill: University of North Carolina Press.

Primm, James N. 1998. *Lion of the Valley: St. Louis, Missouri, 1764–1980*. St. Louis: Missouri Historical Society Press; distributed by University of Missouri Press.

Randolph, Lewis Anthony. 1990. "Development Policy in Four U.S. Cities

(St. Louis, Missouri, Richmond, Virginia, Baltimore, Maryland, Columbus, Ohio)." PhD diss., Ohio State University. Available online at http://www.lib.umi.com/dissertations/fullcit/9105194.

Records of the Governor and Council of the State of Vermont. Vol. 5. 1877. Edited by E. P. Walton. Montpelier: Vermont Historical Society.

Renick, L. W., ed. 1896. *Che-Le-Co-The, Glimpses of Yesterday: A Souvenir of the 100th Anniversary of the Founding of Chillicothe, Ohio.* New York: Knickerbocker Press.

Report of the Joint Special Committee on Selecting a Site for a New State House, 1875. 1875. Providence, RI: Providence Press, Printers to the State, 1875.

Report of the Special Committee on a Site for a New State House, Made to the General Assembly at the January Session, 1873. 1873. Providence, RI: Providence Press, Printers to the State.

Reps, John W. 1969. *Town Planning in Frontier America.* Princeton, NJ: Princeton University Press.

———. 1972. *Tidewater Towns: City Planning in Colonial Virginia and Maryland.* Charlottesville, VA: Colonial Williamsburg Foundation; distributed by the University Press of Virginia.

———. 1981. *The Forgotten Frontier: Urban Planning in the American West before 1890.* Columbia: University of Missouri Press.

———. 1994. *Cities of the Mississippi: Nineteenth-Century Images of Urban Development.* Columbia: University of Missouri Press.

———. 1998. *Bird's Eye Views: Historic Lithographs of North American Cities.* New York: Princeton Architectural Press.

Rice, Bradley R. 1976. "The Rise and Fall of the Galveston-Des Moines Plan: Commission Government in American Cities, 1901–1920." PhD diss., University of Texas, Austin.

Rice, Otis K., and Stephen W. Brown. 1993. *West Virginia: A History.* Lexington: University Press of Kentucky.

Richmond, Robert W. 1972. "Kansas Builds a Capitol." *Kansas Historical Quarterly* 37:249–67.

Rishel, Joseph F., ed. 1992. *American Cities and Towns: Historical Perspectives.* Pittsburgh, PA: Duquesne University Press.

Roberts, O. M. 1898. "The Capitals of Texas." *The Quarterly of the Texas State Historical Association* 2:117–123.

Robertson, David S. 1996. "Oil Derricks and Corinthian Columns: The Industrial Transformation of the Oklahoma State Capitol Grounds." *Journal of Cultural Geography* 16:17–44.

Robinson, Elwyn B. 1966. *History of North Dakota.* Lincoln: University of Nebraska Press.

Rogers, William W., Leah Rawls Atkins, Robert D. Ward, and Wayne Flynt. 1994. *Alabama: The History of a Deep South State.* Tuscaloosa: University of Alabama Press.

Roncayolo, Marcel. 1990. *La ville et ses territoires.* Paris: Gallimard.

———. 2002. *Lectures de villes. Formes et temps*. Marseille: Editions Parenthèses.

Roseberry, Cecil R. 1964. *Capitol Story*. Albany: State of New York.

Rosenthal, Alan. 1989. "The Legislative Institution: Transformed and at Risk." In Van Horn 1989, 65–86.

Ross, Christopher O. 1987. "Organizational Dimensions of Metropolitan Dominance: Prominence in the Network of Corporate Control, 1955–1975." *American Sociological Review* 52:258–67.

Rotenberg, Robert, and Gary W. McDonogh, eds. 1993. *The Cultural Meaning of Urban Space*. Westport, CT: Bergin & Garvey.

Rourke, Francis E. 1964. "Urbanism and American Democracy." *Ethics* 74:255–68.

Rowley, William E. 1971. "The Irish Aristocracy of Albany, 1798–1878." *New York History* 52:275–304.

Sabato, Larry. 1983. *Goodbye to Good-Time Charlie: The American Governorship Transformed*. Washington, DC: CQ Press.

Sawyer, Andrew J., ed. 1916. *Lincoln: The Capital City and Lancaster County, Nebraska, Illustrated*. Vol. 1. Chicago: S. J. Clarke.

Schuler, Harold H. 1989. "In Pursuit of Permanence: A Photographic Essay on the Capital of South Dakota." *South Dakota History* 19:26–55.

———. 1990. *Fort Pierre Chouteau*. Vermillion: University of South Dakota Press.

Schultz, Stanley K. 1989. *Constructing Urban Culture: American Cities and City Planning, 1800–1920*. Philadelphia, PA: Temple University Press.

Schwartz, Philip J. 1979. *The Jarring Interests: New York's Boundary Makers, 1664–1776*. Albany, NY: SUNY Press.

Schwieder, Dorothy. 1996. *Iowa: The Middle Land*. Ames: Iowa State University Press.

Scott, Mel. 1971. *American City Planning Since 1890*. Berkeley and Los Angeles: University of California Press.

Shakel, Paul A. 1993. *Personal Discipline and Material Culture: An Archeology of Annapolis, Maryland, 1695–1870*. Knoxville: University of Tennessee Press.

Shay, Michael. 1983. "Denver's Capitol Hill: Reflections of a Neighborhood." *Colorado Heritage* 4:25–35.

Siegel, Adrienne. 1981. *The Image of the American City in Popular Literature, 1820–1870*. New York: National University Publications; and Port Washington, NY: Kennikat Press.

Simmons, W. H. 1903. "The Selection of Tallahassee as the Capital." *Florida House Journal*; reprinted in *Florida Historical Society Quarterly* 1 (1908): 26–44 and 2 (1908): 18–27.

Smith, Alice E. 1973. *The History of Wisconsin*. Vol. 1. *From Exploration to Statehood*. Madison: State Historical Society of Wisconsin.

Smith, Page. 1966. *As a City upon a Hill: The Town in American History*. New York: Alfred A. Knopf.

Sneddon, Leonard J. 1971. "From Philadelphia to Lancaster: The First Move of Pennsylvania's Capital." *Pennsylvania History* 38:349–60.

Socolofsky, Homer E., and Huber Self. 1988. *Historical Atlas of Kansas*. Norman: University of Oklahoma Press.

"The South. Be Just Ain't Right." 2003. *Economist*, February 15, 48.

Spalding, Burleigh F. 1964. *Constitutional Convention, 1889*. Reprinted in *North Dakota History* 31:151–64.

Spence, Mary L., ed. 1997. *The Arizona Diary of Lily Fremont, 1878–1881*. Tucson: University of Arizona Press.

Stanback, Thomas M., and Thierry J. Noyelle. 1982. *Cities in Transition: Changing Job Structures in Atlanta, Denver, Buffalo, Phoenix, Columbus (Ohio), Nashville, Charlotte*. Totowa, NJ: Allanheld, Osmun.

Stansfield, Charles A., Jr. 1998. *A Geography of New Jersey: The City in the Garden*. New Brunswick, NJ: Rutgers University Press.

Stanton, Mike. 2003. *The Prince of Providence: The True Story of Buddy Cianci, America's Most Notorious Mayor, Some Wiseguys, and the Feds*. New York: Random House.

State of Rhode Island and Providence Plantations. 1899. *Message of Elisha Dyer, Governor of Rhode Island to the General Assembly at Its January Session*. Providence, RI: E. L. Freeman and Sons, Printers to the State.

———. 1901. *Message of Elisha Dyer, Governor of Rhode Island to the General Assembly at Its January Session*. Providence, RI: E. L. Freeman and Sons, Printers to the State.

Stein, Mark. 2008. *How the States Got Their Shapes*. New York: Harper.

Stelter, Gilbert A. 1967. "The Birth of a Frontier Boomtown: Cheyenne in 1867." *Annals of Wyoming* 39:5–35.

———. 1973. "The City and Westward Expansion: A Western Case Study (Cheyenne)." *Western Historical Quarterly* 4:187–202.

Strauss, Anselm L. 1961. *Images of the American City*. New York: Free Press.

Stroble, Paul E., Jr. 1992. *High on the Okaw's Western Bank: Vandalia, Illinois, 1819–1839*. Urbana: University of Illinois Press.

Studer, Jacob H. 1873. *Columbus, Ohio: Its History, Resources, and Progress, with Numerous Illustrations*. Columbus, OH: W. Riches, Engraver.

Third Subscription List of Martyrs to Procure State House at Montpelier, 1857. 1857. Montpelier: Vermont Historical Society MSS 27 / 121.

Thomas, Isabelle. 2000. "La restructuration industrielle de l'Ohio et l'évolution de ses aires métropolitaines." PhD diss., Université Paris IV.

Thompson, John. 1986. *Closing the Frontier: Radical Response in Oklahoma, 1889–1923*. Norman: University of Oklahoma Press.

Tick, Jeffrey L. 1981. "The Albany Connection: National, State and Local Politics in Orange County, New York, 1832–1855." PhD diss., SUNY at Binghampton.

Tocqueville, Alexis de. 2000. *Democracy in America*. 2 vols. Edited and translated by Harvey C. Mansfield and Delba Winthrop. Chicago: University of Chicago Press. Orig. pub. 1840.

Trafzer, Clifford E. 1984. "'Harmony and Cooperation': Robert A. Heffner, Mayor of Oklahoma City." *Chronicles of Oklahoma* 62:70–85.

Turner, Lynn W. 1983.*The Ninth State: New Hampshire's Formative Years.* Chapel Hill: University of North Carolina Press.

U.S. Bureau of the Census. 1998. *Population of Urban Places, 1790 to 1990.* Available online at http://www.census.gov/population/documentation/twps0027/tab01 to 22.txt.

University of Rhode Island. 1996. *"A Most Admirable Public Building": The Rhode Island State House Centennial Exhibition.* Cranston, RI: Lewis Graphics.

Upton, William W. 1990. "A Seat of Government." In Kestenbaum 1990, 389–402.

Van Horn, Carl E., ed. 1989. *The State of the States.* Washington, DC: CQ Press.

Van Zandt, Franklin K. 1976. *Boundaries of the United States and the Several States, Geological Survey Professional Paper 909.* Washington, DC: United States Government Printing Office.

Vance, James E. 1995. *The North American Railroad: Its Origin, Evolution and Geography.* Baltimore, MD: Johns Hopkins University Press.

Vedder, Richard K. 1997. "Capital crimes: Political Centers as Parasite Economies." *USA Today: The Magazine of the American Scene* , September, 1997, 20–22.

Wade, Richard C. 1959. *The Urban Frontier: Pioneer Life in Early Pittsburgh, Cincinnati, Lexington, Louisville and St. Louis.* Chicago: University of Chicago Press.

Walcott, Susan M. 1995. "Niches and Networks: Urban Regional Growth in Indianapolis, 1979–1994." PhD diss., Indiana University.

Walker, David B. 2000. *The Rebirth of Federalism: Slouching Toward Washington.* New York: Chatham House.

Walker, Edwin R. 1929. *A History of Trenton, 1679–1929: Two Hundred and Fifty Years of a Notable Town with Links in Four Centuries.* 2 vols. Princeton, NJ: Princeton University Press.

Wallerstein, Immanuel. 2004. *World-Systems Analysis: An Introduction.* Durham, NC: Duke University Press.

Waug, Elizabeth C., and Editorial Committee. 1967. *North Carolina's Capital, Raleigh.* Chapel Hill: University of North Carolina Press.

Weatherby, James B., and Stephanie L. Witt. 1994. *The Urban West: Managing Growth and Decline.* Westport, CT: Praeger Publishers.

Weber, Adna F. 1899. *The Growth of Cities in the Nineteenth Century: A Study in Statistics.* New York: Macmillan.

Weil, François. 1992. *Naissance de l'Amérique urbaine 1820–1920.* Paris: CDU-SEDES.

Welsh, Carol H. 1982. "Cattle Market for the World: The Oklahoma National Stockyards." *Chronicles of Oklahoma* 60:42–55.

Wenger, Mark R. 1993. "Thomas Jefferson and the Virginia State Capitol." *Virginia Magazine of History and Biography* 101:77–102.

Wheeler, Kenneth W. 1968. *To Wear a City's Crown: The Beginnings of Urban Growth in Texas, 1836–1865.* Cambridge, MA: Harvard University Press.

White, Bruce M. 1999. "The Power of Whiteness: Or, the Life and Times of Joseph Rolette Jr." In *Making Minnesota Territory, 1849–1858*, special issue of *Minnesota History*, edited by Anne R. Kaplan, and Marilyn Ziebarth, 26–49. St. Paul: Minnesota Historical Society.

White, Morton, and Lucia White. 1962. *The Intellectual versus the City: From Thomas Jefferson to Frank Lloyd Wright*. Cambridge, MA: Harvard University Press and MIT Press.

White, William A. 1939. *The Changing West: An Economic Theory about Our Golden Age*. New York: Macmillan.

Widdowfield, Rebekah. 2000. "The place of emotions in academic research." *Area* 32: 199–208.

Williams, Clanton W. 1979. *The Early History of Montgomery and Incidentally of the State of Alabama*. University, AL: Confederate Publishing.

Wilson, Chris. 1997. *The Myth of Santa Fe: Creating a Modern Regional Tradition*. Albuquerque: University of New Mexico Press.

Wilson, William H. 1980. "Harrisburg Successful City Beautiful Movement, 1900–1915." *Pennsylvania History* 47:213–33.

Winslow, Walter C. 1908. "Contests over the Capital of Oregon." *Oregon Historical Society Quarterly* 9:173–78.

Wood, Gordon S. 1991. *La création de la république américaine*. Paris: Belin.

Wood, Joseph S. 1997. *The New England Village*. Baltimore, MD: Johns Hopkins University Press.

Workers of the Writers' Program (WWP) of the Works Progress Administration for Ohio. 1938. *Chillicothe and Ross County*. Chillicothe: Ross County Northwest Territorial Committee.

Workers of the Writers' Program of the Works Progress Administration for the State of Delaware.1937. *New Castle on the Delaware*. Wilmington, DE: Press of W. N. Cann.

Workers of the Writers' Program of the Works Progress Administration for the State of Delaware. 1938. *Delaware: A Guide to the First State*. New York: Viking Press.

Workers of the Writers' Program of the Works Progress Administration for the State of Iowa. 1959. *Iowa: A Guide to the Hawkeye State*. New York: Hastings House.

Workers of the Writers' Program of the Works Progress Administration in the State of Missouri. 1959. *Missouri: A Guide to the "Show Me" State*. New York: Hastings House.

Workers of the Writers' Program of the Works Progress Administration for the State of Nebraska. 1937. *Lincoln, Nebraska*. Lincoln: Woodruff Printing.

Workers of the Writers' Program of the Works Progress Administration for the State of Nebraska. 1939. *Nebraska: A Guide to the Cornhusker State*. New York: Viking Press.

Workers of the Writers' Program of the Works Progress Administration for the

State of New Jersey. 1939. *New Jersey: A Guide to Its Present and Past.* New York: Viking Press.

Workers of the Writers' Program of the Works Progress Administration in the State of Pennsylvania. 1940. *Pennsylvania: A Guide to the Keystone State.* New York: Oxford University Press.

Workers of the Writers' Program of the Works Progress Administration in the State of South Carolina. 1941. *South Carolina: A Guide to the Palmetto State.* New York: Oxford University Press.

Workers of the Writers' Program of the Works Progress Administration in the State of South Dakota. 1952. *South Dakota: A Guide to the State. Second Edition Completely Revised by M. Lisle Reese.* New York: Hastings House.

Workers of the Writers' Program of the Works Progress Administration in the State of Texas. 1940. *Texas: A Guide to the Lone Star State.* New York: Hastings House.

Workers of the Writers' Program of the Works Progress Administration for the State of Utah. 1941. *Utah: A Guide to the State.* New York: Hastings House.

Workers of the Writers' Program of the Works Progress Administration in the State of Virginia. 1940. *Virginia: A Guide to the Old Dominion.* New York: Oxford University Press.

Workers of the Writers' Program of the Works Progress Administration in the State of West Virginia. 1941. *West Virginia: A Guide to the Mountain State.* New York: Oxford University Press.

Workers of the Writers' Program of the Works Progress Administration in the State of Wyoming. 1941. *Wyoming: A Guide to Its History, Highways, and People.* New York: Oxford University Press.

Wright, Deil S. 1988. *Understanding Intergovernmental Relations.* Pacific Grove: Brooks/Cole Publishing Company.

WWP. See entries under "Workers of the Writers' Program of the Works Progress Administration," alphabetized by state.

Zagarri, Rosemarie. 1987. *The Politics of Size: Representation in the United States, 1776–1850.* Ithaca, NY: Cornell University Press.

———. 1988. "Representation and the Removal of State Capitals, 1776–1812." *Journal of American History* 74:1239–56.

Zauner, Phyllis. 1984. *Carson City.* Sonoma, CA: Zanel.

Zielbauer, Paul. 2002. "Poverty in the Land of Plenty: Can Hartford Ever Recover?" *New York Times*, August 26.

Zornow, William F. 1957. *Kansas: A History of the Jayhawk State.* Norman: University of Oklahoma Press.

FURTHER READING

Arthur, T. S., and W. H. Carpenter. 1852. *The History of Kentucky: From its Earliest Settlement to the Present Time.* Philadelphia, PA: Lippincott, Grambo.

Buchta, David L. 2010. *Kentucky's State Capitol.* Charleston, SC: Arcadia.

Cabbage, Henry. 1999. *Tales of Historic Tallahassee.* Tallahassee: Artemis Associates.

"Capital and Capitol History of South Dakota." *South Dakota Historical Collections* 5:110–272.

Clark, Thomas D. 1937. *A History of Kentucky.* New York: Prentice Hall.

Cohan, Zara. 1969. "History of the State House." Draft revision of master's thesis, Newark State College.

Connelly, John L. 1980. "Old Nashville and Germantown." *Tennessee Historical Quarterly* 39:115–48.

Dalzel, Frederick. 1992. "Prudence and the Golden Egg: Establishing the Federal Government in Providence, Rhode Island." *New England Quarterly* 65:355–88.

Debo, Angie. 1944. *Prairie City: The Story of an American Community.* New York: Alfred A. Knopf.

Detzler, Jack J., ed. 1969. *Diary of Howard Stillwell Stanfield (1846–1923), 1864–1865.* Bloomington: Indiana University Press.

Dufour, Charles L. 1967. *Ten Flags in the Wind: The Story of Louisiana.* New York: Harper and Row.

Eldridge, Charles W. 1931. "Journal of a Tour Through Vermont to Montreal and Quebec in 1833." *Proceedings of the Vermont Historical Society* 2:53–82.

Faulk, Odie B. 1970. *Arizona: A Short History.* Norman: University of Oklahoma Press.

Flick, Alexander C., ed. 1933. *History of the State of New York in Ten Volumes.* Vol. 4. New York: Columbia University Press.

Folwell, William W. 1956. *A History of Minnesota.* Vol. 1. St. Paul: Minnesota Historical Society.

Gjerde, Jon. 1997. *The Minds of the West: Ethnocultural Evolution in the Rural Middle West, 1830–1917.* Chapel Hill: University of North Carolina Press.

Henderson, Andrew. 2002. *Images of America: Forgotten Columbus.* Chicago: Arcadia.

Hulse, James W. 1991. *The Silver State: Nevada's Heritage Reinterpreted.* Reno: University of Nevada Press.

Hurst, Harold F. 1981. "The Northernmost Southern Town: A Sketch of Pre–Civil War Annapolis." *Maryland Historical Magazine* 76:240–49.

Hutchinson, W. H. 1969. *California.* Palo Alto, CA: American West.

Judd, Richard W., E. A. Churchill, and Joel W. Eastman, eds. 1995. *Maine: The Pine Tree State, from Prehistory to Present.* Orono: University of Maine Press.

Layton, Charles, and Mary Walton. 1998. "Missing the Story at the State House." *American Journalism Review* 20:42–57.

Markusen, Ann. 1996. "Sticky Places in Slippery Space: A Typology of Industrial Districts." *Economic Geography* 72:293–313.

McReynolds, Edwin C. 1962. *Missouri: A History of the Crossroads State.* Norman: University of Oklahoma Press.

Moore, Albert B. 1934. *History of Alabama*. University, AL: University Supply Store.

Noel, Thomas J., Paul F. Mahoney, and Richard E. Stevens. 1994. *Historical Atlas of Colorado*. Norman: University of Oklahoma Press.

Paullin, Charles O., and John K. Wright. 1932. *Atlas of the Historical Geography of the United States*. Washington, DC: Carnegie Institute and American Geographical Society.

Rafferty, Milton D. 1982. *Historical Atlas of Missouri*. Norman: University of Oklahoma Press.

Remini, Robert. 1958. "The Albany Regency." *New York History* 39:341–55.

Rogers, William W. 1985. "A Great Stirring in the Land: Tallahassee and Leon County in 1860." *Florida Historical Quarterly* 64:148–60.

Teske, Paul, and Renee Johnson. 1994. "Moving Towards an American Industrial Technology Policy." *Political Studies Journal* 22:296–310.

Tregle, Joseph G., Jr. 1999. *Louisiana in the Age of Jackson: A Clash of Culture and Personalities*. Baton Rouge: Louisiana University Press.

Wallace, Christopher E. 1983. "The Opportunity to Grow: Springfield, Illinois, during the 1850s." PhD diss., Purdue University.

Weinard Phil, and K. F. Weinard, eds. 1993. "Phil Weinard Remembers Early Helena, 1880." *Montana* 43:64–8.

Wilson, William E. 1966. *Indiana: A History*. Bloomington: Indiana University Press.

Index

Entries in bold refer to current state capitals. Entries in italics refer to former state capitals. Page numbers followed by *f* or *t* refer to figures or tables, respectively.

224. RUTHERFORD H. PLATT, SHEILA G. PELCZARSKI, AND BARBARA K. BURBANK, eds.,
Cities on the Beach: Management Issues of Developed Coastal Barriers, 1987

223. JOSEPH H. ASTROTH JR.,
Understanding Peasant Agriculture: An Integrated Land-Use Model for the Punjab, 1990

222. MARILYN APRIL DORN,
The Administrative Partitioning of Costa Rica: Politics and Planners in the 1970s, 1989

217–18. MICHAEL P. CONZEN, ed.,
World Patterns of Modern Urban Change: Essays in Honor of Chauncy D. Harris, 1986

216. NANCY J. OBERMEYER,
Bureaucrats, Clients, and Geography: The Bailly Nuclear Power Plant Battle in Northern Indiana, 1989

213. RICHARD LOUIS EDMONDS,
Northern Frontiers of Qing China and Tokugawa Japan: A Comparative Study of Frontier Policy, 1985

210. JAMES L. WESCOAT JR.,
Integrated Water Development: Water Use and Conservation Practice in Western Colorado, 1984

209. THOMAS F. SAARINEN, DAVID SEAMON, AND JAMES L. SELL, eds.,
Environmental Perception and Behavior: An Inventory and Prospect, 1984

207–8. PAUL WHEATLEY,
Nagara and Commandery: Origins of the Southeast Asian Urban Traditions, 1983

206. CHAUNCY D. HARRIS,
Bibliography of Geography, part 2, Regional, volume 1, The United States of America, 1984

194. CHAUNCY D. HARRIS,
Annotated World List of Selected Current Geographical Serials, 4th ed., 1980

186. KARL W. BUTZER,
Recent History of an Ethiopian Delta: The Omo River and the Level of Lake Rudolf, 1971

152. MARVIN W. MIKESELL, ed.,
Geographers Abroad: Essays on the Problems and Prospects of Research in Foreign Areas, 1973

132. NORMAN T. MOLINE,
Mobility and the Small Town, 1900–1930, 1971

127. PETER G. GOHEEN,
Victorian Toronto, 1850 to 1900: Pattern and Process of Growth, 1970